Guerino Mazzola · Gérard Milmeister
Jody Weissmann

Comprehensive Mathematics for Computer Scientists 2

Calculus and ODEs, Splines, Probability,
Fourier and Wavelet Theory,
Fractals and Neural Networks,
Categories and Lambda Calculus

With 114 Figures

 Springer

Guerino Mazzola
Gérard Milmeister
Jody Weissmann

Department of Informatics
University of Zurich
Winterthurerstr. 190
8057 Zurich, Switzerland

The text has been created using LaTeX 2_ε. The graphics were drawn using the open source illustrating software Dia and Inkscape, with a little help from Mathematica. The main text has been set in the Y&Y Lucida Bright type family, the heading in Bitstream Zapf Humanist 601.

Library of Congress Control Number: 2004102307

Mathematics Subject Classification (1998): 00A06

ISBN 3-540-20861-5 Springer Berlin Heidelberg New York

Springer is a part of Springer Science+Business Media

springeronline.com

© Springer-Verlag Berlin Heidelberg 2005
Printed in Germany

Cover design: Erich Kirchner, Heidelberg
Typesetting: Camera ready by the authors
Production: LE-TeX Jelonek, Schmidt & Vöckler GbR, Leipzig

Printed on acid-free paper   40/3142YL - 5 4 3 2 1 0

Preface

This second volume of a comprehensive tour through mathematical core subjects for computer scientists completes the first volume in two regards:

Part III first adds topology, differential, and integral calculus to the topics of sets, graphs, algebra, formal logic, machines, and linear geometry, of volume 1. With this spectrum of fundamentals in mathematical education, young professionals should be able to successfully attack more involved subjects, which may be relevant to the computational sciences.

In a second regard, the end of part III and part IV add a selection of more advanced topics. In view of the overwhelming variety of mathematical approaches in the computational sciences, any selection, even the most empirical, requires a methodological justification. Our primary criterion has been the search for harmonization and optimization of thematic diversity and logical coherence. This is why we have, for instance, bundled such seemingly distant subjects as recursive constructions, ordinary differential equations, and fractals under the unifying perspective of contraction theory.

For the same reason, the entry point to part IV is category theory. The reader will recognize that a huge number of classical results presented in volume 1 are perfect illustrations of the categorical point of view, which will definitely dominate the language of mathematics and theoretical computer science of the decades to come. Categories are advantageous or even mandatory for a thorough understanding of higher subjects, such as splines, fractals, neural networks, and λ-calculus. Even for the specialist, our presentation may here and there offer a fresh view on classical subjects. For example, the systematic usage of categorical limits in neural networks has enabled an original formal restatement of Hebbian learning, perceptron convergence, and the back-propagation algorithm.

However, a secondary, but no less relevant selection criterion has been applied. It concerns the delimitation from subjects which may be very important for certain computational sciences, but which seem to be neither mathematically nor conceptually of germinal power. In this spirit, we have also refrained from writing a proper course in theoretical computer science or in statistics. Such an enterprise would anyway have exceeded by far the volume of such a work and should be the subject of a specific education in computer science or applied mathematics. Nonetheless, the reader will find some interfaces to these topics not only in volume 1, but also in volume 2, e.g., in the chapters on probability theory, in spline theory, and in the final chapter on λ-calculus, which also relates to partial recursive functions and to λ-calculus as a programming language.

We should not conclude this preface without recalling the insight that *there is no valid science without a thorough mathematical culture.* One of the most intriguing illustrations of this universal, but often surprising presence of mathematics is the theory of Lie derivatives and Lie brackets, which the beginner might reject as "abstract nonsense": It turns out (using the main theorem of ordinary differential equations) that the Lie bracket of two vector fields is directly responsible for the control of complex robot motion, or, still more down to earth: to everyday's sideward parking problem. We wish that the reader may always keep in mind these universal tools of thought while guiding the universal machine, which is the computer, to intelligent and successful applications.

Zurich,   *Guerino Mazzola*
August 2004   *Gérard Milmeister*
                                                                   Jody Weissmann

Contents

Topology and Calculus

CHAPTER 27

Limits and Topology

27.1 Introduction

This chapter opens a line of mathematical thought and methods which is quite different from purely set-theoretical, algebraic and formally logical approaches: topology and calculus. Generally speaking this perspective is about the "logic of space", which in fact explains the Greek etymology of the word "topology", which is "logos of topos", i.e., the theory of space. The "logos" is this: We learned that a classical type of logical algebras, the Boolean algebras, are exemplified by the power sets 2^a of given sets a, together with the logical operations induced by union, intersection and complementation of subsets of a (see volume 1, chapter 3). The logic which is addressed by topology is a more refined one, and it appears in the context of convergent sequences of real numbers, which we have already studied in volume 1, section 9.3, to construct important operations such as the n-th root of a positive real number. In this context, not every subset of $\mathbb{R}$ is equally interesting. One rather focuses on subsets $C \subset \mathbb{R}$ which are "closed" with respect to convergent sequences, i.e., if we are given a convergent sequence $(c_i)_i$ having all its members $c_i \in C$, then $l = \lim_{i \to \infty} c_i$ must also be an element of C. This is a useful property, since mathematical objects are often constructed through limit processes, and one wants to be sure that the limit is contained in the same set that the convergent series was initially defined in.

Actually, for many purposes, one is better off with sets complementary to closed sets, and these are called open sets. Intuitively, an open set

O in $\mathbb{R}$ is a set such that with each of its points x, a small interval of points to the left and to the right of x is still contained in O. So one may move a little around x without leaving the open set. Again, thinking about convergent sequences, if such a sequence is outside an open set, then its limit l cannot be in O since otherwise the sequence would eventually approach the limit l and then would stay in the small interval around l within O.

In the sequel, we shall not develop the general theory of topological spaces, which is of little use in our elementary context. We shall only deal with topologies on real vector spaces, and then mostly only of finite dimension. However, the axiomatic description of open and closed sets will be presented in order to give at least a hint of the general power of this conceptualization. There is also a more profound reason for letting the reader know the axioms of topology: It turns out that the open sets of a given real vector space V form a subset of the Boolean algebra 2^V which in its own right (with its own implication operator) is a Heyting algebra! Thus, topology is really a kind of spatial logic, however not a plain Boolean logic, but one which is related to intuitionistic logic. The point is that the double negation (logically speaking) of an open set is not just the complement of the complement, but may be an open set larger than the original. In other words, if it comes to convergent sequences and their limits, the logic involved here is not the classical Boolean logic. This is the deeper reason why calculus is sometimes more involved than discrete mathematics and requires very diligent reasoning with regard to the objects it produces.

27.2 Topologies on Real Vector Spaces

Throughout this section we work with the n-dimensional real vector space $\mathbb{R}^n$. The scalar product $(?, ?)$ in $\mathbb{R}^n$ gives rise to the norm $\|x\| = \sqrt{(x,x)} = \sqrt{\sum_i x_i^2}$ of a vector $x = (x_1, x_2, \ldots x_n) \in \mathbb{R}^n$. Recall that for $n = 1$ the norm of x is just the absolute value of x. Actually, the theory developed here is applicable to any finite-dimensional real vector space which is equipped with a norm, and to some extent even for any infinite-dimensional real vector space with norm, but we shall only on very rare occasions encounter this generalized situation. In the following, we shall use the distance function or metric d defined through the given norm via $d(x, y) = \|x - y\|$, as defined in volume 1, section 24.3. Our first definition introduces the elementary type of sets used in the topology of real vector spaces:

Definition 175 *Given a positive real number ε, and a point $x \in \mathbb{R}^n$, the ε-cube around x is the set*

$$K_\varepsilon(x) = \{y \mid |y_i - x_i| < \varepsilon, \text{ for all } i = 1, 2, \ldots n\},$$

whereas the ε-ball around x is the set

$$B_\varepsilon(x) = \{y \mid d(x, y) < \varepsilon\}.$$

Example 98 To give a geometric intuition of the preceding concepts, consider the concrete situation for real vector spaces of dimensions 1, 2 and 3.

On the real line $\mathbb{R}$ the ε-ball and the ε-cube around x reduce to the same concept, namely the open interval of length 2ε with midpoint x, i.e., $]x - \varepsilon, x + \varepsilon[$.

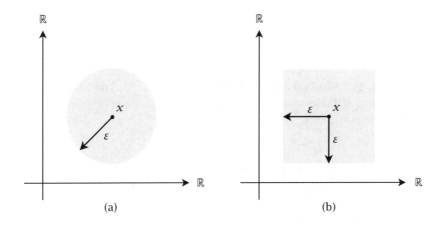

Fig. 27.1. The ε-ball (a) and ε-cube (b) around x in $\mathbb{R}^2$. The boundaries are *not* part of these sets.

On the Euclidean plane $\mathbb{R}^2$, the ε-ball around x is a disk with center x and radius ε. The boundary[1], a circle with center x and radius ε, is *not* part

---

[1] The precise definition of "boundary" is not needed now and will be given in definition 199.

of the disk. The ε-cube is a square with center x with distances from the center to the sides equal to ε. Again, the sides are not part of the square (figure 27.1).

The situation in the Euclidean space $\mathbb{R}^3$ explains the terminology used. In fact, the ε-ball around x is the sphere with center x and radius ε and the ε-cube is the cube with center x, where the distances from the center to the sides are equal to ε, see figure 27.2.

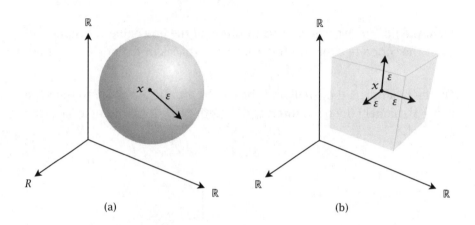

Fig. 27.2. The ε-ball (a) and ε-cube (b) around x in $\mathbb{R}^3$. The boundaries are *not* part of these sets.

The fact that both concepts, considered topologically, are in a sense equivalent, is embodied by the following lemma.

Lemma 230 *For a subset $O \subset \mathbb{R}^n$, the following properties are equivalent:*

(i) *For every $x \in O$, there is a real number $\varepsilon > 0$ such that $K_\varepsilon(x) \subset O$.*

(ii) *For every $x \in O$, there is a real number $\varepsilon > 0$ such that $B_\varepsilon(x) \subset O$.*

Proof Up to translation, it is sufficient to show that for every $\varepsilon > 0$, there is a positive real number δ such that $B_\delta(0) \subset K_\varepsilon(0)$, and conversely, there is a positive real number δ' such that $K_{\delta'}(0) \subset B_\varepsilon(0)$. For the first claim, take $\delta = \varepsilon$. Then $z = (z_1, \ldots z_n) \in B_\delta(0)$ means $\sum_i z_i^2 < \varepsilon^2$, so for every i, $|z_i| < \varepsilon$, i.e., $z \in K_\varepsilon(0)$. For the second claim, take $\delta' = \frac{\varepsilon}{\sqrt{n}}$. Then $z = (z_1, \ldots z_n) \in K_{\delta'}(0)$ means $|z_i| < \frac{\varepsilon}{\sqrt{n}}$, i.e., $\sum_i z_i^2 < n \cdot \frac{\varepsilon^2}{n}$, whence $\|z\| < \varepsilon$, i.e., $z \in B_\varepsilon(0)$. $\qquad\square$

Definition 176 *A subset $O \subset \mathbb{R}^n$ is called* open *(in $\mathbb{R}^n$), iff it has the equivalent properties from definition 230. A subset $C \subset \mathbb{R}^n$ is called* closed *(in $\mathbb{R}^n$), iff its complement $\mathbb{R}^n - C$ is open.*

Example 99 Figure 27.3 shows an open set O in $\mathbb{R}^2$ and illustrates alternative (ii) of lemma 230. Taking an arbitrary point x_1 in the open set, there is an open ball around x_1 (shown in dark gray) that is entirely contained in the open set. Two magnifications exhibit points x_2, x_3 and x_4 increasingly close to the boundary, but always an open ball can be found that lies within O, since the boundary of O is not part of O itself.

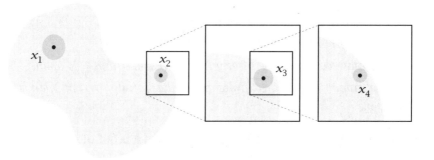

Fig. 27.3. An open set in $\mathbb{R}^2$.

In contrast, figure 27.4 shows the same set, but now it includes its boundary. Again an open ball around x_1 lies within the set, but choosing a point x_2 *on* the boundary, no ε-ball can be found that is entirely contained in the set, however small ε may be. Thus this set cannot be open. In fact, it is closed, as its complement is open.

Note that there are sets that are *both* open and closed. In $\mathbb{R}^n$ the entire set $\mathbb{R}^n$ and the empty set $\varnothing$ are both open and closed. There are also sets that are *neither* open nor closed, for example, in $\mathbb{R}$, the interval $[a, b[$ that includes a, but not b, is neither open nor closed.

Exercise 133 Show that every ball $B_\varepsilon(x)$ and every cube $K_\varepsilon(x)$ is open.

Exercise 134 Use the triangle inequality for distance functions (volume 1, proposition 213) to show that the intersection of any two balls $B_{\varepsilon_x}(x)$, $B_{\varepsilon_y}(y)$ and any two cubes $K_{\varepsilon_x}(x)$, $K_{\varepsilon_y}(y)$ is open.

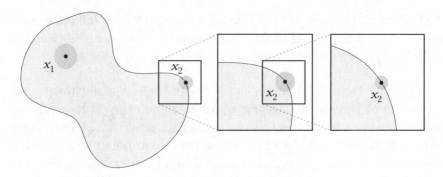

Fig. 27.4. A closed set in $\mathbb{R}^2$.

Sorite 231 *We are considering subsets of $\mathbb{R}^n$. Then:*

(i) *The empty set $\varnothing$ and the total space $\mathbb{R}^n$ are open.*

(ii) *The intersection $U \cap V$ of any two open sets U and V is open.*

(iii) *The union $\bigcup_\iota U_\iota$ of any (finite or infinite) family $(U_\iota)_\iota$ of open sets is open.*

Exercise 135 Use exercises 133 and 134 to give a proof of the properties of sorite 231.

Remark 30 More generally, a *topology* on a set X is a set $\mathcal{T}$ of subsets of X satisfying as axioms the properties of sorite 231.

Example 100 Here is a seemingly exotic, but crucial relation to logical algebras: The set *Open*$(\mathbb{R}^n)$ of open sets in $\mathbb{R}^n$ becomes a Heyting algebra by the following definitions: The maximum and minimum are $\mathbb{R}^n$ and $\varnothing$, respectively, the meet $U \wedge V$ is the intersection $U \cap V$, the join $U \vee V$ is the union $U \cup V$, and the implication $U \Rightarrow V$ is the union $\bigcup_{O \cap U \subseteq V} O$. (Give a proof of the Heyting properties thus defined.)

Classical two-valued logic: For any non-empty set A, consider the topology consisting of the open sets $\bot = \varnothing$ and $\top = A$. With $\vee$ and $\wedge$ as above, define $\neg U = (U \Rightarrow \bot)$. Then $\neg\top = \bigcup_{O \cap \top \subseteq \bot} O = \bot$ and $\neg\bot = \bigcup_{O \cap \bot \subseteq \top} O = \top$. These definitions satisfy the properties of a Boolean algebra.

A three-valued logic: We choose a set A, with the topology consisting of the open sets $\bot = \varnothing$, $\top = A$ and a third set X, with $X \neq \varnothing$ and $X \neq A$. Again $\neg U = (U \Rightarrow \bot)$, and we have: $\neg\top = \bot$, $\neg\bot = \top$ and $\neg X = \bot$. This last equation shows that this logic is not a Boolean algebra, since it is not the case that $x = \neg\neg x$ for all x.

A fuzzy logic: Let $A = [0, 1[$ with the topology of all intervals $I_x = [0, x[\subset A$. We have $I_x \lor I_y = I_{\max(x,y)}$ and $I_x \land I_y = I_{\min(x,y)}$, as well as $\bot = \varnothing$ and $\top = A$. The implication is $I_x \Rightarrow I_y = \top$, if $x \le y$, and $I_x \Rightarrow I_y = I_y$, if $x > y$. This logic is not Boolean either.

The next definition establishes the connection to convergent sequences.

Definition 177 *A sequence $(c_i)_i$ of elements in $\mathbb{R}^n$ is called* convergent *if there is a vector $c \in \mathbb{R}^n$ such that for every $\varepsilon > 0$, there is an index N with $c_i \in B_\varepsilon(c)$ for $i > N$. Equivalently, we may require that for every $\varepsilon > 0$, there is an index M with $c_i \in K_\varepsilon(c)$ for $i > M$. If $(c_i)_i$ converges to c, one writes $\lim_{i \to \infty} c_i = c$. A sequence which does not converge is called* divergent.

A sequence $(c_i)_i$ of elements in $\mathbb{R}^n$ is called a Cauchy *sequence, if for every $\varepsilon > 0$, there is an index N with $c_i \in B_\varepsilon(c_j)$ for $i, j > N$. Equivalently, we may require that for every $\varepsilon > 0$, there is an index M with $c_i \in K_\varepsilon(c_j)$ for $i, j > M$.*

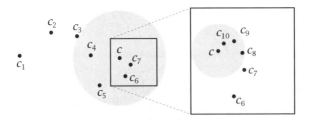

Fig. 27.5. The sequence $(c_i)_i$ converges to c. A given ε-ball around c contains all c_i for $i > 3$. In the magnification, another, smaller, ε-ball contains all c_i for $i > 7$.

Observe that this definition coincides with the already known concept of convergent and Cauchy sequences in the case $n = 1$. For example, because the ε-cube around x corresponds to the interval $]x - \varepsilon, x + \varepsilon[$ in $\mathbb{R}$, the expression $c_i \in K_\varepsilon(c_j)$ corresponds to $c_i \in]c_j - \varepsilon, c_j + \varepsilon[$, which in turn is equivalent to $|c_i - c_j| < \varepsilon$.

Exercise 136 Give a proof of the claimed equivalences in definition 177.

Convergence of a sequence in $\mathbb{R}^n$ is equivalent to the convergence of each of its component sequences:

Proposition 232 *For a sequence $(c_i)_i$ of elements in $\mathbb{R}^n$, and $j = 1, 2, \ldots n$, we denote by $(c_{i,j})_i$ the j-th projection of $(c_i)_i$, whose i-th member $c_{i,j}$ is the j-th coordinate of the vector c_i. Then $(c_i)_i$ is convergent (Cauchy), iff all its projections $(c_{i,j})_i$ for $j = 1, 2, \ldots n$ are so. Therefore, a sequence is convergent, iff it is Cauchy, and then the limit $\lim_{i \to \infty} c_i$ is uniquely determined. It is in fact the vector whose coordinates are the limits of the coordinate sequences, i.e., $(\lim_{i \to \infty} c_i)_j = \lim_{i \to \infty} c_{i,j}$.*

Proof We make use of the characterization in definition 177 of convergent or Cauchy sequences by means of cubes $K_\varepsilon(x)$. In this setting, $y \in K_\varepsilon(x)$ is equivalent to $y_j \in K_\varepsilon(x_j)$ for all projections y_j, x_j of the vectors $y = (y_1, \ldots y_n), x = (x_1, \ldots x_n)$ for $j = 1, \ldots n$. The claims follow immediately from this fact. □

Convergent sequences provide an important characterization of closed sets:

Proposition 233 *For a subset $C \subset \mathbb{R}^n$, the following two properties are equivalent:*

(i) *The set C is closed.*

(ii) *Every Cauchy sequence $(c_i)_i$ with members $c_i \in C$ has its limit $\lim_{i \to \infty} c_i$ in C.*

Proof Suppose that C is closed and assume that the limit $c = \lim_{i \to \infty} c_i$ is in the open complement $D = \mathbb{R}^n - C$. Then there is an open ε-ball $B_\varepsilon(c) \subset D$. But there is an index N such that $i \geq N$ implies $c_i \in B_\varepsilon(c)$, a contradiction to the hypothesis that all c_i are in C. Suppose that C is not closed. Then D is not open. So there is an element $c \in D$ such that for every $i \in \mathbb{N}$, there is an element $c_i \in B_{\frac{1}{i+1}}(c) \cap C$. But then the sequence $(c_i)_i$ converges to c. □

Not every sequence is convergent, but if its members are bounded, we may extract a convergent "subsequence" from it. Boundedness is defined as follows:

Definition 178 *A bounded sequence is a sequence $(c_i)_i$ such that there is a real number R such that for all i, $c_i \in B_R(0)$.*

Intuitively for a bounded sequence, one can find a ball, such that the entire sequence lies within this ball, i.e., members of the sequence do not "grow indefinitely". Here is an important class of bounded sequences:

Lemma 234 *A Cauchy sequence is bounded.*

Proof This is immediate. □

Of course, the converse is false, as can be seen in the trivial example $(c_i = (-1)^i)_i$, whose members all lie in the open interval between -2 and 2. But we may extract parts of bounded sequences which are Cauchy:

Definition 179 *For a sequence $(c_i)_i$, a subsequence $(d_i)_i$ of $(c_i)_i$ is a sequence $(d_i)_i$ defined by an ordered injection $s : \mathbb{N} \to \mathbb{N}$, i.e., $n < m$ implies $s(n) < s(m)$, by means of $d_i = c_{s(i)}$.*

Exercise 137 Show that a subsequence $(e_i)_i$ of a subsequence $(d_i)_i$ of a sequence $(c_i)_i$ is a subsequence of $(c_i)_i$.

Proposition 235 (Bolzano-Weierstrass) *Every bounded sequence $(c_i)_i$ has a convergent subsequence.*

Proof For the proof of this theorem, we need auxiliary closed sets, namely closed cubes. A closed cube is a set of the form $K = \prod_{i=1,2,...n}[a_i, b_i]$ for a sequence $a_i < b_i$ of pairs of real numbers. Such a cube K is the union of 2^n closed subcubes K^j, with $j = 1, 2, \ldots 2^n$, where each cube is defined by either the lower interval $[a_i, (a_i + b_i)/2]$ or the upper interval $[(a_i + b_i)/2, b_i]$ in the i-th coordinate. Clearly, the successive subdivision cubes $K^{j_1, j_2, \ldots j_k}$ are contained in cubes $K_\varepsilon(x)$ for any positive ε as k tends to infinity. Now, since $(c_i)_i$ is bounded, it is contained in a closed cube K. We define our convergent subsequence: Begin by taking $d_0 = c_0$. Then one of the subdivision cubes K^{j_1} contains the c_i for an infinity of indices. Take $d_1 = c_{i_1}$ with the first index $i_1 > 0$ such that $c_{i_1} \in K^{j_1}$. Then at least one of its subdivision cubes K^{j_1, j_2} contains the c_i for an infinity of indexes larger than i_1. Take the first index i_2 such that $c_{i_2} \in K^{j_1, j_2}$ and set $d_2 = c_{i_2}$. Proceeding with this procedure, we thereby define a subsequence $(d_i)_i$ of $(c_i)_i$ which is contained in progressively smaller subdivision cubes. This is a Cauchy sequence, and the proposition is proved. □

Example 101 Figure 27.6 shows a bounded sequence, where the upper and lower bounds are indicated by dashed lines. A convergent subsequence is emphasized through heavy dots.

A sequence contained in a closed set C doesn't necessarily contain any converging subsequence, an example being the sequence $(c_i = i)_i$ of natural numbers, contained in the closed set $\mathbb{R}$. But if the closed set C is bounded, i.e., if there is a radius R such that $x \in B_R(0)$ for all $x \in C$, then a fortiori, any sequence in C is bounded. But then, by the Bolzano-Weierstrass theorem, it has a convergent subsequence and its limit must

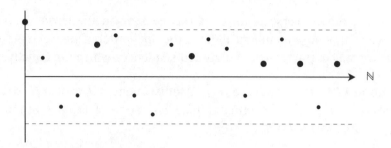

Fig. 27.6. A convergent subsequence (heavy dots) of a bounded sequence.

be an element of C by proposition 233. So every sequence in C has a convergent subsequence which converges within C! This type of closed sets is extremely important in the entire calculus and deserves its own name.

Proposition 236 *For a subset $C \subset \mathbb{R}^n$, the following properties are equivalent:*

(i) *The set C is closed and bounded.*

(ii) *Every sequence $(c_i)_i$ in C has a subsequence which converges to a point in C.*

(iii) *If $(U_i)_i$ is a (finite or infinite) family of open sets such that $C \subset \bigcup_i U_i$ (a so-called* open covering *of C), then there is a finite subfamily $U_{i_1}, \ldots U_{i_k}$ which also covers C, i.e., $C \subset \bigcup_j U_{i_j}$ (a subcovering of $(U_i)_i$).*

Proof (i) implies (ii): Let C be closed and bounded. A sequence $(c_i)_i$ in C has a convergent subsequence by proposition 235. Since C is closed, the limit of the subsequence is in C by proposition 233.

(ii) implies (i): If C is not bounded, then, evidently, there is a sequence $(c_i)_i$ which tends to infinity, so no subsequence can converge. If C is not closed, again by proposition 233, it contains a Cauchy sequence $(c_i)_i$ which has its limit outside C. But then every subsequence of this sequence converges to the same point outside C.

Let us now prove the equivalence of the first and third properties.

(iii) implies (i): If C is not bounded, then the open covering $(U_i = K_{i+1}(0))_i$ of $\mathbb{R}^n$ has no finite subcovering containing C. If C is bounded, but not closed, then let $x = (x_1, \ldots x_n) \notin C$ be a point such that $K_{\frac{1}{2^j}}(x) \cap C \neq \emptyset$ for all $j \in \mathbb{N}$. Take the following open covering of C. Start with the open set $U_0 = \mathbb{R}^n - \prod_i [x_i - 1, x_i + 1]$, complement of the closed cube $\prod_i [x_i - 1, x_i + 1]$. Then take the open set, $U_j = K_2(x) - \prod_i [x_i - \frac{1}{2^j}, x_i + \frac{1}{2^j}]$ for $j = 1, 2, \ldots$. This family of open sets covers $\mathbb{R}^n - \{x\}$, hence also C, but, because of the choice of x, none of its finite subcoverings contains all of C

(i) implies (iii): The converse is more delicate. Suppose that C is closed and bounded. The strategy is this: We first construct a denumerable subcovering $(U_{i_j})_{j \in \mathbb{N}}$ of C. Suppose that no finite subfamily covers C. Then for all finite sub-families $U_{i_0}, U_{i_1}, \ldots U_{i_m}$, there is an element $c_m \in C - \bigcup_{j=0}^m U_{i_j}$. Since C is closed and bounded, we may even suppose that $(c_m)_m$ converges to $c \in C$. But then there is an open set $U_{i_{m_0}}$ which contains c, since $(U_{i_j})_{j \in \mathbb{N}}$ covers C. This means, by construction of $(c_m)_m$, that the members of this convergent sequence stay outside some open cube $K_\varepsilon(c)$ for $m \to \infty$, a contradiction. We now construct a denumerable subcovering of C. Clearly, if U_k is a member of our covering and if $x \in U_k$, then there is an open cube $K = \prod_i]\xi_i, \eta_i[$ which is contained in U_k, contains x, and such that its interval points ξ_i, η_i are all rational numbers. So U_k is covered by a family of open cubes with rational boundary numbers. The denumerable family $(K_r)_{r \in \mathbb{N}}$ of all these cubes, when summed up for all U_k of the given covering, also cover C, and each K_r is contained in an open set $U_{k(r)}$. Therefore the open subcovering $(U_{k(r)})_r$ is denumerable, what was claimed, and we are done. □

Definition 180 *A set $C \subset \mathbb{R}^n$ is called* compact, *iff it has the equivalent properties described in proposition 236.*

Exercise 138 Show that a compact set in $\mathbb{R}$ has a minimum and a maximum.

Proposition 237 *The Cartesian product $X \times Y \subset \mathbb{R}^{m+n}$ of two compact sets $X \subset \mathbb{R}^m$ and $Y \subset \mathbb{R}^n$ is compact.*

Proof First, it is clear that the Cartesian product of two bounded sets is bounded. Next, we show that the complement of $X \times Y$ is open. We have $\mathbb{R}^{m+n} - X \times Y = ((\mathbb{R}^m - X) \times \mathbb{R}^n) \cup (\mathbb{R}^m \times (\mathbb{R}^n - Y))$. We show that $(\mathbb{R}^m - X) \times \mathbb{R}^n$ is open, the other set $X \times (\mathbb{R}^n - Y)$ being then open for the same reason after exchanging left and right factors. Now, let $(x, y) \in (\mathbb{R}^m - X) \times \mathbb{R}^n$. Then there is a cube $K_\varepsilon(x) \subset \mathbb{R}^m - X$ in $\mathbb{R}^m$. Since no conditions are imposed on y, we have $(x, y) \in K_\varepsilon(x, y) \subset (\mathbb{R}^m - X) \times \mathbb{R}^n$, so $(\mathbb{R}^m - X) \times \mathbb{R}^n$ is open. □

Definition 181 *For a real number $\varepsilon > 0$ and $x \in \mathbb{R}^n$, the* closed ball $\overline{B}_\varepsilon(x)$ *is defined by $\overline{B}_\varepsilon(x) = \{y \mid d(x, y) \leq \varepsilon\}$. A* closed cube *in $\mathbb{R}^n$ is a set $[a_1, b_1] \times [a_2, b_2] \times \ldots [a_n, b_n]$ for pairs $a_i \leq b_i, i = 1, 2, \ldots n$. In particular, we have a closed cube $\overline{K}_\varepsilon(x) = [x_1 - \varepsilon, x_1 + \varepsilon] \times [x_2 - \varepsilon, x_2 + \varepsilon] \times \ldots [x_n - \varepsilon, x_n + \varepsilon]$.*

Exercise 139 Show that in $\mathbb{R}^n$ a closed cube $[a_1, b_1] \times \ldots [a_n, b_n]$ for pairs $a_i \le b_i, i = 1, \ldots n$, as well as a closed ball $\overline{B}_\varepsilon(x)$ are compact.

Example 102 The upper half-plane $H^+ = \{(x, y) \in \mathbb{R}^2 \mid y \ge 0\}$ is a closed set in $\mathbb{R}^2$, since its complement $H^- = \{(x, y) \in \mathbb{R}^2 \mid y < 0\}$ is an open set. However H^+ is not compact, since it is not bounded (property (i)). Alternatively, we find a sequence $(c_i)_i = ((0, i))_i$ that has no convergent subsequence in H^+ (property (ii)).

The subset of integers $\mathbb{Z}$ in $\mathbb{R}$ is closed. In fact its complement in $\mathbb{R}$ is $\mathbb{Z}^C = \bigcup_{i=-\infty}^{\infty}]i, i+1[$, which is an open set, since it is the union of open intervals. $\mathbb{Z}$ is not bounded, hence not compact. Also, $U = \bigcup_{i=-\infty}^{\infty}]i - \varepsilon, i + \varepsilon[$, where $\varepsilon < 1$, is an open covering of $\mathbb{Z}$, but U contains no finite subcovering of $\mathbb{Z}$, thus property (iii) is violated.

In contrast, every closed disk $\overline{B}_r(x_0) = \{x \mid d(x_0, x) \le r\}$, and every finite union of such closed disks, is compact.

27.3 Continuity

So far, we have only dealt with topological considerations on all of $\mathbb{R}^n$. In most practical cases, we do not have all of $\mathbb{R}^n$ at hand. For example, a function may be defined only on a closed interval of $\mathbb{R}$, or even only on an interval of type $]0, 1]$, such as $f(x) = 1/x$. When applying topological considerations to such functions, we would like to deal strictly with what happens within their domains. Also, when composing two functions, the specific codomains and domains should coincide, as it is required for the composition of set functions. So we are forced to set up a minimal conceptual environment to apply topology to set functions.[2]

---

[2] This small extra effort will pay off: We obtain a "category" of topological spaces, i.e., topologically reasonable maps, the possibility to compose such maps and to compare topologically specified sets by means of such maps. Compare the category of matrixes, the category of sets and set maps, the category of modules and linear homomorphisms, the category of digraphs, the category of acceptors,... Later in chapter 36, we shall give a systematic account of such a conceptualization. For the moment, you just have to recognize that the present topological considerations are completely integrated within a big program of building categories of mathematical objects in order to obtain a global control of mathematical structures.

Definition 182 *Given a subset $X \subset \mathbb{R}^n$, a subset $U \subset X$ is called* open
(closed) *in X, iff there is an open (closed) set $O \subset \mathbb{R}^n$ such that $U = X \cap O$. The set of open sets in X is also called the* relative topology *of X. In particular, we write $B_\varepsilon^X(x) = B_\varepsilon(x) \cap X$ and $K_\varepsilon^X(x) = K_\varepsilon(x) \cap X$ for the restrictions of the open balls and cubes, respectively, and call these open sets in X the* open ball, *or* open cube, *in X, respectively.*

Exercise 140 Show that for a given subset X of $\mathbb{R}^n$, the properties of sorite 231 are true for the open sets in X, where X plays the role of the "total space". Moreover, show that the closed sets in X are precisely the complements in X of the open sets in X.

Lemma 238 *If $X \subset \mathbb{R}^m$ and $Y \subset \mathbb{R}^n$ are two subsets of Euclidean spaces, and if $f : X \to Y$ is a set map, then the following properties are equivalent.*

(i) *The inverse image $f^{-1}(U)$ of any open set U in Y is open in X.*

(ii) *For any point $x \in X$ and for any positive real number ε, there is a positive real number δ (generally depending on x and on ε) such that $f(B_\delta^X(x)) \subset B_\varepsilon^Y(f(x))$.*

(iii) *For any point $x \in X$ and for any positive real number ε, there is a positive real number δ (generally depending on x and on ε) such that $f(K_\delta^X(x)) \subset K_\varepsilon^Y(f(x))$.*

(iv) *The inverse image $f^{-1}(U)$ of any closed set U in Y is closed in X.*

(v) *For any point $x \in X$ and for any convergent sequence $(c_i)_i$ with $\lim_{i \to \infty} c_i = x$, the image sequence $(f(c_i))_i$ converges to $f(x)$.*

Proof (i) implies (ii): Since $f^{-1}(B_\varepsilon^Y(f(x)))$ is open and contains x, there is an open ball $B_\delta^X(x) \subset f^{-1}(B_\varepsilon^Y(f(x)))$. Therefore $f(B_\delta^X(x)) \subset B_\varepsilon^Y(f(x))$.

(ii) implies (i): Since every open set U is the union of open balls, its inverse image is the union of inverse images of open balls. But by (ii), the inverse image of an open ball is a union of open balls, and, therefore, open, whence (i).

The same argument yields the equivalence of (i) and (iii).

The equivalence of (i) and (iv) results from the set-theoretic fact that complements and inverse images commute.

(ii) implies (v): Let $\varepsilon > 0$. Then there is $\delta > 0$ such that $f(B_\delta^X(x)) \subset B_\varepsilon^Y(f(x))$. So by the convergence of $(c_i)_i$, there is a natural number N such that $i \geq N$ implies $c_i \in B_\delta^X(x)$. Hence $i \geq N$ implies $f(c_i) \in B_\varepsilon^Y(f(x))$, therefore $(f(c_i))_i$ converges to $f(x)$.

(v) implies (ii): Suppose (ii) is false for an $x \in X$. Then there is $\varepsilon_0 > 0$ such that for every $i \in \mathbb{N}$, there is $c_i \in B_{\frac{1}{i+1}}(x)$ with $f(c_i) \notin B_{\varepsilon_0}(f(x))$. But the sequence

$(c_i)_i$ evidently converges to x, while the images $f(c_i)$ stay outside the open ball $B_{\varepsilon_0}(f(x))$, which contradicts (v). □

Definition 183 *If $X \subset \mathbb{R}^m$ and $Y \subset \mathbb{R}^n$ are two subsets of Euclidean spaces, a set map $f : X \to Y$ with the equivalent properties of lemma 238 is called* continuous. *A continuous bijection f such that its inverse f^{-1} is also continuous, is called a* homeomorphism. *The set of continuous maps $f : X \to Y$ is denoted by $Top(X,Y)$. In particular, if $Y = \mathbb{R}$, one writes $C^0(X) = Top(X, \mathbb{R})$. If in lemma 238, the conditions (ii) to (iv) are valid for a specific point x only, f is called* continuous in x. *This means that f is continuous, iff it is continuous in every x of its domain.*

Example 103 To illustrate property (i) of lemma 238, it is best to show a case where the property fails. In figure 27.7, the function f is non-continuous, as is clear by the jump at the argument x. The value at x, $f(x)$, is indicated by the heavy dot. Now, the inverse image $f^{-1}(U)$ of the open interval U, is not open, but a half-open interval, i.e., open at the left and closed at the right with x as the endpoint.

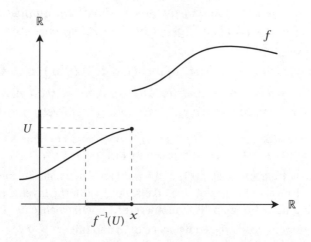

Fig. 27.7. The function f being *non*-continuous, the inverse image of the open set U is not open, in fact it is a half-open interval.

Sorite 239 *Let $X \subset \mathbb{R}^m, Y \subset \mathbb{R}^n, V \subset \mathbb{R}^s, W \subset \mathbb{R}^t, Z \subset \mathbb{R}^l$ be subsets of Euclidean spaces.*

(i) *The identity Id_X is always a homeomorphism.*

(ii) *If $f : X \to Y$ and $g : Y \to Z$ are continuous, then their composition $g \circ f$ is also continuous.*

(iii) *If $f : X \to Y$, $g : Y \to Z$ and $h : Z \to W$ are continuous, then $(h \circ g) \circ f = h \circ (g \circ f) = h \circ g \circ f$.*

(iv) *If $f : X \to Y$ and $u : U \to V$ are continuous, then so is the Cartesian product map $f \times u : X \times U \to Y \times V$.*

(v) *The projections $pr_X : X \times Y \to X$ and $pr_Y : X \times Y \to Y$ are continuous.*

(vi) *If $a : U \to X$ and $b : U \to Y$ are continuous maps, then so is the universal map $(a, b) : U \to X \times Y$ associated with a and b (see volume 1, proposition 57).*

Proof (i) is evident.

(ii) follows from the fact that for an open set $U \subset Z$, we have $(g \circ f)^{-1}(U) = g^{-1}(f^{-1}(U))$, and since $V = f^{-1}(U)$ is an open set so is $g^{-1}(V)$.

(iii) Associativity is clear, since it is true for any set maps.

(iv) It suffices to show that the inverse image of an open cube $K_\varepsilon^{X \times Y}(x, y)$ under $f \times u$ is open. But we have $K_\varepsilon^{X \times Y}(x, y) = K_\varepsilon^X(x) \times K_\varepsilon^Y(y)$, therefore $(f \times u)^{-1}(K_\varepsilon^{X \times Y}(x, y)) = f^{-1}(K_\varepsilon^X(x)) \times u^{-1}(K_\varepsilon^Y(y))$, and this is open.

For (v), observe that the cube $K_\varepsilon^{X \times Y}(x, y)$ is mapped by pr_X into the cube $K_\varepsilon^X(x)$, since cube elements are characterized coordinatewise. Similarly for the second projection.

As to (vi), if at a point $v \in U$, we have $a(K_\delta(v)) \subset K_\varepsilon(a(v))$ and $b(K_\delta(v)) \subset K_\varepsilon(b(v))$, then $(a, b)(K_\delta(v)) \subset K_\varepsilon^X(a(v)) \times K_\varepsilon^Y(b(v)) = K_\varepsilon^{X \times Y}((a, b)(v))$. $\quad\square$

These seemingly innocent general properties of continuous maps have a large number of very important consequences concerning the continuity of functions which are known from the theory of polynomials and from linear geometry. The crucial fact is this:

Lemma 240 *The maps of addition, $+ : \mathbb{R} \times \mathbb{R} \to \mathbb{R}$, and multiplication, $\cdot : \mathbb{R} \times \mathbb{R} \to \mathbb{R}$, are continuous. The inversion $?^{-1} : \mathbb{R}^* \to \mathbb{R}^*$ is continuous.*

Proof By the basic properties of the real number arithmetic, we have, for the addition, $+(K_{\varepsilon/2}(x, y)) \subset K_\varepsilon(x + y)$.

For the product $x \cdot y$, we have $|(x + v) \cdot (y + \mu) - x \cdot y| \le |x||\mu| + |y||v| + |\mu||v|$. If $xy \ne 0$, take $\delta = \min\{\varepsilon/3|x|, \varepsilon/3|y|, \sqrt{\varepsilon/3}\}$. If $x = 0$ and $y \ne 0$, then take $\delta = \min\{\varepsilon/2|y|, \sqrt{\varepsilon/2}\}$. If $x \ne 0$ and $y = 0$, then take $\delta = \min\{\varepsilon/2|x|, \sqrt{\varepsilon/2}\}$. If $x = y = 0$, take $\delta = \sqrt{\varepsilon}$. We then obtain $(K_\delta(x, y)) \subset K_\varepsilon(x \cdot y)$.

The third statement is left as an exercise for the reader. $\quad\square$

Proposition 241 *A polynomial function* $P : \mathbb{R}^n \to \mathbb{R}$ *defined by a polynomial* $P \in \mathbb{R}[X_1, \ldots X_n]$ *of* n *variables* $X_1, \ldots X_n$ *is continuous.*

Proof If $P = a \in \mathbb{R}$ is a constant, the polynomial function $P : \mathbb{R}^n \to \mathbb{R}$ is constant, and this is evidently continuous. In general, if $f_i : \mathbb{R}^n \to \mathbb{R}$, for $i = 1, \ldots k$, are continuous, then, by proposition 240, their sum $\sum_i f_i : \mathbb{R}^n \to \mathbb{R} : x \mapsto \sum_i f_i(x)$ and their product $\prod_i f_i : \mathbb{R}^n \to \mathbb{R} : x \mapsto \prod_i f_i(x)$ are continuous since we have $\sum_i f_i = \sum \circ (f_1, \ldots f_k)$ and $\prod_i f_i = \prod \circ (f_1, \ldots f_k)$, where $(f_1, \ldots f_k) : \mathbb{R}^n \to \mathbb{R}^k$ is the universal map of Cartesian products, and where $\sum$ and $\prod$ are the continuous k-fold sum and product maps. But the polynomial function P is the sum of its monomials, so it is continuous if the monomials are so. Further, each monomial $aX^{n_1} \ldots X^{n_t}$ in P is a product of the constant a and the projection functions X_j, which are all continuous by sorite 239, hence P is continuous. $\qquad \square$

Exercise 141 Give a proof of proposition 241 for the polynomial $P = 2X_1^2 - X_2 \cdot X_3 + 1.5$ using sorite 239 and lemma 240.

Lemma 242 *The maps of addition and multiplication,* $+ : \mathbb{C} \times \mathbb{C} \to \mathbb{C}$ *and* $\cdot : \mathbb{C} \times \mathbb{C} \to \mathbb{C}$ *are continuous for the complex numbers, where we interpret* $\mathbb{C}$ *as the real vector space* $\mathbb{R}^2$ *to define its topology. Conjugation* $\overline{?} : \mathbb{C} \to \mathbb{C}$ *of complex numbers is a homeomorphism.*

Proof This follows immediately since these operations, when rewritten in real coordinates, are polynomial functions. So proposition 241 applies. $\qquad \square$

Using the above general facts from sorite 239, we deduce the following theorem about continuity of matrix operations. This requires that matrixes $M \in \mathbb{M}_{m,n}(\mathbb{R})$ are viewed as vectors in some Euclidean space. We do this in the usual way by the well-known identification of $\mathbb{M}_{m,n}(\mathbb{R})$ with $\mathbb{R}^{mn}$, the Euclidean structure on $\mathbb{M}_{m,n}(\mathbb{R})$ being induced from the Euclidean structure on $\mathbb{R}^{mn}$. For example, the norm of a matrix $M = (M_{i,j})$ is $\|M\| = \sqrt{\sum_{i,j} M_{i,j}^2}$.

Proposition 243 *The following maps are all continuous:*

(i) *Addition* $+ : \mathbb{M}_{m,n}(\mathbb{R}) \times \mathbb{M}_{m,n}(\mathbb{R}) \to \mathbb{M}_{m,n}(\mathbb{R}) : (M, N) \mapsto M + N$,

(ii) *Multiplication* $\cdot : \mathbb{M}_{l,m}(\mathbb{R}) \times \mathbb{M}_{m,n}(\mathbb{R}) \to \mathbb{M}_{l,n}(\mathbb{R}) : (M, N) \mapsto M \cdot N$,

(iii) *Scalar product* $(?, ?) : \mathbb{R}^n \times \mathbb{R}^n \to \mathbb{R}$,

(iv) *Scalar multiplication* $\mathbb{R} \times \mathbb{M}_{m,n}(\mathbb{R}) \to \mathbb{M}_{m,n}(\mathbb{R}) : (t, M) \mapsto t \cdot M$,

(v) *Determinant function* $\det : \mathbb{M}_{n,n}(\mathbb{R}) \to \mathbb{R}$,

(vi) *Matrix transposition* $\tau : \mathbb{M}_{m,n}(\mathbb{R}) \to \mathbb{M}_{n,m}(\mathbb{R})$,

(vii) *Matrix adjunction* $Ad : \mathbb{M}_{n,n}(\mathbb{R}) \to \mathbb{M}_{n,n}(\mathbb{R})$.

Proof All the claims of this proposition are immediate from the polynomial character of the involved functions and their combinations, following sorite 239 and proposition 241. We leave the details to the reader as a useful exercise. □

A central fact about continuous maps is

Proposition 244 *The image $f(X) \subset \mathbb{R}^m$ of a compact set $X \subset \mathbb{R}^n$ under a continuous map $f : X \to \mathbb{R}^m$ is compact.*

Proof Let $(U_i)_i$ be an open covering of $f(X)$. Then the inverse image of $f(X) = \bigcup_i U_i$ is $X = \bigcup_i f^{-1}(U_i)$, an open covering of X. So there is a finite subcovering $\bigcup_{j=1}^{J} f^{-1}(U_{i_j})$ of X. Therefore $f(X) = f(\bigcup_{j=1}^{J} f^{-1}(U_{i_j})) = \bigcup_{j=1}^{J} f(f^{-1}(U_{i_j})) \subset \bigcup_{j=1}^{J} U_{i_j} \subset f(X)$, so we obtain the finite subcovering $\bigcup_{j=1}^{J} U_{i_j}$ of $f(X)$. □

In particular, by exercise 138, if we are given a continuous function $f : X \to \mathbb{R}$ on a compact set X, there are two arguments $x, y \in X$ such that $f(x) \le f(z) \le f(y)$ for all $z \in X$, i.e., the minimum and maximum of $f(X)$ are obtained as function values. But we do not know whether all intermediate values are obtained. This property is guaranteed by the famous *intermediate value theorem (Zwischenwertsatz)* first proved by the German mathematician Bernhard Bolzano in 1817.

Proposition 245 (Bolzano) *If $K = [a_1, b_1] \times [a_2, b_2] \times \ldots [a_n, b_n]$ for pairs $a_i \le b_i, i = 1, 2, \ldots n$ is a closed cube in $\mathbb{R}^n$, and if $f : K \to \mathbb{R}$ is continuous, then $\mathrm{Im}(f)$ is a closed interval $[a, b]$, i.e., for each value c between the minimum $a = f(x)$ and maximum $b = f(y)$ of $\mathrm{Im}(f)$, there is an argument $z \in K$ such that $c = f(z)$.*

Proof By proposition 244, $f(K)$ is compact, i.e., closed and bounded by proposition 236. Therefore $b = \sup(f(K))$ is finite. But taking a sequence $(c_i)_i$ in $f(K)$ which converges to b, closedness of $f(K)$ implies that $b \in f(K)$. A similar argument works with $a = \inf(f(K))$, the infimum[3] of $f(K)$. Therefore there is a maximal and a minimal value for $f(K)$. Now, let $r = f(u) < s = f(v)$ for $u, v \in K$ be any two values in $f(K)$. We claim that any value $c \in [r, s]$ is taken by an argument $z \in K$, i.e., $c = f(z)$. Consider the map $y : [0, 1] \to K : \xi \mapsto \xi \cdot u + (1 - \xi) \cdot v$. This is evidently a continuous map, since it is even affine. The composition $g = f \cdot y : [0, 1] \to \mathbb{R}$ is continuous and we have $g(0) = r < g(1) = s$. So we have reduced the problem to a one-dimensional cube $[0, 1]$. Suppose that there is $x \in T = [r, s] - g([0, 1])$. Take the supremum y of T. Since $s \notin T$,

---

[3] The infimum of a non-empty set $A \subset \mathbb{R}$, which is bounded from below, i.e., there is $l < x$, for all $x \in A$, is the number $\inf(A) = -\sup(-A)$, where $-A = \{-x \mid x \in A\}$.

the supremum is an element of the interval $[r, s]$, and smaller than s. Take a sequence $(c_i)_i$, $c_i \in]y, s]$ which converges to y. By construction of y, there is a sequence $(d_i)_i$ in $[0, 1]$ with $f(d_i) = c_i$, for all i. But $[0, 1]$ is compact, so there is even a convergent subsequence $(e_i)_i$ of $(d_i)_i$, converging to $e \in [0, 1]$, say. But then, by continuity, $f(e) = f(\lim_{i \to \infty} e_i) = \lim_{i \to \infty} f(e_i) = y$, a contradiction. $\quad \square$

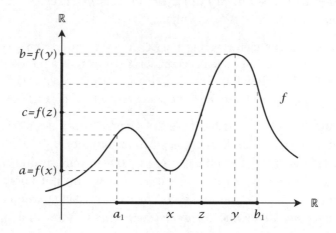

Fig. 27.8. Intermediate value theorem.

Recall that we used this result to prove proposition 219 in chapter 25.3 of volume 1.

Corollary 246 *For a polynomial $P \in \mathbb{R}[X]$ of odd degree, there is an argument $x \in \mathbb{R}$ such that $P(x) = 0$.*

Proof Since P is continuous, it suffices to find arguments $a, b \in \mathbb{R}$ such that $P(a) < 0$ and $P(b) > 0$. Let $P(x) = a_{2n+1}x^{2n+1} + a_{2n}x^{2n} + \ldots a_0$. We may evidently suppose $a_{2n+1} = 1$, since the general case follows immediately from this special case. For $x \neq 0$, we write $P(x) = x^{2n+1}(1 + \frac{a_{2n}}{x} + \ldots \frac{a_0}{x^{2n+1}})$. Consider positive natural numbers $x = i$ as arguments of P. If $i \to \infty$, then the summands $\frac{a_{2n}}{i} \ldots \frac{a_0}{i^{2n+1}}$ converge to 0. Therefore the factor $1 + \frac{a_{2n}}{i} + \ldots \frac{a_0}{i^{2n+1}}$ converges to 1. This implies that the product $P(i) = i^{2n+1}(1 + \frac{a_{2n}}{i} + \ldots \frac{a_0}{i^{2n+1}})$ tends to ∞ as $i \to \infty$. For integers $i < 0$, if $i \to -\infty$, then $P(i) = i^{2n+1}(1 + \frac{a_{2n}}{i} + \ldots \frac{a_0}{i^{2n+1}})$ tends to $i^{2n+1} < 0$, so we have positive and negative values and then, by proposition 245, there is an x such that $P(x) = 0$. $\quad \square$

This last result was used in chapter 25.1 of volume 1.

27.4 Series

This section introduces a more systematic study of sequences, and in particular sequences deduced from partial sums of given sequences. These series play a central role in the construction of basic continuous functions, but also, as we shall see later in this book under the title of Taylor series, in the reconstruction of quite general functions in terms of convergent sequences of polynomial functions.

To begin with, consider the real vector space $Sequ(\mathbb{R}, n) = (\mathbb{R}^n)^{\mathbb{N}}$ of sequences $(c_i)_i$ with values in $\mathbb{R}^n$. Recall that the sum and scalar multiplication are defined coordinatewise, i.e., $(c_i)_i + (d_i)_i = (c_i + d_i)_i$, and $\lambda(c_i)_i = (\lambda c_i)_i$ for $\lambda \in \mathbb{R}$. Denote by $C(\mathbb{R}, n)$ the subset of Cauchy or, equivalently, convergent sequences in $Sequ(\mathbb{R}, n)$.

Lemma 247 *The set $C(\mathbb{R}, n)$ is a vector subspace of $Sequ(\mathbb{R}, n)$. The map $\lim_{i \to \infty} : C(\mathbb{R}, n) \to \mathbb{R}^n : (c_i)_i \mapsto \lim_{i \to \infty} c_i$ is linear and its kernel is the sub-vector space $\mathcal{O}(\mathbb{R}, n)$ of zero sequences.*

Exercise 142 Give a proof of lemma 247. Check in particular that the statement of linearity of the map $\lim_{i \to \infty}$ is equivalent to the fact that limits of sums of sequences are the sums of their limits, whereas the product of a constant λ with the members of a convergent sequence converges to the scaling of the sequence limit by λ.

Definition 184 *Consider the following two linear endomorphisms Σ and Δ of $Sequ(\mathbb{R}, n)$:*

$$\Sigma : Sequ(\mathbb{R}, n) \to Sequ(\mathbb{R}, n) : (c_i)_i \mapsto \Sigma(c_i)_i,$$

the i-th member of $\Sigma(c_i)_i$ being $\Sigma(c_i) = \Sigma_{j=0}^{i} c_j$. The image sequence $\Sigma(c_i)_i$ is called the (associated) series *of $(c_i)_i$. And*

$$\Delta : Sequ(\mathbb{R}, n) \to Sequ(\mathbb{R}, n) : (c_i)_i \mapsto \Delta(c_i)_i,$$

the i-th member of $\Delta(c_i)_i$ being $\Delta(c_i) = c_i - c_{i-1}$ for positive i and $\Delta(c_0) = c_0$. The image $\Delta(c_i)_i$ is called the (associated) difference *of $(c_i)_i$.*

Lemma 248 *The endomorphisms Σ and Δ are automorphisms and inverses of each other, i.e.,*

$$\Delta \circ \Sigma = \Sigma \circ \Delta = Id_{Sequ(\mathbb{R}, n)}.$$

Proof This is immediate by a straightforward calculation, which we leave to the reader. □

This means that the inverse image $\Delta^{-1}C(\mathbb{R}, n)$ is the vector space of sequences having convergent series. If a series $\Sigma(c_i)_i$ converges, we write $\Sigma_{i=0}^{\infty}c_i$ for its limit. By the identification of Cauchy and convergent sequences, we have:

Proposition 249 *If a series $\Sigma(c_i)_i$ converges, then $(c_i)_i$ is a zero sequence, i.e,*

$$\Delta(C(\mathbb{R}, n)) \subset \mathcal{O}(\mathbb{R}, n).$$

Proof This follows from the Cauchy condition $|\Sigma(c_{i+1}) - \Sigma(c_i)| < \varepsilon$ for subsequent partial sums of sufficiently high index i of the series. □

Example 104 Given a real number q, the sequence $(q^i)_i$ gives rise to the *geometric series* $\Sigma(q^i)_i$ with general member

$$\Sigma(q^i) = 1 + q + q^2 + \dots q^i.$$

For $q \neq 1$, one has the formula $1 + q + q^2 + \dots q^i = \frac{1-q^{i+1}}{1-q}$. Since we have $\Sigma(q^i) = \frac{1}{1-q} - \frac{q^{i+1}}{1-q}$, convergence is a linear map, and the second summand converges to zero for $|q| < 1$, we have the very important formula

$$\Sigma_{i=0}^{\infty}q^i = \frac{1}{1-q}$$

for $|q| < 1$. Try to understand this result geometrically for the intuitive special value $q = \frac{1}{2}$.

But there are zero sequences without converging associated series:

Example 105 The *harmonic series* $\Sigma\left(\frac{1}{i+1}\right)_i$, with partial sums $\Sigma\left(\frac{1}{i+1}\right) = 1 + \frac{1}{2} + \dots \frac{1}{i+1}$, is divergent. Nonetheless, the very similar *alternating series* $\Sigma\left(\frac{(-1)^i}{i+1}\right)_i$ is convergent. This is a special case of the following *Leibniz criterion.*

Proposition 250 *If $(c_i)_i \in Sequ(\mathbb{R}, 1)$ is a zero sequence which is monotonously decreasing, i.e., $c_i \geq c_{i+1}$ for all i, then the alternating series $\Sigma((-1)^i c_i)_i$ converges.*

Proof We are given a series with $c_0 \geq c_1 \geq \ldots$ which converges to 0. Let us show by induction on N that the partial sums $S_N = \sum_{i=0}^{N} (-1)^i c_i$ satisfy $0 \leq S_N \leq c_0$. This is true for $N = 0, 1, 2$ by immediate check. In general, if N is even, we have $S_N = S_{N-2} - c_{N-1} + c_N$, whence $S_N \leq S_{N-2} \leq c_0$, but also $S_N = S_{N-1} + c_N \geq S_{N-1} \geq 0$. If N is odd, then $S_N = S_{N-2} + c_{N-1} - c_N$, whence $S_N \geq S_{N-2} \geq 0$, but also $S_N = S_{N-1} - c_N \leq S_{N-1} \leq c_0$. Now, Cauchy's criterion for convergence requires $|S_N - S_M| < \varepsilon$ for N, M sufficiently large. But $S_N - S_M$ is just a partial sum of such an alternating series starting from $m = \min(M, N)$. If this minimum is sufficiently large, by the above, the difference is limited by c_m, which converges to 0, so we are done. □

A partially converse criterion is the famous *criterion of absolute convergence*.

Definition 185 *A series $\Sigma(c_i)_i \in Sequ(\mathbb{R}, n)$ is said to be absolutely convergent if the series $\Sigma(\|c_i\|)_i \in Sequ(\mathbb{R}, 1)$ converges.*

Proposition 251 *An absolutely convergent series $\Sigma(c_i)_i \in Sequ(\mathbb{R}, n)$ is convergent.*

Proof Let $\Sigma(c_i)_i$ be absolutely convergent. Then for two indexes $N \leq M$, the triangle inequality in $\mathbb{R}^n$ yields $\|\Sigma(c_M) - \Sigma(c_N)\| = \|\sum_{i=N+1}^{M} c_i\| \leq \sum_{i=N+1}^{M} \|c_i\|$, and the latter is smaller than any positive ε for M, N sufficiently large by the absolute convergence hypothesis. Therefore the Cauchy criterion yields convergence of the series. □

The next criterion gives us a large variety of absolutely convergent series at hand:

Proposition 252 *If a series $\Sigma(c_i)_i \in Sequ(\mathbb{R}, n)$ is based on a sequence $(c_i)_i$ with non-zero members such that there is a real number $0 < q < 1$ with this property: There is a natural N such that $\frac{\|c_{i+1}\|}{\|c_i\|} \leq q$ for all $i > N$, then $\Sigma(c_i)_i$ is absolutely convergent.*

Proof Since the initial portion of a sequence is irrelevant for its convergence, we may suppose that $\frac{\|c_{i+1}\|}{\|c_i\|} \leq q$ for all $i \geq 0$. Then we have $\|c_i\| \leq q^i \|c_0\|$, all $i \in \mathbb{N}$. Therefore $\Sigma(\|c_i\|) \leq \|c_0\| \cdot (1 + q + q^2 + \ldots q^i)$ which is a convergent geometric series. □

For the next result, we again interpret complex numbers as vectors in $\mathbb{R}^2$ and accordingly consider sequences with members in $\mathbb{C}$ as series in the Euclidean space $\mathbb{R}^2$.

Corollary 253 *Given a complex number $z \in \mathbb{C}$, the* power series *(involving powers of z)* $\Sigma \left(\frac{z^k}{k!} \right)_k$ *is absolutely convergent. We therefore define the complex-valued function*

$$\exp(z) = \sum_{k=0}^{\infty} \frac{z^k}{k!}$$

which is called the exponential function.

Proof The absolute convergence follows immediately from the ratio

$$\frac{\frac{z^{k+1}}{(k+1)!}}{\frac{z^k}{(k)!}} = \frac{z}{k+1},$$

which tends to 0 for $k \to \infty$, and the proposition 252 applies. $\square$

27.4.1 Fundamental Properties of the Exponential Function

In this subsection, we want to deal with some technical aspects which are of general interest, but which are also crucial for the establishment of fundamental properties of the exponential function. In particular, we want to calculate the value $\exp(w + z)$, and since this involves the powers $(w + z)^k$ as functions of w and z, we need to calculate polynomials $(X + Y)^k \in \mathbb{Z}[X, Y]$ first. To this end, we need a formula for the coefficients of such polynomials. These coefficients will also play an important role in the calculus of probability, to name but one example. They are in fact omnipresent in mathematics as soon as it comes to the calculation of any combinatorial quantities.

Definition 186 *Let $0 \le k \le n$ be natural numbers. Then one sets*

$$\binom{n}{k} = \frac{n!}{k!(n-k)!} = \frac{n(n-1)(n-2)\ldots(n-k+1)}{k!}$$

(with the special value $0! = 1$) and calls this rational number the binomial coefficient *n over k.*

Here is the basic result which allows the inductive calculation of binomial coefficients:

Lemma 254 *For natural numbers $0 \le k < n$, we have*

$$\binom{n}{k} + \binom{n}{k+1} = \binom{n+1}{k+1}.$$

In particular, by induction on n, and observing that $\binom{n}{0} = 1$, it follows that binomial coefficients are integers.

Proof We have

$$\binom{n}{k} + \binom{n}{k+1} = \frac{n \cdot (n-1) \cdot \ldots (n-k+1) \cdot (k+1)}{k!(k+1)} +$$
$$\frac{n \cdot (n-1) \cdot \ldots (n-k+1) \cdot (n-k)}{k!(k+1)}$$
$$= \frac{n \cdot (n-1) \cdot \ldots (n-k+1)}{(k+1)!}((k+1) + (n-k))$$
$$= \binom{n+1}{k+1}.$$

□

The *Pascal triangle* (figure 27.9) is a graphical representation of the above result: We represent the binomial coefficients for a given n on a row and develop the coefficients from $n = 0$ on downwards. Observe the vertical symmetry axis in the triangle, which stems from the obvious fact that $\binom{n}{k} = \binom{n}{n-k}$.

$$
\begin{array}{ccccccccccccc}
& & & & & & 1 & & & & & & \\
& & & & & 1 & & 1 & & & & & \\
& & & & 1 & & 2 & & 1 & & & & \\
& & & 1 & & 3 & & 3 & & 1 & & & \\
& & 1 & & 4 & & 6 & & 4 & & 1 & & \\
& 1 & & 5 & & 10 & & 10 & & 5 & & 1 & \\
1 & & 6 & & 15 & & 20 & & 15 & & 6 & & 1 \\
\end{array}
$$

1   7   21   35   35   21   7   1

$\ldots$

Fig. 27.9. The Pascal triangle.

This yields the coefficients of $(X + Y)^n$ as follows:

Proposition 255 *If $n \in \mathbb{N}$, then the polynomial $(X + Y)^n \in \mathbb{Z}[X, Y]$ has this representation in terms of monomials:*

$$(X + Y)^n = \sum_{k=0}^{n} \binom{n}{k} X^{n-k} \cdot Y^k = X^n + nX^{n-1} \cdot Y + \ldots nX \cdot Y^{n-1} + Y^n.$$

Proof One proves the proposition by induction on n using the recursive formula from lemma 254. This is just an exercise in reindexing sums, we therefore omit it and refer to [14]. □

This allows us to regard the expression $\exp(w + z)$ as a series of the following products:

$$\exp(w + z) = \sum_{n=0}^{\infty} \sum_{k=0}^{n} \frac{1}{n!} \binom{n}{k} w^{n-k} z^k = \sum_{n=0}^{\infty} \sum_{k=0}^{n} \frac{1}{(n-k)!} w^{n-k} \frac{1}{k!} z^k.$$

So we are confronted with the problem of whether a product of series is the series of the products of their summands. This is precisely what the following proposition guarantees:

Proposition 256 *Identifying $C(\mathbb{C}, 1)$ with the vector space $C(\mathbb{R}, 2)$ over the Euclidean space $\mathbb{R}^2$, if $\Sigma(c_i)_i$ and $\Sigma(d_i)_i$ are absolutely convergent series in $C(\mathbb{C}, 1)$, then we have the Cauchy product formula*

$$\left(\sum_{i=0}^{\infty} c_i \right) \cdot \left(\sum_{i=0}^{\infty} d_i \right) = \sum_{i=0}^{\infty} \sum_{k=0}^{i} c_{i-k} d_k.$$

This is a special case of a formula guaranteeing that a series is absolutely convergent, iff it is "unconditionally" convergent, which means that it converges to the same limit for any permutation of the summation. We cannot delve into those details and refer to [14].

Proposition 256 implies the following result.

Proposition 257 *The map*

$$\exp : \mathbb{C} \to \mathbb{C}^*$$

is a surjective continuous group homomorphism from the additive group of complex numbers to the multiplicative group of non-zero complex numbers, i.e., $\exp(0) = 1$ and $\exp(w + z) = \exp(w) \cdot \exp(z)$ for all $w, z, \in \mathbb{C}$. There is a number $\pi = 3.1415926\ldots$ such that

$$Ker(\exp) = i2\pi\mathbb{Z}.$$

In particular, $\mathbb{C}/i2\pi\mathbb{Z} \xrightarrow{\sim} \mathbb{C}^*$. The inverse image of the unit circle subgroup $U \subset \mathbb{C}^*$ is the additive group $i \cdot \mathbb{R}$, in particular,

$$U \xrightarrow{\sim} i \cdot \mathbb{R}/i2\pi\mathbb{Z} \xrightarrow{\sim} \mathbb{R}/\mathbb{Z}.$$

This combines to the group isomorphism $\mathbb{R} \times U \overset{\sim}{\to} \mathbb{C}^* : (r, u) \mapsto \exp(r) \cdot u$, *which is called the* polar coordinate representation *of (non-zero) complex numbers. The uniquely determined angle* $-\pi < \theta \leq \pi$, *such that* $u = \exp(i \cdot \theta) \in U$ *in the polar coordinate representation* $z = \exp(r) \cdot u$ *is denoted by* $\arg(z)$. *The Euler formula* $\exp(i \cdot \theta) = \cos(\theta) + i\sin(\theta)$ *established in proposition 210 in volume 1 (we used the symbol $A(\theta)$ for $\exp(i \cdot \theta)$ there), implies the representation of the sine and cosine functions, which are both continuous, in terms of power series:*

$$\cos(\theta) = 1 - \frac{\theta^2}{2!} + \frac{\theta^4}{4!} - \frac{\theta^6}{6!} + \cdots (-1)^n \frac{\theta^{2n}}{(2n)!} + \cdots$$

$$\sin(\theta) = \theta - \frac{\theta^3}{3!} + \frac{\theta^5}{5!} - \frac{\theta^7}{7!} \cdots (-1)^n \frac{\theta^{2n+1}}{(2n+1)!} + \cdots$$

The Euler formula implies this alternative definition of the sine and cosine functions:

$$\cos(\theta) = \frac{1}{2}(\exp(i \cdot \theta) + \exp(-i \cdot \theta)),$$

$$\sin(\theta) = \frac{1}{2i}(\exp(i \cdot \theta) - \exp(-i \cdot \theta)).$$

The restriction $\exp|_{\mathbb{R}} : \mathbb{R} \to \mathbb{R}_+$ *is continuous and ordered[4] isomorphism of the additive group of $\mathbb{R}$ onto the multiplicative group $\mathbb{R}_+$ of positive real numbers. Its inverse* $\log : \mathbb{R}_+ \to \mathbb{R}$ *is called the* (natural) logarithm. *In particular,* $\log(1) = 0$, *and* $\log(x \cdot y) = \log(x) + \log(y)$ *for all* $x, y \in \mathbb{R}_+$.
The number $e = \exp(1) = \sum_{k=0}^{\infty} \frac{1}{k!} = 2.7182818\ldots$ *is called the* Euler number; *it is also equal to* $\lim_{n \to \infty}(1 + \frac{1}{n})^n$. *For a rational number* $\frac{p}{q}$ *with* $q > 0$, *we have* $\exp(\frac{p}{q}) = e^{\frac{p}{q}} = (\sqrt[q]{e})^p$. *The general value* $\exp(z)$ *for* $z \in \mathbb{C}$ *is therefore also written as* e^z.

Proof By proposition 256, exp is a group homomorphism, i.e., for all $w, z \in \mathbb{C}$, $\exp(w + z) = \exp(w) \cdot \exp(z)$. In particular, $1 = \exp(0) = \exp(w + (-w)) = \exp(w) \cdot \exp(-w)$, whence $\exp : \mathbb{C} \to \mathbb{C}^*$ is a group homomorphism into the multiplicative group of non-zero complex numbers. Moreover, exp is continuous. In fact, for any $w \in \mathbb{C}$, we have $\exp(w + z) - \exp(w) = \exp(z)(\exp(w) - 1)$. So we have to show that $\exp(w) - 1 \to 0$ if $w \to 0$. But $\| \exp(w) - 1 \| \leq \|w\| \cdot \sum_k \frac{\|w\|^k}{(k+1)!} \leq \|w\| \cdot \sum_k \frac{\|w\|^k}{k!} \leq \|w\| \cdot \sum_k \|w\|^k = \frac{\|w\|}{1 - \|w\|}$ for $\|w\| < 1$, which evidently converges

---

[4] This means that $x < y$ implies $\exp(x) < \exp(y)$. In calculus this is also called a *strictly monotonous* map.

to 0 as $\|w\| \to 0$. Now, clearly $\exp(\overline{z}) = \overline{\exp(z)}$. Therefore, for $\theta \in \mathbb{R}$, we have $\frac{1}{\exp(i \cdot \theta)} = \exp(-i \cdot \theta) = \overline{\exp(i \cdot \theta)}$, which means that we have a group homomorphism $\exp : i \cdot \mathbb{R} \to U$. Setting the Euler equation $\exp(i \cdot \theta) = \cos(\theta) + i \cdot \sin(\theta)$ for the real an complex parts of $\exp(i \cdot \theta)$, we have $\cos(\theta)^2 + \sin(\theta)^2 = 1$, and the alternative definitions of $\cos(\theta)$ and $\sin(\theta)$ in terms of the exponential function follow immediately. The series for $\cos(\theta)$ and $\sin(\theta)$ are also visible from the real and imaginary contributions in the series expansion of the exponential function.

We now want to calculate the kernel of $\exp$. To this end, observe that for $k \geq 2$ and $0 < \theta \leq 3$, we have $\frac{\theta^k}{k!} > \frac{\theta^{k+1}}{(k+1)!}$. Therefore, under these conditions, the series for the cosine converges by the Leibniz criterion proposition 250. Coming back to that proposition's proof, we recognize that $1 - \frac{\theta^2}{2} < \cos(\theta) < 1 - \frac{\theta^2}{2} + \frac{\theta^4}{24}$ for $0 < \theta < 3$. But then, $u = \sqrt{2}$ is the smallest zero of $1 - \frac{\theta^2}{2}$ while $v = \sqrt{6 - 2\sqrt{3}}$ is the smallest zero of $1 - \frac{\theta^2}{2} + \frac{\theta^4}{24}$ in the interval $0 < \theta < 3$. Therefore, by proposition 245, there is a zero of $\cos(\theta)$ in the interval $]u, v[$. Since $\cos(\theta)$ is continuous by the way it is derived from $\exp(i \cdot \theta)$, and since $\cos(0) = 1$, there is a smallest zero of $\cos$, lying between u and v. Call it $\frac{\pi}{2}$. Therefore all values in $[0, 1]$ are taken for arguments θ between 0 and $\frac{\pi}{2}$ by $\cos(\theta)$. So $\exp(i \cdot \frac{\pi}{2}) = i, \exp(i \cdot \pi) = -1, \exp(i \cdot \frac{3\pi}{2}) = -i$, and $\exp(i \cdot 2\pi) = 1$. Therefore, the cosine takes all values between 1 and -1. This implies that $\exp(i \cdot \theta)$ is onto U. The goniometric addition theorem from proposition 210 in volume 1 is a consequence of the group homomorphism property of $\exp$. For $0 \leq \theta < \theta + \eta < \frac{\pi}{2}$, it yields $\cos(\theta + \eta) = \cos(\theta)\cos(\eta) - \sin(\theta)\sin(\eta) < \cos(\theta)\cos(\eta) < \cos(\theta)$, so the cosine function is strictly monotonously decreasing. So for every $x \in [0, 1]$, there is exactly one $\theta \in [0, \frac{\pi}{2}]$ such that $\cos(\theta) = x$. By $\cos(\theta)^2 + \sin(\theta)^2 = 1$, the sine function is monotonously increasing from 0 to 1 as θ moves from 0 to $\frac{\pi}{2}$. Again, by the addition theorem for the cosine function, we have $\cos(\theta + \frac{\pi}{2}) = -\sin(\theta)$. This gives us the values for $\cos(\theta)$ for the arguments in $[\frac{\pi}{2}, \pi]$: The values $\cos(\theta)$ decrease monotonously from 0 to -1 as θ moves from $\frac{\pi}{2}$ to π. By the same argumentation, from π to 2π, $\cos(\theta)$ increases monotonously from -1 to 1. All this together proves that $i \cdot 2\pi\mathbb{Z}$ is the kernel of $\exp$.

Let us finally concentrate on the real arguments in $\exp$. Since $e = \exp(1) > 1$, there are arbitrary large real numbers $\exp(n) = e^n$ for real arguments, and by $\exp(-n) = \frac{1}{\exp(n)}$ also arbitrary small real values for real arguments. By proposition 245, every positive real value is taken by $\exp(x)$ for $x \in \mathbb{R}$. Now, every complex number $z \neq 0$ can be written as $z = \|z\|u, u \in U$. Therefore there are $x, \theta \in \mathbb{R}$, such that $\exp(x) = \|z\|$ and $\exp(i \cdot \theta) = u$. This means that $z = \|z\|u = \exp(x)\exp(i \cdot \theta) = \exp(x + i \cdot \theta)$, and we have shown that $\exp : \mathbb{C} \to \mathbb{C}^*$ is surjective. For $x \in \mathbb{R}$, we have $\exp(-x) = \frac{1}{\exp(x)}$. But for positive $x \in \mathbb{R}$, $\exp(x) > 1$. So $\exp(x) > 0$ for all $x \in \mathbb{R}$. Moreover, for real numbers $x < y$, we have $\exp(x) < \exp(x)\exp(y - x) = \exp(y)$, whence $\exp|_{\mathbb{R}}$ is strictly monotonous onto the multiplicative group $\mathbb{R}_+$ of positive real numbers. The statements about the logarithm are now immediate. The statements about the coincidence $\exp(\frac{p}{q}) = e^{\frac{p}{q}}$ are left as an exercise. For the equation $e = \lim_{n\to\infty}(1 + \frac{1}{n})^n$, we refer to [14]. □

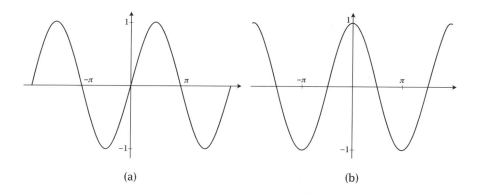

(a)   (b)

Fig. 27.10. The sine (a) and cosine (b) functions, with their domains restricted to $\mathbb{R}$.

Definition 187 *If $a \in \mathbb{R}_+$, one defines the* exponential function for basis a *by $a^x = \exp_a(x) = \exp(x \cdot \log(a))$. If moreover $a \neq 1$, one also defines the* logarithm function for basis a *by $\log_a(x) = \frac{1}{\log(a)}\log(x)$. In older literature, $\log$ is also denoted by $\ln$ (logarithmus naturalis), while one uses the notation $\log$ for $\log_{10}$ and calls that the* decadic *logarithm, but we refrain from such atavisms.*

Sorite 258 *The logarithm $\log_a$ for basis $a \in \mathbb{R}_+$ has the following properties. Let $x, y \in \mathbb{R}$.*

(i) *If $b \in \mathbb{R}_+$ is a second basis, we have $\log_b(x) = \log_b(a) \cdot \log_a(x)$,*

(ii) *$\log_b(a) \cdot \log_a(b) = 1$,*

(iii) *$\log_a(b^x) = x \cdot \log_a(b)$,*

(iv) *if $x \in \mathbb{Q}$, then the exponential function a^x and the rational powers defined earlier, denoted by the same signs, coincide,*

(v) *$a^{x+y} = a^x \cdot a^y$, and $(a^x)^y = a^{x \cdot y}$.*

(vi) *If $b \in \mathbb{R}_+$ is a second basis, we have $b^x = a^{\log_a(b) \cdot x}$.*

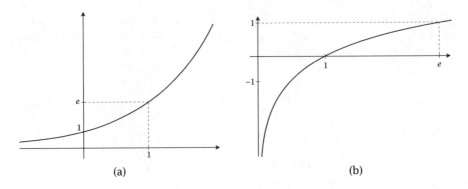

Fig. 27.11. The exponential (a) and logarithm (b) functions, with their domains restricted to ℝ.

Proof The proof of this sorite is left as an exercise, using the straightforward definition of the logarithm, i.e., applying the exponential function as the inverse isomorphism to log to verify the specific claims.   □

27.5 Proof of Euler's Formula for Polyhedra and Kuratowski's Planarity Theorem

Recall from chapter 13 of volume 1 that a skeletal graph $\Gamma : A \to {}^2V$ is a graph without multiple edges or loops. A drawing of a skeletal graph Γ as defined in definition 84, chapter 13, volume 1, is intuitively a family of (continuous) curves $c_a : [0,1] \to \mathbb{R}$ such that $c_a(]0,1[)$ is disjoint of the image of all other curves. Recall also that a drawing may also be defined on the unit sphere $S^2 \subset \mathbb{R}^3$, instead of $\mathbb{R}^2$. The *Northpole* is the top point with coordinates $(0,0,1)$. By the *stereographic projection* $\tau : S^2 - Northpole \overset{\sim}{\to} \mathbb{R}^2$, which is a homeomorphism of topological spaces, every drawing on S^2 induces one in $\mathbb{R}^2$, and conversely. Here is the definition of τ (see figure 27.12). We write $x = (h,v) \in \mathbb{R}^2 \times \mathbb{R}$ for a point in S^2.

$$\tau(h,v) = \frac{1}{1-v}h.$$

Exercise 143 Show that inverse map is $\tau^{-1}(z) = \left(\frac{2}{\|z\|^2+1} \cdot z, \frac{\|z\|^2-1}{\|z\|^2+1} \right)$. Use propositions 240 and 241 to show that these maps are continuous.

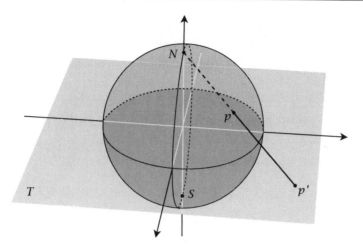

Fig. 27.12. A stereographic projection of a sphere onto the plane T through the equator. The point p is mapped to p' on T, where p, p' and the northpole N are collinear.

We called a polyhedron a drawing of a connected skeletal graph Γ on S^2, but we also use this terminology for a drawing D of Γ in $\mathbb{R}^2$, too. The elements of the finite set C of connected components of the drawn graph $D(\Gamma)$ are called the *faces of $D(\Gamma)$*. In the drawing on $\mathbb{R}^2$ there is one face which is not bounded, this one corresponds to the face on S^2 which includes the *Northpole*. It is called the *exterior face*, the others are called the *interior faces* of the drawing. We now want to prove Euler's formula for polyhedra from proposition 108, volume 1. Recall from that proposition that $\varepsilon = card(V)$, $\phi = card(A)$, $\sigma = card(C)$.

Proof The proof is by induction on the number $\xi = \varepsilon + \phi$. For $\xi = 1$, there is a single point and no edge, whence Euler's formula for polyhedra $\varepsilon - \phi + \sigma = 1 + 0 + 1 = 2$. Suppose that the drawing D has a "bridge", i.e., an edge line c_a such that the drawing minus this line is no more connected (see figure 27.13 (a)). Then the drawing of the remainder of the graph after omitting a decomposes into a disjoint union of two connected subdrawings D', D''. These subdrawings obviously each have a ξ which is smaller than that of the drawing D. Therefore, Euler's formula for polyhedra holds for both D' and D'': $\varepsilon' - \phi' + \sigma' = 2$ and $\varepsilon'' - \phi'' + \sigma'' = 2$. Now let us express ε, ϕ, and σ of D in terms of the values for D' and D'':

$$\varepsilon = \varepsilon' + \varepsilon''$$
$$\phi = \phi' + \phi'' + 1 \quad \text{(the bridge)}$$
$$\sigma = \sigma' + \sigma'' - 1 \quad \text{(both share the exterior face)}$$

So we have

$$\varepsilon - \phi + \sigma = (\varepsilon' + \varepsilon'') - (\phi' + \phi'' + 1) + (\sigma' + \sigma'' - 1)$$
$$= \varepsilon' - \phi + \sigma' + \varepsilon'' - \phi'' + \sigma'' - 1 - 1$$
$$= 2 + 2 - 2$$
$$= 2$$

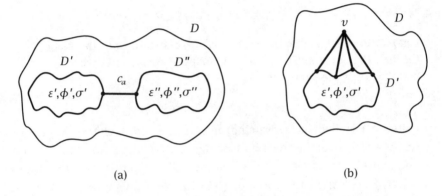

(a)   (b)

Fig. 27.13. Reducing the drawing D of a graph to drawings of smaller graphs: (a) by removing a connecting edge, (b) by removing a vertex and the edges connected to it.

If there is no bridge, take a vertex v which is on the boundary of the exterior face. The k lines terminating at v define a total of k faces containing v in their boundaries (see figure 27.13 (b)). Omitting the point v and all lines terminating in v in the drawing D defines the drawing D' of a connected graph with $k - 1$ less faces, since the interior faces around v are now united to the exterior face of D. Again, Euler's formula for polyhedra holds for D', and we have

$$\varepsilon = \varepsilon' + 1 \quad \text{(the vertex } v\text{)}$$
$$\phi = \phi' + k \quad (k \text{ edges connecting to } v)$$
$$\sigma = \sigma' + (k - 1) \quad (k - 1 \text{ new faces})$$

This yields
$$\varepsilon - \phi + \sigma = (\varepsilon' + 1) - (\phi' + k) + (\sigma' + k - 1) = 2,$$

and the proof is complete. $\square$

Corollary 259 *The graphs K_5 and $K_{3,3}$ are not planar.*

Proof In fact, suppose that we have a drawing D of K_5. Then every three vertexes define a triangular face, so $\sigma = \binom{5}{3} = 10$, but then $\varepsilon - \phi + \sigma = 5 - 10 + 10 = 5 \neq 2$. For $K_{3,3}$, the faces are defined by rectangular cycles through 4 vertexes each. There are 9 such cycles (choose 2 upper and 2 lower points and connect them to a cycle), so we have $\varepsilon - \phi + \sigma = 6 - 9 + 9 = 6 \neq 2$ □

Now, Kuratowski's theorem attributes a central role to the two graphs K_5 and $K_{3,3}$ in that planarity of any skeletal graph Γ is based on the non-inclusion of essentially one of these non-planar graphs. "Essentially" means one of these alternatives: (1) There is a subgraph $\Gamma' \subset \Gamma$ which has a contraction isomorphic to K_5 or $K_{3,3}$ (see definition 85, chapter 13, vol 1). (2) There is a subgraph $\Gamma' \subset \Gamma$ which results from K_5 or $K_{3,3}$ by a succession of subdivisions of their edges. A subdivisions of an edge $x \xrightarrow{\;a\;} y$ is the addition of one more vertex v to V and the replacement of a by two edges $x \xrightarrow{\;a_x\;} v$ and $v \xrightarrow{\;a_y\;} y$.

We have these auxiliary facts:

Lemma 260 *Suppose that we can prove the special case that Γ is planar iff it contains no subgraph which is a subdivision of a graph isomorphic to K_5 or $K_{3,3}$. Then the theorem follows.*

Proof If Γ has no subgraph which can be contracted to a graph isomorphic to K_5 or $K_{3,3}$, then in particular, it has no subgraph, which is a subdivision of a graph isomorphic to K_5 or $K_{3,3}$, since subdivisions can be contracted to the original graphs. By the assumption made in lemma 260 it then can be concluded that Γ is planar. Conversely, if Γ is planar and there is a contraction to a graph isomorphic to K_5 or $K_{3,3}$, then there is a sequence of elementary contractions, which define this contraction. The idea is this: If it can be shown that an elementary contraction preserves the planarity of a graph, then K_5 or $K_{3,3}$ must be planar, which a contradiction. So, if a drawing of a planar graph Γ is given, an elementary contraction of the line $x \xrightarrow{\;a\;} y$ can be performed by isolating a small tubular neighborhood around the drawing of a and then piping the lines ending at x within that tubular neighborhood to y (see figure 27.14). Obviously, this construction conserves planarity. Thus the lemma is proved.   □

Kuratowski's theorem

So one is left with the proof of the subdivision version of Kuratowski's theorem. Now, we already know that a graph containing a subdivision of

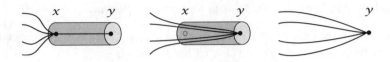

Fig. 27.14. The graphical process of an elementary contraction conserves planarity.

a copy of K_5 or of $K_{3,3}$ can not be planar, since the contractions yielding K_5 or $K_{3,3}$ would yield drawings of graphs containing drawings of K_5 or of $K_{3,3}$, which is impossible by corollary 259. So we are left with the proof of the other implication, i.e., that a non-planar graph must necessarily contain a subdivision of drawings of K_5 or of $K_{3,3}$.

Suppose there is a Γ which, being non-planar, contains no subgraph which is a subdivision of a copy of K_5 or of $K_{3,3}$. Take one with a minimal number of edges. It cannot have a bridge line, since then it is easily seen that one of the subgraphs connected by this bridge would be non-planar and therefore would contain a subdivision of one of the two critical graphs. Moreover, it cannot contain points x with $deg(x) = 1, 2$, since the non-planarity would be conserved omitting these points. So all points have $deg(x) \geq 3$. Then the omission of an arbitrary line $x \overset{l}{\rule{1cm}{0.4pt}} y$ in Γ yields a smaller graph Φ which does not contain a subdivision of K_5 or of $K_{3,3}$ and therefore is planar.

The proof idea is to show that, under these assumptions, one can find a subgraph of Φ which is isomorphic to K_5 or $K_{3,3}$, and this would contradict the assumption that Γ does not contain any of these subgraphs. To do so, one first shows that there is a cycle Z in Φ containing the points x, y defined above. One then makes a drawing of Φ such that there is a maximum of faces interior to the drawing of Z. One considers the components of the subgraph of Φ induced on the vertexes outside the drawing of Z and then defines *outer pieces* as those subgraphs of Φ which are either induced on outer components, plus the points on Z which they are connected to, or else which are outer edges of the drawing of Z connecting two points of Z. Inner components and inner pieces are defined in an analogous way. For a pair of points u, v on Z, one looks for inner or outer pieces such that they contain points $\neq u, v$ on both walks on Z (in clockwise orientation, say) between u and v. These pieces are called *(u–v)-separating*.

One can find an inner piece H and four points u_0, u_1, v_0, v_1 such that

$$Z = u_0 \quad\text{---}\quad \ldots u_1 \quad\text{---}\quad \ldots v_0 \quad\text{---}\quad \ldots v_1 \quad\text{---}\quad \ldots u_0,$$

and such that H is $(u_0\text{-}v_0)$- and $(u_1\text{-}v_1)$-separating. Thus H meets the clockwise walks $u_0 \quad\text{---}\quad \ldots u_1$, $u_1 \quad\text{---}\quad \ldots v_0$, $v_0 \quad\text{---}\quad \ldots v_1$, and $v_1 \quad\text{---}\quad \ldots u_0$ in four points q, r, s, t, all different from u_0, u_1, v_0, v_1. The proof now closes with an analysis of four cases of possible positions of the points q, r, s, t on the cycle Z, and where each case yields a subgraph isomorphic to K_5 or $K_{3,3}$. This is a contradiction to the assumption that the original graph Γ (of which Φ is a subgraph) does not contain a subgraph isomorphic to K_5 or $K_{3,3}$. The details of the proof are described in [12]. It goes back to Gabriel Andrew Dirac and Seymour Schuster, A theorem of Kuratowski. Nderl. Akad. Wetensch. Proc. Ser. A 57, 1954.

CHAPTER 28

Differentiability

28.1 Introduction

Differentiation is probably the single most influential concept in the history of modern science. It is at the basis of virtually all of the physical theories which have changed our lives and ideas so fundamentally. Isaac Newton's (1643–1727) principles of mechanics and gravitation and James Clerk Maxwell's (1831–1897) equations of electrodynamics cannot even be stated without differentiation as a basic language. It was indeed Galileo Galilei (1564–1642) who recognized in his creation of mathematical physics that nature is like a book which we can only read if we learn the language and the symbols in which it is written, and that this language is mathematics. At Galileo's times, this language was still not sufficiently developed to be able to control mechanical phenomena. This discrepancy became irritating when Galileo was performing physical experiments with balls moving down an inclined plane, since these experiments suggested a concept of instantaneous velocity of a body being constantly accelerated. While this concept became manageable by Galileo's experimental access to phenomena, there was at that time no concept of instantaneous velocity except for a constant velocity being maintained for a finite time. In the Medieval scholastic tradition, such as described by the French scientist Nicholas Oresme (1323–1382), an accelerated movement was only conceivable as a succession of locally constant movements, so that a ball would move down the inclined plane in step-wise portions of successively increasing, but at every time constant, velocities.

While Galileo's understanding of instantaneous velocity was experimentally tractable, it took a few decades more until the the problem of defining the concept of a velocity having no duration whatsoever was finally settled. The solution came from what we now call "(infinitesimal) calculus". It was a parallel and independent discovery by Newton (published in detail under the title of a "fluxion method" the first time in an appendix to his book "Opticks" 1704) and by Leibniz (in 1684, he published the first description of the differential dx in the journal "Acta Eruditorum", and in 1686, in the same journal, the first description of the now (in)famous integral sign $\int$). Although these scientists had a perfect, though somewhat mystical, intuition of what these "infinitary" concepts were about, the first precise description of their basics was given by Jean Le Rond d'Alembert in his article about limits in the famous "Encyclopédie" published in 1751:

> *One magnitude is said to be the limit of another magnitude when the second may approach the first within any given magnitude, however small, though the second magnitude may never exceed the magnitude it approaches.*

This definition is not very operational, but we have already learned how the limit of a Cauchy sequence or the continuity of a function make this idea precise. Fortunately, the concepts of infinitesimal calculus are just one more application of limits. Before starting with the technical discussion of calculus, we should mention that the concept of a limit has also been generalized to the end that a majority of fundamental constructions in virtually all important fields of mathematics, such as algebra, the theory of machines, differential geometry, for example, are special cases of constructions by limits. We shall learn more about this approach, which is also fundamental in theoretical computer science, in chapter 36.

In this section, we shall introduce the theory of differentiation of functions $f : U \rightarrow V$ having domains $U \subset \mathbb{R}^n$ and codomains $V \subset \mathbb{R}^m$ in the standard Euclidean spaces. We shall give the fundamental propositions about such functions and thereby solve the Medieval problem of instantaneous velocity.

We begin with the basic definition of a limit for a given function:

Definition 188 *Given a open set $U \subset \mathbb{R}^n$ and $x \in U$, let $f : U - \{x\} \rightarrow \mathbb{R}^m$ be a function defined on the punctured open set $U - \{x\}$. Given a point*

$z \in \mathbb{R}^m$, *one says that* the limit of f at x is z *iff for every real $\varepsilon > 0$ there is a $\delta > 0$ such that $y \in B_\delta(x)$ implies that $f(y) \in B_\varepsilon(z)$. In symbols:* $\lim_{y \to x} f(y) = z$, *or:* $f(y) \to z$ *if $y \to x$.*

Exercise 144 Show that the limit property of a function f from definition 188 is equivalent to the following: For every sequence $(x_i)_i$ in $U - \{x\}$ which converges to x, the sequence $(f(x_i))_i$ converges to z.

Observe that the limit property and its equivalent from exercise 144 is a substitute for continuity of f in x if f were defined in x, too!

28.2 Differentiation

As we are only interested in the local behavior of functions, we first need to make precise what it means to speak of the behavior of a function on "arbitrary small neighborhoods" of a given argument x.

Definition 189 *If $f : U \to \mathbb{R}^m$ and $g : V \to \mathbb{R}^m$ are two functions defined in open neighborhoods U, V of $0 \in \mathbb{R}^n$, then they are called* equivalent *if there is a neighborhood $N \subset U \cap V$ of 0 such that $f|_N = g|_N$. This relation is an equivalence relation, and the equivalence class $[f]$ of a function f is called the* germ *of f (at 0).*

The set F_0 of germs of functions $f : U \to \mathbb{R}^m$ with $f(0) = 0$ is a real vector space as follows: (1) The sum of germs is $[f] + [g] = [f|_W + g|_W]$, $W = U \cap V$ being the intersection of the domains U and V of f and g representing the germs $[f]$ and $[g]$, and (2) the scalar multiplication is $\lambda[f] = [\lambda f]$.

Exercise 145 Show that the vector space structure defined on the set of function germs F_0 at $0 \in \mathbb{R}^n$ in definition 189 is well defined.

Lemma 261 *The canonical linear map $Lin_\mathbb{R}(\mathbb{R}^n, \mathbb{R}^m) \to F_0 : f \mapsto [f]$ is injective.*

Proof According to proposition 116, chapter 15, volume 1, we have to show that the kernel of $Lin_\mathbb{R}(\mathbb{R}^n, \mathbb{R}^m)$ is trivial. But $[f] = 0$ means that there is an open cube $K_\varepsilon(0)$ with $f|_{K_\varepsilon(0)} = 0$. Evidently the basis $(\varepsilon/2 \cdot e_i)_i$ for the canonical basis vectors $e_1 = (1, 0, \ldots 0), \ldots e_n = (0, \ldots 0, 1)$, is contained in $K_\varepsilon(0)$, therefore the linear map f vanishes. $\square$

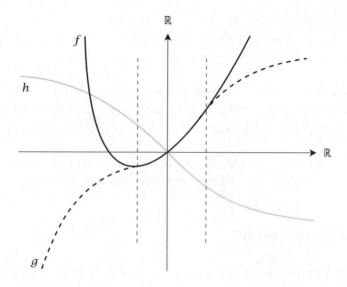

Fig. 28.1. The functions f and g are in the same germ $[f] = [g]$, both functions are identical on the interval delimited by the dashed vertical lines. The third function h is not in $[f]$, it intersects f and g in a single point, the origin.

This means that linear maps are determined by their germs, i.e., their behavior in any vicinity of 0. In other words, for linear maps, local and global behavior is the same. This is the missing link in the Medieval struggle between constant and instantaneous velocity.

Proposition 262 *The subset DF_0 of those germs $[f] \in F_0$ such that $\frac{\|f(z)\|}{\|z\|} \to 0$ if $z \to 0$ is a vector subspace $DF_0 \subset F_0$ such that $DF_0 \cap Lin_\mathbb{R}(\mathbb{R}^n, \mathbb{R}^m) = 0$.*

Proof If $\delta > 0$ is such that for two germs $[f], [g] \in DF_0$ we have $\frac{\|f(z)\|}{\|z\|} < \frac{\varepsilon}{2}$ and $\frac{\|g(z)\|}{\|z\|} < \frac{\varepsilon}{2}$ for $\|z\| < \delta$, then by the triangle inequality, $\frac{\|(f+g)(z)\|}{\|z\|} \leq \frac{\|f(z)\|}{\|z\|} + \frac{\|g(z)\|}{\|z\|} < \varepsilon$. If for $[f] \in DF_0$ and a real number $\lambda \neq 0$, $\delta > 0$ is such that $\frac{\|f(z)\|}{\|z\|} < \frac{\varepsilon}{\lambda}$, for $\|z\| < \delta$, then $\frac{\|(\lambda \cdot f)(z)\|}{\|z\|} < \varepsilon$; the case $\lambda = 0$ is trivial. If $f \in Lin_\mathbb{R}(\mathbb{R}^n, \mathbb{R}^m) \cap DF_0$, then for a canonical basis vector e_i and $\lambda \neq 0$, one has $\frac{\|f(\lambda \cdot e_i)\|}{\|\lambda \cdot e_i\|} = \frac{|\lambda| \cdot \|f(e_i)\|}{|\lambda| \|e_i\|} = \frac{\|f(e_i)\|}{\|e_i\|}$, a constant. If $\lambda \to 0$, this must tend to 0, i.e., it is 0. Whence $f(e_i) = 0$ for all i. $\square$

Definition 190 *A function $f : U \to \mathbb{R}^m$ which is defined in an open neighborhood U of a point $x \in \mathbb{R}^n$ is differentiable in x, iff there is a linear* *map $D \in Lin_{\mathbb{R}}(\mathbb{R}^n, \mathbb{R}^m)$ such that (the germ of) $\Delta_x f - D \in DF_0$, where $\Delta_x f(z) = f(x + z) - f(x)$. By proposition 262, D is uniquely determined and is denoted by Df_x.*

The matrix of Df_x is called the Jacobian matrix *of f at x, its coefficient in the standard bases at row i and column j is denoted by $\partial f_i / \partial x_j(x)$.*

The function $f : U \to V \subset \mathbb{R}^m$ is called differentiable, *iff it is differentiable in every $x \in U$. The set of these differentiable functions is denoted by $Diff(U, V)$. The* derivative *of $f \in Diff(U, V)$ is the function $Df : U \to Lin_{\mathbb{R}}(\mathbb{R}^n, \mathbb{R}^m) : x \mapsto Df_x$ where the Euclidean space $Lin_{\mathbb{R}}(\mathbb{R}^n, \mathbb{R}^m)$ is identified with $\mathbb{R}^{nm}$ as usual.*

Differentiability is stronger than continuity:

Lemma 263 *If $f : U \to \mathbb{R}^m$ is differentiable in $x \in U$, then it is continuous in x. Therefore we have $Diff(U, V) \subset Top(U, V)$ for any open sets $U \subset \mathbb{R}^n$ and $V \subset \mathbb{R}^m$.*

Exercise 146 Give a proof of lemma 263 using these facts:

1. The limit condition on $\frac{\|f(z+x) - f(x) - D(z)\|}{\|z\|}$ implies that $f(z+x) - f(x) - D(z)$ is continuous at x.

2. The linear function D is continuous at x.

3. The sum of continuous functions is continuous.

4. The constant function with value $f(x)$ is continuous.

Intuitively, differentiability of f in x does not only mean that the value of f at x and the values of f at neighboring points $x + z$ around x differ much like the linear map $D(z)$, but also that the difference between the f differences and the linear values "tends faster to zero" than the argument z does. In fact, the statement $\Delta_x f - D \in DF_0$ means

$$\frac{\|f(z + x) - f(x) - D(z)\|}{\|z\|} \to 0 \text{ if } z \to 0.$$

The classical one-dimensional case may give further evidence to what we have defined so far. In this case, $n = m = 1$, $Lin_{\mathbb{R}}(\mathbb{R}^1, \mathbb{R}^1) \xrightarrow{\sim} \mathbb{R}$ and the derivative Df_x identifies with the number $f'(x)$ which describes a linear map $Df_x : \mathbb{R} \to \mathbb{R} : z \mapsto f'(x) \cdot z$. This number is precisely the slope of the line which is tangent to the graph of the function $f : U \to \mathbb{R}$ in the point $(x, f(x))$, see figure 28.2. In other words, we have

$$f'(x) = (pr_2 \circ Tf \circ \Delta)(x),$$

where $\Delta : x \mapsto (x, 1)$.

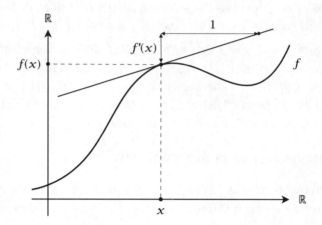

Fig. 28.2. The derivative $f'(x)$ at x is equal to the slope of the tangent to the curve at the point $(x, f(x))$.

The calculation of the derivative of a function can be quite intricate. There is however an efficient method, called chain rule, which we shall expose in the following, to the effect that using this rule we may calculate the derivatives of a large number of common functions. The most elegant formulation of the chain rule requires a tiny extra effort, but we contend that it is worth the reward which we shall draw from it. Coming back to the derivative $Df : U \to Lin_{\mathbb{R}}(\mathbb{R}^n, \mathbb{R}^m)$ of a function, the meaning of the derivative is somewhat hidden in this form. In fact, the meaning of differentiability is that we may not only consider values $f(x + z)$ but also values of the linear map $Df_x(z)$. So we are dealing with two arguments: x and z, and with two values, $f(x)$ and $Df_x(z)$. This entails a natural restatement of the derivative: For any open set $U \subset \mathbb{R}^n$, the *tangent bundle over U* is the Cartesian product $TU = U \times \mathbb{R}^n$. An element $(x, t) \in TU$ is called a *tangent of U at x*.

Definition 191 *Given $f \in Diff(U, V)$ as above, the* tangent map *of f is defined by*

$$Tf : TU \to TV : (x, t) \mapsto Tf(x, t) = (f(x), Df_x(t)).$$

This means that for a fixed $x \in U$, the map $t \mapsto Df_x(t)$ is linear from the fiber $\{x\} \times \mathbb{R}^n$ to the fiber $\{f(x)\} \times \mathbb{R}^m$.

Here is the famous chain rule in terms of tangent bundles.[1]

Proposition 264 *Given differentiable functions $f \in \text{Diff}(U, V)$ and $g \in \text{Diff}(V, W)$, their (usual set-theoretic) composition $g \circ f$ is differentiable and we have*

$$T(g \circ f) = Tg \circ Tf.$$

In other words, if $(x, t) \in TU$, then $D(g \circ f)_x(t) = (Dg_{f(x)} \circ Df_x)(t)$, or

$$D(g \circ f)_x = Dg_{f(x)} \circ Df_x,$$

the classical chain rule statement. Moreover, $T Id_U = Id_{TU}$.

Proof Observe that for a linear map $l : \mathbb{R}^n \to \mathbb{R}^m$, there is a constant $\lambda \geq 0$ such that $\|l(x)\| \leq \lambda \cdot \|x\|$. In fact, taking the canonical basis $(e_i)_i$ of $\mathbb{R}^n$, if $x = \sum_i \xi_i \cdot e_i$, we have $\|l(x)\| = \|\sum_i \xi_i \cdot l(e_i)\| \leq \sum_i |\xi_i| \cdot \|l(e_i)\|$. Let $\lambda = n \cdot \max_i(\|l(e_i)\|)$. Then we have $\sum_i |\xi_i| \cdot \|l(e_i)\| \leq \frac{\lambda}{n} \sum_i |\xi_i| \leq \frac{\lambda}{n} \sum_i \|x\| = \lambda \cdot \|x\|$. Now consider these functions:

$$\phi(t) = f(x + t) - f(x) - Df_x(t),$$
$$\psi(t) = g(f(x) + t) - g(f(x)) - Dg_{f(x)}(t),$$
$$\sigma(t) = g(f(x + t)) - g(f(x)) - Dg_{f(x)}(Df_x(t)),$$

with

$$\frac{\|\phi(t)\|}{\|t\|} \to 0 \quad \text{and} \quad \frac{\|\psi(t)\|}{\|t\|} \to 0$$

for $t \to 0$. We have to show that also $\frac{\|\sigma(t)\|}{\|t\|} \to 0$ if $t \to 0$. But

$$\begin{aligned}
\sigma(t) &= g(f(x + t)) - g(f(x)) - Dg_{f(x)}(Df_x(t)) \\
&= g(f(x + t)) - g(f(x)) - Dg_{f(x)}(f(x + t) - f(x) - \phi(t)) \\
&= \psi(f(x + t) - f(x)) + Dg_{f(x)}(\phi(t)).
\end{aligned}$$

So now, by the above limitation for linear maps, there exists a $\lambda \geq 0$ with $\|Dg_{f(x)}(s)\| \leq \lambda \cdot \|s\|$, and therefore

$$\begin{aligned}
\frac{\|\sigma(t)\|}{\|t\|} &\leq \frac{\|\psi(f(x + t) - f(x))\|}{\|t\|} + \frac{\|Dg_{f(x)}(\phi(t))\|}{\|t\|} \\
&\leq \frac{\|\psi(f(x + t) - f(x))\|}{\|t\|} + \frac{\|\phi(t)\|}{\|t\|} \cdot \lambda.
\end{aligned}$$

---

[1] This is a modern statement of so-called functorial character, i.e., the construction $U \mapsto TU$ "commutes with the composition of differentiable functions". We shall discuss functoriality in a precise way in chapter 36. Check that we have quite often encountered functorial behavior in previous chapters!

So the second term tends to 0 if $t \to 0$. As to the first term, we know that $\|\psi(f(x+t) - f(x))\| \le \|f(x+t) - f(x)\| \cdot \varepsilon$ if $\|f(x+t) - f(x)\| < \delta'$ for an adequate $\delta' > 0$. Since f is continuous at x by lemma 263, there is a $\delta > 0$ such that $\|t\| < \delta$ implies $\|f(x+t) - f(x)\| < \delta'$. Therefore, $\|t\| < \delta$ implies $\frac{\|\psi(f(x+t) - f(x))\|}{\|t\|} < \varepsilon$, and we are done. $\qquad\qquad\square$

Exercise 147 Show that in the one-dimensional case of proposition 264, we have the well-known chain rule

$$(g \circ f)'(x) = g'(f(x)) \cdot f'(x).$$

The chain rule applies in this sense: If one is given a function which can be written as a composition of functions whose derivatives are known, then the chain rule allows the calculation of the derivative of the composed function.

Sorite 265 *If $U \subset \mathbb{R}^n$ and $V \subset \mathbb{R}^m$, then we have a canonical bijection $T(U \times V) \overset{\sim}{\to} TU \times TV : ((u,v),(t,s)) \mapsto ((u,t),(v,s))$, which we use without special mention in the following statements:*

(i) *If $f_1 : U_1 \to V_1$ and $f_2 : U_2 \to V_2$ are differentiable, then so is $f_1 \times f_2$, and we have $T(f_1 \times f_2) = Tf_1 \times Tf_2$.*

(ii) *The projections $pr_U : U \times V \to U$ and $pr_V : U \times V \to V$ are differentiable, and we have $Tpr_U = pr_{TU}$ and $Tpr_V = pr_{TV}$, where $pr_{TU} : TU \times TV \to TU$ and $pr_{TV} : TU \times TV \to TV$ are the canonical projections of the tangent bundles.*

(iii) *If $f : U \to V$ and $g : U \to W$ are differentiable, then so is the universal map $(f,g) : U \to V \times W$, and we have $T(f,g) = (Tf, Tg)$, the universal map of the tangent maps.*

(iv) *If $f : U \to V$ is constant in a neighborhood of $x \in U$, then $Df_x = 0$.*

(v) *If $f : U \to V$ is the restriction of a linear map $f = L|_U$, then $Df(x) = L$ for all $x \in U$.*

(vi) *The product function $\mu : \mathbb{R}^2 \to \mathbb{R} : (x,y) \mapsto x \cdot y$ is differentiable and we have $D\mu_{(x,y)} = (y,x)$.*

(vii) *If $f, g : U \to \mathbb{R}^m$ are differentiable, then so is $f \cdot g : U \to \mathbb{R} : u \mapsto (f(u), g(u))$, the standard scalar product (see volume 1, exercise 122), and for $u \in U$, we have $D(f \cdot g) = f^\tau \cdot Dg + g^\tau \cdot Df$, the latter evaluating to the product of matrixes at each argument $u \in U$. In particular, if $m = n = 1$, we have the classical product rule for derivatives of functions, i.e., $(f \cdot g)' = f \cdot g' + g \cdot f'$.*

(viii) *If $f, g \in Diff(U, \mathbb{R})$, and if for $u \in U$, $g(u) \neq 0$, then $f/g : W \to \mathbb{R} : u \mapsto f(u)/g(u)$ is defined in a neighborhood W of u, is differentiable in u, and we have*

$$D(f/g)_u = \frac{g(u)Df_u - f(u)Dg_u}{g(u)^2}.$$

Proof Claim (i): The Cartesian product $f_1 \times f_2$ at a point (x_1, x_2) such that f_i is differentiable at x_i for $i = 1, 2$, involves two functions $\phi_i(t_i) = f_i(x_i + t_i) - f(x_i) - Df_{x_i}(t_i)$. We know that $\frac{\|\phi_i(t_i)\|}{\|t_i\|} \to 0$ for $t_i \to 0$. Therefore, for $\varepsilon > 0$, there is $\delta > 0$ such that $\|x_i\| < \delta$ implies $\|\phi_i(t_i)\| < \|t_i\| \cdot \frac{\varepsilon}{2}$. But then

$$\frac{\|(\phi_1(t_1), \phi_2(t_2))\|}{\|(t_1, t_2)\|} \leq \frac{\|t_1\|}{\|(t_1, t_2)\|} \cdot \frac{\varepsilon}{2} + \frac{\|t_2\|}{\|(t_1, t_2)\|} \cdot \frac{\varepsilon}{2} < \varepsilon.$$

But this means that the coordinatewise derivatives yield the derivative of the Cartesian product, whence the claim (i).

Claim (v) is an easy exercise.

Claim (ii) is a special case of (v) since projections are linear.

Claim (iii) results from the fact that $(f, g) = (f \times g) \circ \Delta$, where $\Delta_U : U \to U \times U : x \mapsto (x, x)$ is the linear diagonal map, so, by the chain rule and claim (v), we have $T(f, g) = (Tf \times Tg) \circ \Delta_{TU} = (Tf, Tg)$.

Claim (iv) is clear, and (vi) is an easy exercise.

Claim (vii) results from the decomposition

$$f \cdot g : U \xrightarrow{(f,g)} \mathbb{R}^m \times \mathbb{R}^m \xrightarrow{\mu^m} \mathbb{R}^m \xrightarrow{+} \mathbb{R},$$

where μ^m is the m-fold Cartesian product of the product map μ, and where $+$ is the linear sum map $(x_i) \mapsto \sum_i x$. In fact, we then apply the chain rule and the statements (iii), (v), and (vi).

As to the last claim (viii), if f/g is differentiable, we have $f = g \cdot f/g$, and the formula from claim (vii) yields the formula in (viii). So we have to test its validity in the definition of differentiability. This is a routine calculation and is left to the reader. □

This powerful sorite allows us to differentiate a large number of functions. But we first need to consider derivatives of the single most important function of mathematics, exp, together with its "satellites" cos and sin:

Proposition 266 *The functions* exp, cos, sin : $\mathbb{R} \to \mathbb{R}$ *are differentiable on all of $\mathbb{R}$ and we have* exp$'$ = exp, cos$'$ = $-$ sin *and* sin$'$ = cos.

Proof We have

$$\frac{\exp(x+t) - \exp(x) - \exp(x) \cdot t}{t} = \exp(x)\frac{\exp(t) - 1}{t} - \exp(x).$$

So it suffices to prove $\frac{\exp(t)-1}{t} \to 1$ for $t \to 0$. Now, $\frac{\exp(t)-1}{t} = 1 + \frac{t}{2!} + \frac{t^2}{3!} + \dots$, which clearly tends to 1 as $t \to 0$. As to the derivative of sin, observe that by the power series representation of sin and cos in proposition 257, we have $\frac{\cos(t)-1}{t} \to 0$, $\frac{\sin(t)}{t} \to 1$ as $t \to 0$. Now, using the addition theorem for the sine function, we get

$$\frac{\sin(x+t) - \sin(x)}{t} = \sin(x) \cdot \frac{\cos(t) - 1}{t} + \cos(x)\frac{\sin(t)}{t},$$

which converges to $\cos(x)$ as $t \to 0$, therefore $\sin' = \cos$. Since we have $\cos(x) = \sin(x + \frac{\pi}{2})$, the chain rule yields $\cos' = -\sin$. □

Here is a list of derivatives whose calculation we leave as an exercise.

Exercise 148 Examples of derivatives of functions calculated by use of the above rules.

1. For an integer n, if $f(x) = x^n$ then $f'(x) = nx^{n-1}$. More precisely: If $n = 0$, then f is constant and f' is the zero function on all of $\mathbb{R}$. If $n < 0$, then $f(x)$ is not defined at $x = 0$, but the formula is valid for all other arguments. Give a proof of this statement by induction on $n > 0$ and using proposition 264 and sorite 265.

2. Let $f(x)$ be the real function associated with a polynomial $f(X) = \sum_{i=0}^{n} a_i X^i$. Then the derivative $f'(x)$ is the function associated with the polynomial $f'(X) = \sum_{i=1}^{n} i a_i X^{i-1}$.

3. For a positive basis number $a \neq 1$, we have $\exp'_a = \log(a) \cdot \exp$.

4. Let $f(x) = x^{\frac{1}{p}}$, $p \neq 0$, $x > 0$. Use the chain rule and the fact that $(x^{\frac{1}{p}})^p = x$ to show that $f'(x) = \frac{1}{p} x^{\frac{1}{p}-1}$, in particular, $(\sqrt{x})' = \frac{1}{2\sqrt{x}}$.

5. Use the chain rule to show that $\log'(x) = \frac{1}{x}$, for $x > 0$.

6. Let $f(x) = x^r$, $r \neq 0$, $x > 0$. Use the chain rule and the fact that $x^r = e^{r\log(x)}$ to show that $f'(x) = rx^{r-1}$.

Refer to [7] for derivatives of frequently occurring functions.

Exercise 149 Define the domains and calculate the derivatives of these functions:

1. $f(x, y, z) = x^z$

2. $f(x, y) = \cos(xy)$

3. $f(x, y) = \det \begin{pmatrix} \cos(x) & \sin(y) \\ -2x^2 & y \cdot \cos(3x) \end{pmatrix}$

4. $f(x, y) = \begin{pmatrix} 2 & 5 \\ -2 & 3 \end{pmatrix} \cdot \begin{pmatrix} y^2 \\ x^3 \end{pmatrix}.$

Example 106 Consider the function $f : U \to \mathbb{R}^3 : (x, y) \mapsto (x, y, g(x, y))$, where

$$g(x, y) = 3 + \sqrt{1 - \frac{1}{4}x^2 - y^2},$$

and choose $U = \{(x, y) \in \mathbb{R}^2 : \frac{1}{4}x^2 - y^2 \le 1\}$, i.e., an ellipse two units long and one unit wide. This choice of U ensures that the square root in g always yields real numbers. The graph of f is the upper half of an ellipsoid, see figure 28.3.

Calculate the Jacobian matrix of f at x by applying the chain rule for the differentiation of g: Writing g as $g = s \circ t$, where $t(x, y) = 1 - \frac{1}{4}x^2 - y^2$, and $s(z) = 3 + \sqrt{z}$ results in

$$Dg_{(x,y)} = \left(-\frac{1}{4} \frac{x}{\sqrt{(1 - \frac{1}{4}x^2 - y^2)}}, -\frac{y}{\sqrt{(1 - \frac{1}{4}x^2 - y^2)}} \right),$$

since

$$Dt_{(x,y)} = \left(-\frac{1}{2}x, -2y \right)$$

and

$$Ds_z = \frac{1}{2}z^{-\frac{1}{2}}.$$

Therefore, the differential of f at (x, y) is:

$$Df_{(x,y)} = \begin{pmatrix} 1 & 0 \\ 0 & 1 \\ -\frac{1}{4}\frac{x}{\sqrt{(1-\frac{1}{4}x^2-y^2)}} & -\frac{y}{\sqrt{(1-\frac{1}{4}x^2-y^2)}} \end{pmatrix}$$

This enables the calculation of the tangential plane at any point of the ellipsoid: Select a point $p = (x, y) \in U$ and a vector $t \in \mathbb{R}^2$. Calculate the image p_E of p on the ellipsoid, $p_E = (x, y, g(x, y))$. Now use the tangent map Tf to get the tangent $Df_p(t))$. To make this example concrete, assume $p = (1.4, 0.3)$, and $t = (-0.5, 0.1)$. Then $p_E \approx (1.4, 0.3, 3.64807)$, and the Jacobian matrix is

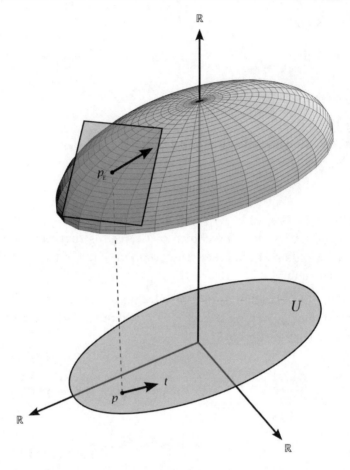

Fig. 28.3. Tangential plane at a point on the ellipsoid from example 106.

$$Df_p \approx \begin{pmatrix} 1 & 0 \\ 0 & 1 \\ -0.540062 & -0.46291 \end{pmatrix},$$

so

$$Df_p(t) \approx \begin{pmatrix} 1 & 0 \\ 0 & 1 \\ -0.540062 & -0.46291 \end{pmatrix} \begin{pmatrix} -0.5 \\ 0.1 \end{pmatrix} \approx \begin{pmatrix} -0.5 \\ 0.1 \\ 0.22374 \end{pmatrix}.$$

The tangent calculated here is the vector starting at p_E and ending at $p_E + Df_p(t)$.

The derivative $Df : U \to Lin_{\mathbb{R}}(\mathbb{R}^n, \mathbb{R}^m) \xrightarrow{\sim} \mathbb{R}^{nm}$ is a function which may again be differentiable. We write $D^2 f = D(Df)$ in that case. More generally, for $r \in \mathbb{N}$, we define $D^r f$, if it exists, recursively by $D^0 f = f$ and $D^{r+1} f = D(D^r f)$, and call $D^r f$ the r-th derivative of f. We may also recursively extend the tangent operator $f \mapsto Tf$ to higher powers if the r-th derivative of the involved functions exists, by $T^r f = T(T^{r-1} f)$.

Exercise 150 Show that if the functions $f : U \to V$ and $g : V \to W$ are r times differentiable, then we have $T^r(g \circ f) = T^r g \circ T^r f$.

Definition 192 *For open sets $U \subset \mathbb{R}^n$ and $V \subset \mathbb{R}^m$, the set of functions $f : U \to V$ such that all derivatives $D^s f$ for $s = 0, 1, \ldots r$ exist and are continuous (check that this is always the case if they exist, except for the last derivative $D^r f$), is denoted by $C^r(U, V)$. Such a function is called r times continuously differentiable. In particular, $C^0(U, V) = Top(U, V)$, and $Diff(U, V) \subset C^1(U, V)$. The set $\bigcap_{r=0,1,\ldots} C^r(U, V)$ is denoted by $C^\infty(U, V)$, and its elements are called C^∞ functions. C^1 functions are also called continuously differentiable.*

Example 107 All polynomial functions as well as exp, sin, and cos are C^∞ functions.

The function

$$f(x) = \begin{cases} x^2 & \text{if } x \geq 0, \\ -x^2 & \text{if } x < 0 \end{cases}$$

is continuous on $\mathbb{R}$. Its derivative exists, is continuous, and is defined by

$$f'(x) = \begin{cases} 2x & \text{if } x \geq 0, \\ -2x & \text{if } x < 0. \end{cases}$$

The second derivative f'', however, does not exist at $x = 0$, where there is a jump from -2 to 2. Therefore f is in C^1, but not in C^2.

Let us close with the very important *mean value theorem*.

Proposition 267 (Mean Value Theorem) *If for real numbers $a < b$, a function $f : [a, b] \to \mathbb{R}$ is continuous and also differentiable in $]a, b[$, then there is a point $x \in]a, b[$ such that $f'(x) = \frac{f(b)-f(a)}{b-a}$.*

Proof Replacing $f(x)$ by $g(x) = f(x) - \frac{f(b)-f(a)}{b-a}(x-a)$, we have $g(a) = g(b) = f(a)$. If we can prove the proposition for g, we find a $\xi \in]a, b[$ with $g'(\xi) = 0$.

But $g'(x) = f'(x) - \frac{f(b)-f(a)}{b-a}$. Whence $f'(\xi) = \frac{f(b)-f(a)}{b-a}$, and we are done. So suppose that $f(a) = f(b)$. If f is constant, everything is clear. If not, then the closed interval $f([a,b])$, image of the compact set $[a,b]$ under the continuous map f according to proposition 245, has a maximum or a minimum $\neq f(a)$ for an argument $\xi \in \,]a,b[$. Suppose that it is a maximum, the minimum case works alike. Then suppose that $f'(\xi) > 0$ (the case $f'(\xi) < 0$ is similar). Then there is $\delta > 0$ such that $0 \le t < \delta$ implies $|f(\xi+t) - f(\xi) - t \cdot f'(\xi)| < t \cdot f'(\xi)$. But this means that $f(\xi+t) > f(\xi)$, a contradiction. So taking $x = \xi$ yields $f'(x) = 0$. $\square$

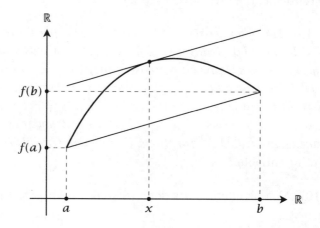

Fig. 28.4. Mean value theorem.

28.2.1 Partial Derivatives

The existence of the derivative of a function is not evident in general, but there is an important additional information that gives necessary and sufficient conditions for the existence of derivatives. This information is provided by partial derivatives. The idea is completely natural: In order to understand the behavior of a function f on an open set $U \subset \mathbb{R}^n$, one does not evaluate the function at every point of U but rather considers the restriction of f to special curves $c_i : U_i \to U$ defined on open sets $U_i \subset \mathbb{R}$. Under good conditions for these restrictions, we can tell a lot about the original function f. Here is the technical setup:

Lemma 268 *If $U \subset \mathbb{R}^n$ is an open set, if $j = 1, 2, \ldots n$ is an index, and if*

$$\alpha = (\alpha_1, \alpha_2, \ldots, \alpha_{j-1}, \alpha_{j+1}, \ldots \alpha_n) \in \mathbb{R}^{n-1},$$

then the set $U_j^\alpha = \{x \mid x \in \mathbb{R}, (\alpha_1, \alpha_2, \ldots, \alpha_{j-1}, x, \alpha_{j+1}, \ldots \alpha_n) \in U\}$ is open. The curve $u_j^\alpha : U_j^\alpha \to U : x \mapsto (\alpha_1, \alpha_2, \ldots, \alpha_{j-1}, x, \alpha_{j+1}, \ldots \alpha_n)$ is an injective C^∞ function.

Proof Since open sets in $\mathbb{R}^n$ are unions of open cubes $K_\varepsilon(x)$, the set U_j^α is clearly open. The curve u_j^α is an injection, and we have

$$u_j^\alpha(x) = (\alpha_1, \alpha_2, \ldots, \alpha_{j-1}, 0, \alpha_{j+1}, \ldots \alpha_n) + x \cdot e_j,$$

i.e., the sum of a constant and a linear function, so by sorite 265, the curve is C^∞, since constants, addition, and linear functions are so. □

If $f : U \to V$, and if j, α are as in lemma 268, then we may consider the compositions ${}_j^\alpha f = f \circ u_j^\alpha : U_j^\alpha \to V$. If ${}_j^\alpha f$ is differentiable in $x \in U_j^\alpha$, we have the derivative $D_j^\alpha f(x)$, which is denoted by $D_j f(a)$ with $a = (\alpha_1, \alpha_2, \ldots, \alpha_{j-1}, x, \alpha_{j+1}, \ldots \alpha_n) \in U$. This is called the j-th partial derivative of f in a. Often, if the j-th variable is known by the name v, say, then one writes $D_v f(a)$ instead of $D_j f(a)$.

Proposition 269 A function $f : U \to V, U \subset \mathbb{R}^n, V \subset \mathbb{R}^m$ open, with component functions f_i, for $i = 1, \ldots m$, is continuously differentiable iff all partial derivatives $D_j f_i, j = 1, \ldots n$, exist and are continuous. For $x \in U$, we then have $D_j f_i(x) = \partial f_i / \partial x_j(x)$, i.e., the coefficients of the Jacobian matrix of f are the partial derivatives of the component functions.

Proof By sorite 265, f is C^1 if the components f_i are so. If these are C^1, then by lemma 268, the partial derivatives $D_j f^i$ exist and are continuous. Moreover, the derivative of a curve u_j^α is the column vector $e_j^\top$. Therefore, by the chain rule, for $x \in U$, the entry $\partial f_i / \partial x_j$ equals $D_j f_i(x)$. The converse, i.e., that f is C^1 if its partial derivatives are all continuous, follows by a standard estimation using the mean value theorem 267 and then calculating the difference $f(x + t) - f(x)$ by a decomposition

$$f(x + t) - f(x) = (f(x + t) - f(x + t^{n-1})) +$$
$$(f(x + t^{n-1}) - f(x + t^{n-2})) + \ldots (f(x + t^1) - f(x)),$$

where the step vector $t = (t_1, t_2, \ldots)$ is replaced by a sequence of step vectors $t^1 = (t_1, 0, \ldots 0), t^2 = (t_1, t_2, 0 \ldots 0), \ldots t^{n-1} = (t_1, t_2, \ldots t_{n-1}, 0)$. Thereby, each successive difference $f(x + t^i) - f(x + t^{i-1})$ has a difference in its arguments, which relates only to one coordinate i. To the above decomposition, one then applies the mean value theorem for the partial derivatives in the i-th coordinates and estimates the total difference. We leave the details as an exercise. □

Example 108 Consider the function f from $\mathbb{R}^2$ to $\mathbb{R}^2$ defined by

$$f(x,y) = \left(\cos\left(\frac{x}{2}\right) + y - \frac{1}{2}, x - \sin(y) + \frac{1}{2}\right).$$

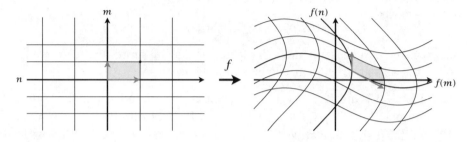

Fig. 28.5. The mapping $f(x,y) = (\cos(\frac{x}{2}) + y - \frac{1}{2}, x - \sin(y) + \frac{1}{2})$ from $\mathbb{R}^2$ to $\mathbb{R}^2$.

The transformation defined by f is illustrated in figure 28.5. Here the grid on the left side, consisting of lines parallel to the axes through $x = -2, -1, 0, 1, 2$ and $y = -2, -1, 0, 1, 2$ respectively, is transformed to the corresponding curves on the right side. The origin $(0,0)$ is mapped to $(\frac{1}{2}, \frac{1}{2})$, the unit rectangle $0 \le x \le 1$, $0 \le y \le 1$, shown in light gray, is mapped to the corresponding patch, and the point $(1, 1)$ is mapped to the corresponding point $(\frac{1}{2} + \cos(\frac{1}{2}), \frac{3}{2} - \cos(1))$. If, in addition, we consider the axes m and n and their images by f, it is obvious that the mapping also changes orientation.

The Jacobian matrix of f at (x, y) is

$$Df_{(x,y)} = \begin{pmatrix} -\frac{1}{2}\sin(\frac{x}{2}) & 1 \\ 1 & -\cos(y) \end{pmatrix}.$$

Its determinant, also simply called Jacobian, is

$$\det(Df_{(x,y)}) = \frac{1}{2}\cos(y)\sin\left(\frac{x}{2}\right) - 1.$$

Partial derivatives are useful in giving necessary conditions for maxima or minima of functions:

Proposition 270 *If a differentiable function $f : U \to \mathbb{R}$ has a maximal or minimal value at $x \in U$, then $D_j f(x) = 0$ for all j.*

Proof We know from the proof of the mean value theorem 267 that a maximal or minimal value $f(x)$ of a continuous function $f : [a, b] \to \mathbb{R}$, which is differentiable in the interior $]a, b[$, has $f'(x) = 0$ at an interior point $x \in]a, b[$. Now, the partial derivative $D_j f$ is the derivative of the composition of f with a curve u_j^{α}, and therefore that argument applies. $\qquad\square$

Attention, the conditions of proposition 270 are not sufficient for a maximum, as is shown by the example $f(x, y) = x^2 - y^2$. Its first partial derivatives are $D_x f(x, y) = 2x - y^2$ and $D_y f(x, y) = x^2 - 2y$. At $(x, y) = (0, 0)$ the said conditions are fulfilled, however f has neither maximum nor minimum, but a so-called *saddle point* at this value, see figure 28.6.

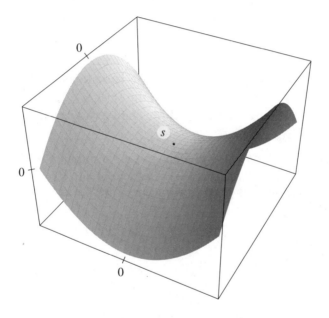

Fig. 28.6. The graph of $f(x, y) = x^2 - y^2$. At the saddle point $s = (0, 0)$, $D_x f(s) = 0$ and $D_y f(s) = 0$, but s is neither a maximum nor a minimum.

28.3 Taylor's Formula

We have defined the exponential function $\exp(z) = \sum_{k=0}^{\infty} \frac{z^k}{k!}$ as a convergent series limiting the sequence $e_n(z) = \sum_{k=0}^{n} \frac{z^k}{k!}$ of polynomial functions $e_n(z)$ of the complex variable z. Taylor's formula inverses this fact: it provides us with the representation of quite general functions of a real variable as limits of polynomial functions. The special form of such representation is motivated by this observation: Let $f = \sum_{k=0}^n a_k X^k$ be a polynomial $f \in \mathbb{R}[X]$. The polynomial function $f : \mathbb{R} \to \mathbb{R} : x \mapsto f(x)$ is in $C^\infty(\mathbb{R})$, and we have $D^k f(0) = k! a_k$ for $k = 0, 1, \dots n$. This means that

$$f(x) = \sum_{k=0}^n \frac{D^k f(0)}{k!} x^k.$$

In other words, the polynomial function is represented by an expression, whose coefficients are determined from calculus: they are derivatives of the given function and do not directly refer to its polynomial character! Comparing this result to the representation $\exp(x) = \sum_{k=0}^\infty \frac{x^k}{k!}, x \in \mathbb{R}$, we recognize that the coefficient of the k-th power of x is also $\frac{D^k f(0)}{k!}$ with $f = \exp$, in view of the fact that $D^k \exp = \exp$ and $\exp(0) = 1$. So the conjecture is that under adequate conditions, any differentiable function f should admit a representation $f(x) = \sum_{k=0}^\infty \frac{D^k f(0)}{k!} x^k$. This is the famous Taylor expansion, which we shall now discuss in more detail.

To begin with, we generalize the reference argument $x = 0$ in the above evaluation of the derivatives to a general x_0 and then obtain the expression $g(h) = f(x_0 + h)$ for $h \in \mathbb{R}$. Then by applying the chain rule (exercise 147), $D^k g(0) = D^k f(x_0)$. Therefore,

$$f(x_0 + h) = \sum_{k=0}^n \frac{D^k f(x_0)}{k!} h^k,$$

or, setting $x = x_0 + h$,

$$f(x) = \sum_{k=0}^n \frac{D^k f(x_0)}{k!} (x - x_0)^k.$$

In this form, suppose that $f \in C^\infty(I)$, where $I =]a, b[$ is an open interval in $\mathbb{R}$. Then for $x_0 \in I$ and $n \in \mathbb{N}$, the *n-th Taylor polynomial of f in x_0* is the polynomial function

$$Taylor^n_{x_0} f(x) = \sum_{k=0}^n \frac{D^k f(x_0)}{k!} (x - x_0)^k.$$

The only critical point is to investigate the conditions under which the *n-th remainder*

$$R_n(x - x_0) = f(x) - Taylor^n_{x_0} f(x)$$

converges to zero. If we have convergence, we obtain *Taylor's formula*

$$f(x) = Taylor_{x_0} f(x) = \sum_{k=0}^{\infty} \frac{D^k f(x_0)}{k!} (x - x_0)^k.$$

There are several useful formulas for the n-th remainder, one of which is particularly elegant:

Proposition 271 *Let $f \in C^\infty(I)$ as above, then there is $\rho \in \,]0, 1[$ such that*

$$R_n(x - x_0) = \frac{1}{(n + 1)!} D^{n+1} f(x_0 + \rho(x - x_0))(x - x_0)^{n+1}.$$

For the Taylor polynomial we therefore have

$$f(x) = f(x_0) + Df(x_0)(x - x_0) + \frac{1}{2!} D^2 f(x_0)(x - x_0)^2 + \ldots$$
$$+ \frac{1}{n!} D^n f(x_0)(x - x_0)^n$$
$$+ \frac{1}{(n + 1)!} D^{n+1} f(x_0 + \rho(x - x_0))(x - x_0)^{n+1}.$$

Proof Consider the C^∞-function in the closed interval $[x_0, x]$

$$\Delta(z) = f(x) - f(z) - Df(z)(x - z) - \frac{1}{2!} D^2 f(z)(x - z)^2 - \ldots$$
$$- \frac{1}{n!} D^n f(z)(x - z)^n - \frac{1}{(n + 1)!} \cdot d \cdot (x - z)^{n+1}.$$

for a constant d, which is chosen such that $\Delta(x_0) = 0$. Then we have $\Delta(x_0) = \Delta(x) = 0$. Therefore, by the mean value theorem 267, there is $0 < \rho < 1$ such that for $\delta = x_0 + \rho(x - x_0)$, we have $\Delta'(\delta) = 0$. But

$$\Delta'(z) = -\frac{1}{n!} D^{n+1} f(z)(x - z)^n + \frac{1}{n!} \cdot d \cdot (x - z)^n,$$

and therefore $d = D^{n+1} f(\delta)$. □

In other words, we have a finite Taylor formula under the condition that the last term takes the derivative of f not exactly at x_0, but somewhere between x_0 and x.

There are several propositions which guarantee the zero convergence of the remainder. We shall present a frequently used criterion here, see [14] for more refined criteria:

Lemma 272 *If $f \in C^\infty(I)$ and there are positive real constants A, B such that for all $x \in I$ and all $n \in \mathbb{N}$, we have*

$$|D^n f(x)| \le A \cdot B^n$$

then the Taylor formula representation $f(x) = Taylor_{x_0} f(x)$ holds for all $x \in I$.

Proof By proposition 271, we have to show that the remainder term

$$R_n(x) = \frac{1}{(n+1)!} D^{n+1} f(x_0 + \rho(x - x_0))(x - x_0)^{n+1}$$

tends to zero as $n \to \infty$. But the supposed estimation implies

$$|R_n(x)| \le \frac{A \cdot (B|x - x_0|)^{n+1}}{(n+1)!},$$

which converges to 0, as we know.   □

Example 109 We look at the first Taylor polynomials in 0 of the function

$$f(x) = \cos(x) + \sin(2x).$$

Derivatives of f must be calculated first:

$$D^0 f(x) = f(x),$$
$$D^1 f(x) = -\sin(x) + 2\cos(2x),$$
$$D^2 f(x) = -\cos(x) - 4\sin(2x),$$
$$D^3 f(x) = \sin(x) - 8\cos(2x),$$
$$D^4 f(x) = \cos(x) + 16\sin(2x).$$

For the Taylor expansion of f in 0, these derivatives must be evaluated at 0:

$$D^0 f(0) = 1,$$
$$D^1 f(0) = 2,$$
$$D^2 f(0) = -1,$$
$$D^3 f(0) = -8,$$
$$D^4 f(0) = 1.$$

Now the Taylor polynomials, according to the Taylor formula, can be calculated:

$$Taylor_0^0 f(x) = t_0 = \frac{1}{0!},$$

$$Taylor_0^1 f(x) = t_1 = \frac{1}{0!} + \frac{2}{1!}x,$$

$$Taylor_0^2 f(x) = t_2 = \frac{1}{0!} + \frac{2}{1!}x + \frac{-1}{2!}x^2,$$

$$Taylor_0^3 f(x) = t_3 = \frac{1}{0!} + \frac{2}{1!}x + \frac{-1}{2!}x^2 + \frac{-8}{3!}x^3,$$

$$Taylor_0^4 f(x) = t_4 = \frac{1}{0!} + \frac{2}{1!}x + \frac{-1}{2!}x^2 + \frac{-8}{3!}x^3 + \frac{1}{4!}x^4.$$

Thus, evaluating, $t_0 = 1$, a constant function and $t_1 = 2x + 1$, an affine function. Further, $t_2 = -\frac{1}{2}x^2 + 2x + 1$, $t_3 = -\frac{4}{3}x^3 - \frac{1}{2}x^2 + 2x + 1$ and $t_4 = \frac{1}{24}x^4 - \frac{4}{3}x^3 - \frac{1}{2}x^2 + 2x + 1$.

The polynomials t_1 to t_4, and further ones up to t_{10}, are shown in figure 28.7. The backmost curve is the exact function f.

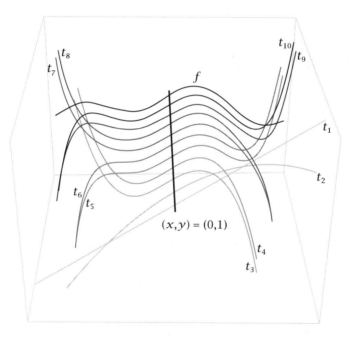

Fig. 28.7. At the back is the function $f(x) = \cos(x) + \sin(2x)$. From front to back, t_1 to t_{10} are increasingly exact Taylor approximations of f. The straight line in the middle indicates that all curves have the value 1 at 0.

CHAPTER 29

Inverse and Implicit Functions

29.1 Introduction

If a function $f : U \to \mathbb{R}$ defined on $U \subset \mathbb{R}$ is continuously differentiable and has $f'(x) > 0$ for a point $x \in U$, then by continuity of f', $f'(y) > 0$ for all y in an open ball $U_\varepsilon(x)$ around x. Therefore, by the mean value theorem 267, if y_1 and y_2, with $y_1 < y_2$, are in $U_\varepsilon(x)$, then there is a z with $y_1 < z < y_2$, such that $f(y_2) - f(y_1) = f'(z)(y_2 - y_1)$, thus $f(y_1) < f(y_2)$, and f is injective on $U_\varepsilon(x)$. Its image $f(U_\varepsilon(x))$ is evidently also an open interval, and there is an inverse function $f^{-1} : f(U_\varepsilon(x)) \to U_\varepsilon(x)$ (see figure 29.1).

Suppose that f^{-1} is differentiable. Then, if $y = f(x)$, $x = f^{-1}(y)$, and, since $f^{-1}(f(x)) = x$, we have

$$(f^{-1}(f(x)))' = (x)'$$
$$(f^{-1})'(f(x)) \cdot f'(x) = 1,$$

by the chain rule, therefore,

$$(f^{-1})'(f(x)) = \frac{1}{f'(x)},$$

and finally, replacing x by $f^{-1}(y)$ again,

$$(f^{-1})'(y) = \frac{1}{f'(f^{-1}(y))},$$

for $y \in f(U_\varepsilon(x))$. The existence of the derivative is easily proved from its very definition.

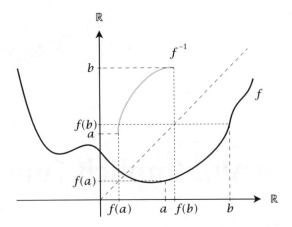

Fig. 29.1. The inverse function f^{-1} (in gray) of f (black) on the interval $]a,b[$.

29.2 The Inverse Function Theorem

The inverse function theorem is the generalization of this fact to n-dimensional continuously differentiable maps. It is easy to guess that the role of the slope $f'(x)$ will be played by the Jacobian matrix, and the fact that the slope is non-zero carries over to the fact that the Jacobian matrix is invertible. We only need a single and easy technical lemma to deal with the general case:

Lemma 273 *Let $K = [a_1, b_1] \times [a_2, b_2] \times \ldots [a_n, b_n]$ be a closed cube in $\mathbb{R}^n$. Suppose that a function $f = (f_1, \ldots f_n) : K \to \mathbb{R}^n$ is continuous and its restriction $f|_{K^o}$ to the open cube $K^o =]a_1, b_1[\times]a_2, b_2[\times \ldots]a_n, b_n[$ is continuously differentiable such that there is a number L with $|D_j f_i(x)| \le L$ for all $x \in K^o$. Then for all $x, y \in K$, we have $\|f(x) - f(y)\| \le n^2 L \|x - y\|$.*

Proof For $x = (x_1, \ldots x_n), y = (y_1, \ldots y_n) \in K$, and for any index i, we have

$$f_i(y) - f_i(x) = \sum_{j=1,\ldots n} (f_i(y_1, \ldots y_j, x_{j+1}, \ldots x_n) - f_i(y_1, \ldots y_{j-1}, x_j, \ldots x_n))$$

and then, by the mean value theorem 267 on each coordinate,

$$|f_i(y_1, \ldots y_j, x_{j+1}, \ldots x_n) - f_i(y_1, \ldots y_{j-1}, x_j, \ldots x_n)| = |y_j - x_j| \cdot |D_j f_i(w_{ij})|$$

for some vector w_{ij}. Supposing $|D_j f_i(x)| \le L$, we get

$$|f_i(y) - f_i(x)| \leq \sum_j |y_j - x_j| \cdot L \leq n\|x - y\|L$$

and therefore

$$|f(x) - f(y)| \leq \sum_i |f_i(y) - f_i(x)| \leq n^2\|x - y\|L.$$

$\square$

Proposition 274 (Inverse Function Theorem) *Let $f \in C^1(U, V)$ for two subsets $U, V \subset \mathbb{R}^n$. Suppose that $\det(Df(x)) \neq 0$ for a point $x \in U$. Then there is an open set O with $x \in O$ such that $W = f(O)$ is open, $f|_O$ is bijective onto W and $(f|_O)^{-1} \in C^1(W, O)$. In particular, $f|_O$ is an open map (i.e., images of open sets $Q \subset O$ are open), and its inverse is open.*

Proof Suppose that we found open sets O and W as claimed. Then by the chain rule, since the inverse f^{-1} has inverse Jacobian matrixes, the determinants of the Jacobian matrixes must be invertible on all $x \in O$. Therefore the theorem applies to any of these points, and the map and (mutatis mutandis the arguments) its inverse are open.

To begin with, the theorem is trivially true for affine automorphisms of $\mathbb{R}^n$, where the Jacobian matrix is the linear part of those automorphisms. So we may assume from the beginning that $a = f(a) = 0$, and that the Jacobian matrix at the origin is $Df_0 = Id$. Now, suppose $f(h) = f(0) = 0$. Then $\frac{\|f(h) - f(0) - Id(h)\|}{\|h\|} = 1$, but we also have $\frac{\|f(h) - f(0) - Id(h)\|}{\|h\|} \to 0$ if $\|h\| \to 0$. So $f(h) \neq f(0)$ for all h in a closed cube $\overline{K(0)}$ around 0. Choosing a sufficiently small $\overline{K(0)}$, we may also assume that $\det(Df_x) \neq 0$, and that $|D_j f_i(x) - D_j f_i(0)| < \frac{1}{2n^2}$ for all i and j and all $x \in \overline{K(0)}$, since the partial derivatives are continuous functions. Applying lemma 273 to $k = f - Id$, and using the inequality $|D_j k(x)| = |D_j f_i(x) - D_j f_i(0)| < \frac{1}{2n^2}$, one sees that

$$\|k(x) - k(y)\| = \|f(x) - x - (f(y) - y)\| \leq \tfrac{1}{2}\|x - y\|$$

for $x, y \in \overline{K(0)}$, and, because $\|x - y\| - \|f(x) - f(y)\| \leq \|f(x) - x - (f(y) - y)\|$, we get

$$\|x - y\| \leq 2\|f(x) - f(y)\|$$

for any $x, y \in \overline{K(0)}$. Further, since f is continuous, the image $f(\partial K(0))$ of the compact set $\partial K = \overline{K(0)} - K(0)$ is compact and disjoint from $0 = f(0)$. Therefore, there is $\delta > 0$ such that $\|f(x)\| \geq \delta$ for all $x \in \partial K(0)$. Clearly then $\|y - f(x)\| > \|y - f(0)\| = \|y\|$ for all $y \in B_{\delta/2}(0)$ and $x \in \partial K(0)$. Now we show that for every $y \in B_{\delta/2}(0)$, there is a unique $x \in K(0)$ such that $y = f(x)$. In fact, consider the function $u : \overline{K(0)} \to \mathbb{R} : x \mapsto \|y - f(x)\|^2$. This is a continuous function with a minimum on the compact set $\overline{K(0)}$. But since on $x \in \partial K(0)$, we have $\|y - f(x)\|^2 > \|y\|^2$, the minimum occurs not on $\partial K(0)$. Therefore the minimum x occurs in the open cube $K(0)$, and there, by proposition 270, all partial derivatives $D_j u(x)$ of u vanish. But this means that

$$\sum_i 2(y_i - f_i(x)) \cdot D_j f_i(x) = 0,$$

for all j. But the Jacobian matrix is invertible, i.e., we have $y_i = f_i(x)$ for all i, i.e., $y = f(x)$. But we know that $\|x - y\| \le 2\|f(x) - f(y)\|$ for any $x, y \in \overline{K(0)}$, so the x is unique. Therefore, in the intersection $V = K(0) \cap f^{-1}(B_{\delta/2})$, there is an inverse function f^{-1}, and, again, by rewriting $\|x - y\| \le 2\|f(x) - f(y)\|$ as $\|f^{-1}(w) - f^{-1}(z)\| \le 2\|w - z\|$, we see that $f^{-1} : W = K(0) \to V$ is a continuous inverse of $f : V \to W$. So we are left with the proof that f^{-1} is differentiable. Take $M = Df(x)$ for $x \in V$. It must be shown that for $y = f(x)$, $Df^{-1}(y) = M^{-1}$. We set $\phi(t) = f(x + t) - f(x) - Df_x(t)$ with $\frac{\|\phi(t)\|}{\|t\|} \to 0$ for $t \to 0$. This implies

$$M^{-1}(f(x + t) - f(x)) = t + M^{-1}(\phi(t)).$$

But since f is a bijection between V and W, setting $s = f(x + t) - f(x)$, this equation may be rewritten as

$$M^{-1}(s) = f^{-1}(y + s) - f^{-1}(y) + M^{-1}(\phi(f^{-1}(y + s) - f^{-1}(y))),$$

whence we must prove that

$$\frac{\|M^{-1}(\phi(f^{-1}(y + s) - f^{-1}(y)))\|}{\|s\|} \to 0$$

for $s \to 0$. But we know from the proof of proposition 264 that for the linear map $M : \mathbb{R}^n \to \mathbb{R}^n$, there is a constant $\lambda \ge 0$ such that $\|M(x)\| \le \lambda \cdot \|x\|$. Therefore we only have to show that

$$\frac{\|\phi(f^{-1}(y + s) - f^{-1}(y))\|}{\|s\|} \to 0$$

for $s \to 0$. But we have

$$\frac{\|\phi(f^{-1}(y+s) - f^{-1}(y))\|}{\|s\|} = \frac{\|\phi(f^{-1}(y+s) - f^{-1}(y))\|}{\|f^{-1}(y+s) - f^{-1}(y)\|} \cdot \frac{\|f^{-1}(y+s) - f^{-1}(y)\|}{\|s\|},$$

where the second factor is less than 2 by the above estimation $\|f^{-1}(w) - f^{-1}(z)\| \le 2\|w - z\|$, and the first factor tends to zero as $s \to 0$, since then also $\phi(f^{-1}(y + s) - f^{-1}(y))$ tends to 0 by the continuity of f^{-1} and by the defining property of ϕ. □

Exercise 151 Consider the function $f : \mathbb{R}^2 \to \mathbb{R}^2 : (x, y) \mapsto (x^2 - y^2, x^2 + y^2)$. Discuss its behavior at the origin $x = (0, 0)$ and the failure of the claim in proposition 274.

Example 110 The function $f(x) = x^3$ is a bijection on $\mathbb{R}$, but $f'(0) = 0$, and the inverse $f^{-1}(x) = \sqrt[3]{x}$ is not differentiable in $x = 0$.

Bijections $f \in C^r(O, W)$ between open sets $O, W \subset \mathbb{R}^n$ such that their inverse maps f^{-1} are in $C^r(W, O)$ are called C^r-*diffeomorphisms*, or simply *diffeomorphisms*, if r is clear. Diffeomorphisms play the role of "isomorphisms" in differential calculus. The inverse function theorem guarantees the local existence of diffeomorphisms for non-vanishing Jacobians.

Corollary 275 *If $f : U \to V$ is a continuously differentiable bijection between open sets of $\mathbb{R}$ such that $f'(x) \neq 0$ for all $x \in U$, then f is a C^1-diffeomorphism.*

Proof By the proposition 274, f has its inverse which is C^1 in a neighborhood of every $x \in U$. So f^{-1} is C^1. □

Example 111 The exponential function $\exp : \mathbb{R} \xrightarrow{\sim} \mathbb{R}_+^*$ and its inverse function $\log : \mathbb{R}_+^* \xrightarrow{\sim} \mathbb{R}$ are C^∞-diffeomorphisms.

Example 112 The restriction $\sin : \left]-\frac{\pi}{2}, \frac{\pi}{2}\right[\xrightarrow{\sim} \left]-1, 1\right[$ is a C^∞-diffeomorphism, whose inverse is denoted by arcsin (figure 29.2). The chain rule yields $\arcsin'(x) = \frac{1}{\sqrt{1-x^2}}$. Similarly, the restriction $\cos : \left]0, \pi\right[\xrightarrow{\sim} \left]-1, 1\right[$ is a C^∞-diffeomorphism, whose inverse is denoted by arccos. The chain rule yields $\arccos'(x) = \frac{-1}{\sqrt{1-x^2}}$.

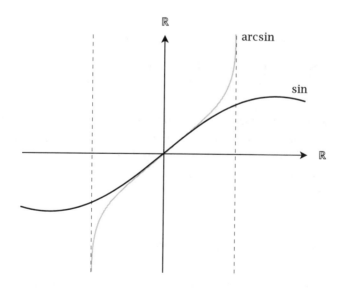

Fig. 29.2. The sine function and its inverse arcsin.

Exercise 152 For $x \in \left]-\frac{\pi}{2}, \frac{\pi}{2}\right[$, we define the *tangent* function $\tan(x) = \frac{\sin(x)}{\cos(x)}$. Show that this is a C^∞-diffeomorphism onto $\mathbb{R}$ with derivative $\tan'(x) = \frac{1}{\cos^2(x)}$ and inverse $\tan^{-1} = \arctan$ (figure 29.3), whose derivative is $\arctan'(x) = \frac{1}{1+x^2}$. For $x \in \left]0, \pi\right[$, we have the *cotangent* function $\cot(x) = \frac{\cos(x)}{\sin(x)}$. Show hat this is a C^∞-diffeomorphism onto $\mathbb{R}$ with derivative $\cot'(x) = \frac{-1}{\sin^2(x)}$ and inverse $\cot^{-1} = \text{arccot}$, whose derivative is $\text{arccot}'(x) = \frac{-1}{1+x^2}$.

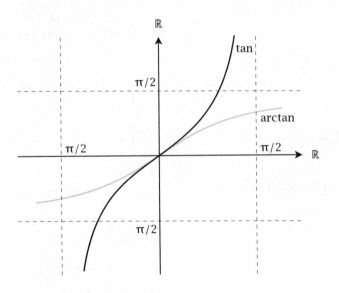

Fig. 29.3. The tangent function and its inverse arctan.

We now turn to a theorem which is intimately related to the inverse function theorem: the implicit function theorem. In some textbooks the inverse function theorem is even interpreted as a special case of the implicit function theorem, but we refrain from this approach in our modest environment.

29.3 The Implicit Function Theorem

The implicit function theorem arises from the intuitively evident fact that many non-functional graphs "locally" look like graphs of functions, i.e., if we do not vary too much the arguments and values, then the graph behaves like a function. For example, take the function $f : \mathbb{R}^3 \to \mathbb{R}$ given by $f(x, y, z) = 3(x^2 + y^2) + 4z^2 - 1$. Then consider the zero fiber $V(f) = f^{-1}(0)$. See figure 29.4 for the visualization of the elliptic shape $V(f)$.

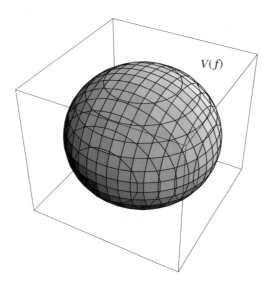

Fig. 29.4. The zero fiber $V(f) = f^{-1}(0)$, i.e., the set of the solutions (x, y, z) of $3(x^2 + y^2) + 4z^2 - 1 = 0$.

If we select a solution $(r, s, t) \in V(f)$ such that $r^2 + s^2 \leq \frac{1}{3}$, then there is an open neighborhood $U(r, s)$ of (r, s) and an open neighborhood $W(t)$ of t such that $V(f) \cap (U(r, s) \times W(t))$ is a functional graph. The function is evidently given by $t = \frac{1}{2}\sqrt{1 - 3(r^2 + s^2)}$ or $t = -\frac{1}{2}\sqrt{1 - 3(r^2 + s^2)}$, depending on the condition $t > 0$ or $t < 0$ for our solution. See figure 29.5 for the graphs of both functions.

Such a functional dependence is said to be given *implicitly* by the equation $f(r, s, t) = 0$. What are the general properties of such a function f and of points $x \in V(f)$ such that a local functional graph can be found in $V(f)$? The next proposition gives the general answer.

Proposition 276 *Given three open sets* $U \subset \mathbb{R}^n, W, T \subset \mathbb{R}^m$, *let* $f : U \times W \to T$ *be a* C^1 *function. We set* $V(f) = f^{-1}(0)$ *and suppose that* $(r, s) \in U \times W$ *is in* $V(f)$. *If the matrix* $D = (D_{n+j}f_i(r, s))_{1 \leq i, j \leq m}$ *has* $\det(D) \neq 0$, *then there are open neighborhoods* $U(r) \subset U$ *of* r *and* $W(s) \subset W$ *of* s,

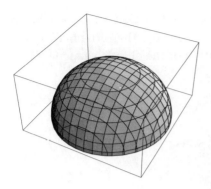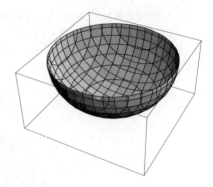

Fig. 29.5. The functional graphs $t = \frac{1}{2}\sqrt{1 - 3(r^2 + s^2)}$ (left) and $t = -\frac{1}{2}\sqrt{1 - 3(r^2 + s^2)}$ (right) on $U = \{(r, s) \mid r^2 + s^2 \leq \frac{1}{3}\}$.

such that $V(f) \cap U(r) \times W(s)$ defines a function $g : U(r) \to W(s)$, i.e., the solutions of $V(f) \cap U(r) \times W(s)$ are exactly the pairs $(x, g(x)), x \in U(r)$. Moreover, the function g is differentiable.

Proof Let $F : U \times W \to \mathbb{R}^n \times T$ be defined by $F(x, y) = (x, f(x, y))$. Clearly $\det(DF_{(r,s)}) = \det(D) \neq 0$. so by proposition 274, there are open sets $V, W \subset \mathbb{R}^{n+m}$, such that $(r, s) \in V, (r, 0) \in W$ and $F : V \overset{\sim}{\to} W$ is a C^1-diffeomorphism. We may even take $V = R \times S$, a product of open sets $R \subset \mathbb{R}^n, S \subset \mathbb{R}^m$. Then we have $F^{-1}(x, y) = (x, q(x, y))$, for a C^1 function q, and

$$f(x, q(x, y)) = f \circ F^{-1}(x, y) = pr_y \circ F \circ F^{-1}(x, y) = y,$$

whence $f(x, q(x, 0)) = 0$, i.e., $(x, q(x, 0)) \in V(f)$. Now, set $g(x) = q(x, 0)$, and we are done. $\qquad\square$

Example 113 A *cardioid* is a member of the family of curves in $\mathbb{R}^2$ satisfying

$$(x^2 + y^2 - 2ax)^2 = 4a^2(x^2 + y^2),$$

i.e, the solution set $V(f_a)$ of the equation $f_a(x, y) = 0$, where

$$f_a(x, y) = (x^2 + y^2 - 2ax)^2 - 4a^2(x^2 + y^2).$$

From now on, we consider the cardioid for $f = f_1$ (figure 29.6).

With the notation of proposition 276, we have $U \subset \mathbb{R}$, $W \subset \mathbb{R}$ and $T \subset \mathbb{R}$. Then the curve is the fiber of 0, $V(f) = f^{-1}(0)$. The matrix

$$D = (D_y f(r, s)) = (4s(r^2 + s^2 - 2r) - 8s)$$

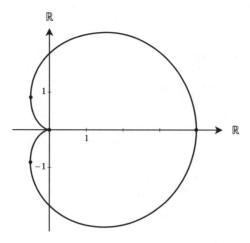

Fig. 29.6. The cardioid $V(f) = V((x^2 + y^2 - 2x)^2 - 4(x^2 + y^2))$.

has determinant $\det(D) = 4s(r^2 + s^2 - 2r) - 8s$. Wanted are the points where $\det(D) = 0$ for $(r, s) \in V(f)$. They are the solutions S of the system of polynomial equations

$$(r^2 + s^2 - 2r)^2 - 4(r^2 + s^2) = 0$$
$$4s(r^2 + s^2 - 2r) - 8s = 0$$

We will not go into the details of solving such equations, we simply state the result:

$$S = \{(0,0), (4,0), (-\tfrac{1}{2}, \tfrac{\sqrt{3}}{2}), (-\tfrac{1}{2}, -\tfrac{\sqrt{3}}{2})\}$$

Now we have to find U_i and V_i, such that $S \cap (U_i \times V_i) \neq 0$, i.e., rectangles covering the curve $V(f)$ except at the "singular" points in S. Looking at figure 29.6, where the singular points are marked in black, we can divide the curve into four arcs, each starting and ending at a singular point. The four arcs are covered by the following open sets

$$O_1 = \,]-\tfrac{1}{2}, 4[\, \times\,]\tfrac{\sqrt{3}}{2}, \infty[\, \cup\,]0, 4[\, \times\, \mathbb{R}$$
$$O_2 = \,]-\tfrac{1}{2}, 4[\, \times\,]-\tfrac{\sqrt{3}}{2}, -\infty[\, \cup\,]0, 4[\, \times\, \mathbb{R}$$
$$O_3 = U_3 \times V_3 = \,]-\tfrac{1}{2}, 0[\, \times\,]0, \tfrac{\sqrt{3}}{2}[$$
$$O_4 = U_4 \times V_4 = \,]-\tfrac{1}{2}, 0[\, \times\,]0, -\tfrac{\sqrt{3}}{2}[$$

Proposition 276 assures us that for each $U_1 = U_2 = \,]-\tfrac{1}{2}, 4[$, and $U_3 = U_4 = \,]-\tfrac{1}{2}, 0[$ we can find an explicit function g_i defined on U_i, whose graph is the arc of the curve which is contained in O_i:

$$g_1(x) = \sqrt{2 + 2x - x^2 + 2\sqrt{1 + 2x}}$$

$$g_2(x) = -\sqrt{2 + 2x - x^2 + 2\sqrt{1 + 2x}}$$

$$g_3(x) = \sqrt{2 + 2x - x^2 - 2\sqrt{1 + 2x}}$$

$$g_4(x) = -\sqrt{2 + 2x - x^2 - 2\sqrt{1 + 2x}}$$

Exercise 153 In example 113, the functions $y = g_i(x)$ result from a consideration of the matrix $D_y f(r, s)$. Similarly, functions $x = h_i(y)$ can be found by considering $D_x h(r, s)$. Perform the analogous procedure to determine the h_i.

Knowing that g in proposition 276 is differentiable, it is no problem to effectively find its derivative. We have $f_i(x, g(x)) = 0$ for $x \in U(r)$. Then, taking the partial derivative with respect to the j-th coordinate of x, and using the chain rule, yields

$$0 = D_j f_i(x, g(x)) + \sum_{t=1,\dots m} D_{n+t} f_i(x, g(x)) \cdot D_j g_t(x).$$

For each $j = 1, 2, \dots n$, we have m linear equations for $i = 1, 2, \dots m$ and m unknowns $D_j g_t(x), t = 1, 2, \dots m$. Since the determinant $\det(D) \neq 0$, this system of m linear equations (for fixed j) in the partial derivatives $D_j g_t(x)$ has a unique solution.

Example 114 Let us come back to the above example $f(x, y, z) = 3(x^2 + y^2) + 4z^2 - 1$ defined on $\mathbb{R}^2 \times \mathbb{R}$, so $n = 2$ and $m = 1$. We have to take the two partial derivatives for $j = 1, 2$:

$$0 = D_1 f(x, y, g(x, y)) + D_3 f(x, y, g(x, y)) \cdot D_1 g(x, y)$$
$$= 6x + 8g(x, y) \cdot D_1 g(x, y)$$
$$0 = D_2 f(x, y, g(x, y)) + D_3 f(x, y, g(x, y)) \cdot D_2 g(x, y)$$
$$= 6y + 8g(x, y) \cdot D_2 g(x, y).$$

Here $\det(D)$ is equal to $8g(x, y)$ and does not vanish for $x^2 + y^2 < \frac{1}{3}$. We have

$$D_1 g(x, y) = \frac{-3x}{4g(x, y)} = \frac{\mp 3x}{2\sqrt{1 - 3(x^2 + y^2)}}$$

and

$$D_2g(x,y) = \frac{-3y}{4g(x,y)} = \frac{\mp 3y}{2\sqrt{1 - 3(x^2 + y^2)}}$$

depending on the solution

$$g(x,y) = \frac{1}{2}\sqrt{1 - 3(x^2 + y^2)}$$

or

$$g(x,y) = -\frac{1}{2}\sqrt{1 - 3(x^2 + y^2)}.$$

29.3.1 A Remark on Global Coordinates and Manifolds

The implicit function theorem 29.3 gives rise to a fascinating generalization of the concept of coordinate spaces. Consider the simple example
$f(x,y,z) = 3(x^2 + y^2) + 4z^2 - 1$ from above and the associated solution set $V(f) = \{(x,y,z) \mid f(x,y,z) = 0\}$. We have seen that if the
Jacobian matrix Df has the third partial derivative $D_z f(x,y,z) \neq 0$,
we may find an open neighborhood B of (x,y,z) such that $V(f) \cap B$
is the graph of a function $g : U \to \mathbb{R}$. This means that the projection
$pr_{x,y} : V(f) \cap B \to U$ is a bijection. So around $(x,y,z) \in V(f)$, $V(f)$
is bijective with an open set of $\mathbb{R}^2$. But what happens on points where
$D_z f(x,y,z) = 0$? Now, the equation $f = 0$ does not actually stress
the *third* coordinate, we just need to consider *one* of the three coordinates $w = x,y,z$ with $D_w f(x,y,z) \neq 0$. Then a permutation of the
coordinates yields such a local parametrization by an open set in $\mathbb{R}^2$.[1]
In fact, we have $df = (6x, 6y, 8z)$, and this number never vanishes on
$V(f)$, since $(0,0,0) \notin V(f)$. So every point $v \in V(f)$ has a neighborhood $U(v)$ such that a projection $pr : U(v) \cap V(f) \to W_v \subset \mathbb{R}^2$ is a local
parametrization, however not always with the same projections! One calls
such a local bijection a *chart* for $V(f)$. So we are interested in the com-
patibility of such charts, more precisely, if $v, w \in V(f)$, then we have
two charts $pr_v : U(v) \cap V(f) \to W_v$ and $pr_w : U(w) \cap V(f) \to W_w$.
Consider now the restrictions $pr_v|_{U(v) \cap U(w)} : U(v) \cap U(w) \cap V(f) \to W_v$
and $pr_w|_{U(v) \cap U(w)} : U(v) \cap U(w) \cap V(f) \to W_w$, and their images, which
we denote by $W_v|w$ and $W_w|v$, respectively, which are in fact open sets.
Then, by composing the inverse of $pr_v|_{U(v) \cap U(w)}$ with $pr_w|_{U(v) \cap U(w)}$, we
have a bijection $q_{v,w} : W_v|w \xrightarrow{\sim} W_w|v$ of open sets, which in fact is a
diffeomorphism.

---

[1] This procedure has been hinted at in exercise 153.

So we have this situation: The solution set $V(f)$ is not globally in bijection with an open set of $\mathbb{R}^2$, but it is so locally in the neighborhood of every point. And moreover, if we have two such local parametrizations by open sets of $\mathbb{R}^2$, they are compatible insofar as on their intersections, the parametrizations are related to each other by a diffeomorphism. This is the birth of what Bernhard Riemann has inaugurated in his habilitation talk in 1854 ("Über die Hypothesen, welche der Geometrie zugrunde liegen"): the theory of differentiable manifolds. In a first description, a *differentiable manifold of dimension k* is a subset $M \subset \mathbb{R}^n$ which is covered by a family $(U_i)_i$ of open sets of $\mathbb{R}^n$ such that every intersection set $M_i = U_i \cap M$ is bijective to an open set $K_i \subset \mathbb{R}^k$, $f_i : M_i \xrightarrow{\sim} K_i$, and such that for any pair i, j of indexes, the restrictions $f_{ij} = f_i|_{M_i \cap M_j} \xrightarrow{\sim} K_{ij} = Im(f_{ij})$ are open in K_i and such that all compositions $f_{ji} \circ f_{ij}^{-1} : M_i \cap M_j \xrightarrow{\sim} M_j \cap M_i$ are diffeomorphisms. The system $(M_i, f_i)_i$ is called an *atlas of M*. Intuitively, this means that M is a patchwork of subsets, which are glued together in a differentiable way from open charts in $\mathbb{R}^k$. Famous examples of manifolds are the unit circle $S^1 = V(X^2 + y^2 - 1)$, the unit sphere $S^2 = V(X^2 + y^2 + z^2 - 1)$, or the torus $T = \{(\cos(t) - \frac{1}{2}\cos(s)\cos(t), \sin(t) - \frac{1}{2}\cos(s)\sin(t), \frac{1}{2}\sin(s)) \mid 0 \le s, t \le 2\pi\}$ (figure 29.7).

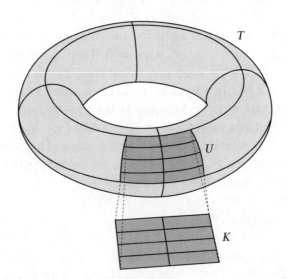

Fig. 29.7. A torus $T \subset \mathbb{R}^3$ is made up of "patches" that are in bijection with open squares in $\mathbb{R}^2$.

Example 115 The curve M in figure 29.8 is a manifold of dimension 1. It is covered by open sets U_i, for $1 \leq i \leq 6$. Every arc $M_i = U_i \cap M$ is bijective to an open interval K_i by a bijection f_i. Thus $(M_i, f_i)_i$ form an atlas of M.

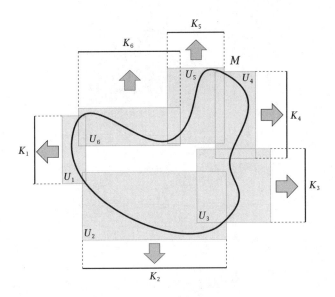

Fig. 29.8. An atlas of M consisting of the open sets $M_i = U_i \cap M$ together with the bijections $M_i \xrightarrow{\sim} K_i$, where K_i are open intervals.

CHAPTER 30

Integration

30.1 Introduction

This chapter is a short introduction to the integral calculus, as it appears from the point of view of surface and volume calculation. In the next chapter, we shall relate the present theory to the differential calculus by the fundamental theorem of calculus (proposition 285), which interprets the calculation of surfaces as an inverse procedure to differentiation. Surface and volume calculation has a very long tradition, which had its first climax in the work of the Greek mathematician Archimedes (287-212 BC), including the calculation of the volume of cylinder and sphere. Only in the seventeenth century did Johannes Kepler reconsider, in his work "Nova stereometria doliorum vinariorum" ("New volume measurement of vine casks") from 1615, the method of infinite summation method introduced by Archimedes. In the infinitesimal calculus of Leibniz, the infinite summation gained its modern shape, also by the introduction of the integral sign ∫ in 1676 as a stylization of the sum sign S. Leibniz called his method "calculus summatorius", but Johann Bernoulli changed that title to the nowadays current "calculus integralis". It is remarkable that Newton did not mention his contribution to the integral calculus (which he called "methodus fluxionum") in his monumental work "Principia", since he did not want to be criticized for something not so certain. With Augustin Cauchy and then Bernhard Riemann in his famous habilitation talk "Über die Hypothesen, welche der Geometrie zugrunde liegen" from 1854, in which he absorbed Cauchy's approach, the Riemann integral was created, giving the infinite summation a precise meaning in terms of limits. Later, when successively more "pathological" functions were considered for integration, more powerful integral concepts based on Henri-Léon Lebesgue's measure theory were created in 1902. However we refrain from these theories, since in an introductory book, they cannot be treated in due detail, and since the special case of a Riemann integral is sufficient for all the examples which we shall deal with.

30.2 Partitions and the Integral

For the theory of integration, we need some easy preliminary structures. The first of these is concerned with the approximation of a function by step functions on a tiling of the domain of the function which one would like to integrate.

Definition 193 *Given a pair (a, b) of real numbers with $a < b$, the set $Part(a, b)$ is the set of all finite subsets $P \subset [a, b]$ with $\{a, b\} \subset P$. The elements of $Part(a, b)$ are called the* partitions *of the interval $[a, b]$. If $P \in Part(a, b)$, then its elements can be put into increasing order $a = x_0 < x_1 < \ldots x_k = b$. Whenever we enumerate the members of P, we mean this ordering.*

By definition, the set $I(P)$ consists of all closed intervals $[x_i, x_{i+1}]$ delimited by successive elements of P and is called the interval set *of P. On $Part(a, b)$, one has the refinement relation which is the set inclusion, i.e., one says that Q refines P, iff $P \subset Q$, which is denoted by $P \to Q$. Since $Part(a, b)$ is closed under intersection and union, there is, for any pair P and P' of partitions of $[a, b]$, a refinement Q such that $P \to Q$ and $P' \to Q$. Such a refinement Q is called a* common refinement *of P and P'. For $P \in Part(a, b)$ and $I = [x_i, x_{i+1}] \in I(P)$, we set $vol(I) = x_{i+1} - x_i$ and call it the* volume *of I.*

Starting with these one-dimensional partitions one introduces n-dimensional partitions as follows.

Definition 194 *For $n \in \mathbb{N}, n > 1$, and a sequence*

$$(a., b.) = (a_1, a_2, \ldots a_n, b_1, b_2, \ldots b_n) \in \mathbb{R}^{2n}$$

with $a_i < b_i, i = 1, 2, \ldots n$, we define the partition set

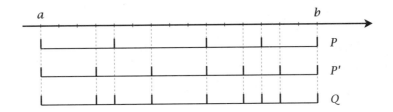

Fig. 30.1. Q is a common refinement of the partitions P and P' of $[a, b]$.

$$Part(a., b.) = \prod_i Part(a_i, b_i)$$

as the Cartesian product of the one-dimensional partition sets $Part(a_i, b_i)$. On $Part(a., b.)$, one considers the product relation of refinements in the n factors, i.e., if $P = (P_i)_i, Q = (Q_i)_i \in Part(a., b.)$, then $P \to Q$, iff $P_i \to Q_i$ for all $i = 1, 2, \ldots n$. A common refinement Q of two partitions $P, P' \in Part(a., b.)$ *always exists and is denoted by $P \to Q$ and $P' \to Q$ in analogy with the 1-dimensional case.*

For $P = (P_i)_i \in Part(a., b.)$, the set of cubes $I(P)$ consists of all closed cubes $K = \prod_i K_i$, for $K_i \in I(P_i)$, i.e., of the Cartesian products of n intervals (the one-dimensional cubes). We denote by $K(a., b.)$ the cube $\prod_i [a_i, b_i]$ defined by the minimal sets $P_i = \{a_i, b_i\}$. In accordance with the elementary concept of a volume, we set $vol(K) = \prod_i vol(K_i)$.

Example 116 In the Euclidean space $\mathbb{R}^3$, we consider partitions of the cuboid $\{(x, y, z) \mid \frac{1}{2} \le x \le 4, \frac{1}{2} \le y \le 3, 1 \le z \le 3\}$. In this case, we have $a_1 = \frac{1}{2}, b_1 = 4, a_2 = \frac{1}{2}, b_2 = 3$ and $a_3 = 1, b_3 = 3$. The set of partitions is given by

$$Part(\tfrac{1}{2}, \tfrac{1}{2}, 1, 4, 3, 3) = Part(\tfrac{1}{2}, 4) \times Part(\tfrac{1}{2}, 3) \times Part(1, 3).$$

The particular partition P in figure 30.2 is

$$P = (\{\tfrac{1}{2}, \tfrac{3}{4}, 3, 4\}, \{\tfrac{1}{2}, 1, \tfrac{3}{2}, \tfrac{5}{2}, 3\}, \{1, \tfrac{7}{4}, 3\})$$

The elements of the three sets are of course unordered, but we enumerate them here in increasing order to emphasize that these sets are partitions. To complete this example, we pick out one K, namely, the dark gray one in the figure, $K = [3, 4] \times [1, \tfrac{3}{2}] \times [1, \tfrac{7}{4}]$, its volume is $vol(K) = (4-3) \cdot (\tfrac{3}{2} - 1) \cdot (\tfrac{7}{4} - 1) = \tfrac{3}{8}$. The volume of the whole cuboid is $vol(K(\tfrac{1}{2}, \tfrac{1}{2}, 1, 4, 3, 3)) = (4 - \tfrac{1}{2}) \cdot (3 - \tfrac{1}{2}) \cdot (3 - 1) = \tfrac{35}{2}$.

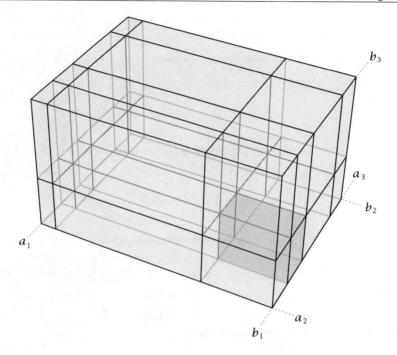

Fig. 30.2. A particular partition in $Part(a_1, a_2, a_3, b_1, b_2, b_3)$.

Let $(a., b.)$ be a sequence of real numbers as in definition 194 and let $K \in I(P)$. Take a bounded function $f : K \to \mathbb{R}$, i.e., there is $b \in \mathbb{R}$ with $|f(x)| \leq b$, for all $x \in K$. Then we denote $m_K(f) = \inf\{f(x) \mid x \in K\}$ and $M_K(f) = \sup\{f(x) \mid x \in K\}$. The *infimum* $\inf(X)$ *of a set* X *of real numbers is by definition* $\inf(X) = -\sup(-X)$, where $-X = \{-x \mid x \in X\}$. Clearly, if a cube K' is such that $K' \subset K$, then $m_{K'}(f) \geq m_K(f)$ and $M_{K'}(f) \leq M_K(f)$.

Definition 195 *Let* $(a., b.)$ *be a sequence of real numbers as in definition 194. If* $f : K(a., b.) \to \mathbb{R}$ *is a bounded function, and if* $P \in Part(a., b.)$, *we define the* lower sum *of* f *over* P *by*

$$L(f, P) = \sum_{K \in I(P)} m_K(f) \cdot vol(K)$$

and the upper sum *of* f *over* P *by*

$$U(f, P) = \sum_{K \in I(P)} M_K(f) \cdot vol(K)$$

Example 117 Figure 30.3 shows the upper and lower sums for the function

$$f : \mathbb{R} \to \mathbb{R} : x \mapsto \tfrac{8}{3}x^3 - \tfrac{41}{5}x^2 + \tfrac{9}{2}x + \tfrac{33}{10}$$

for a partition $P = \{0, \tfrac{1}{2}, 1, \tfrac{3}{2}, 2, \tfrac{5}{2}\}$.

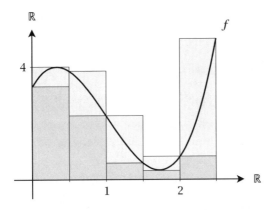

Fig. 30.3. The lower sum (area of the dark gray rectangles) and upper sum (dark gray and light gray rectangles) of f.

The minima of the partition intervals are

$$m_{[0,\frac{1}{2}]} = \tfrac{33}{10}, \quad m_{[\frac{1}{2},1]} = \tfrac{34}{15},$$
$$m_{[1,\frac{3}{2}]} = \tfrac{3}{5}, \quad m_{[\frac{3}{2},2]} = \tfrac{26029-781\sqrt{781}}{12000},$$
$$m_{[2,\frac{5}{2}]} = \tfrac{5}{6}.$$

Since $vol(K) = \tfrac{1}{2}$, for $K \in I(P)$, the lower sum is

$$L(f,P) = \sum_{K \in I(P)} m_K \cdot vol(K) = \frac{1}{2} \sum_{K \in I(P)} m_K \approx 5.49.$$

For the maxima of the partition intervals, we have

$$M_{[0,\frac{1}{2}]} = \tfrac{26029+781\sqrt{781}}{12000}, \quad M_{[\frac{1}{2},1]} = \tfrac{23}{6},$$
$$M_{[1,\frac{3}{2}]} = \tfrac{34}{15}, \quad M_{[\frac{3}{2},2]} = \tfrac{5}{6},$$
$$M_{[2,\frac{5}{2}]} = \tfrac{149}{30},$$

and the upper sum is

$$U(f,P) = \sum_{K \in I(P)} M_K \cdot vol(K) = \frac{1}{2} \sum_{K \in I(P)} M_K \approx 7.94.$$

Figure 30.4 shows the upper and lower sums for the function

$$g : \mathbb{R}^2 \to \mathbb{R} : (x,y) \mapsto \frac{4}{x + 2y + 1}$$

for a partition $P = (\{0,1,2\}, \{0,1,2\})$.

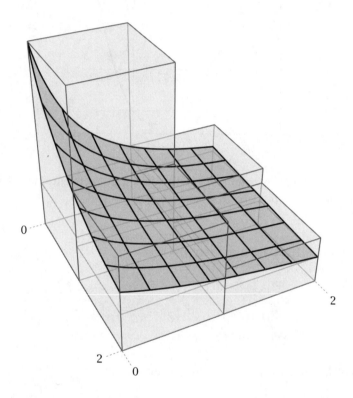

Fig. 30.4. The lower and upper sums of the surface g.

Lemma 277 *If $P \to Q$ is a refinement with respect to the sequence $(a., b.)$, and if $f : K(a., b.) \to \mathbb{R}$ is a bounded function, then $L(f,P) \le L(f,Q)$ and $U(f,Q) \le U(f,P)$.*

Proof Since every cube $K \in I(P)$ is the union of a number $K_1, \ldots K_r$ of cubes of Q, and since different cubes in $I(P)$ are unions of disjoint sets of cubes of Q, it suffices to show the lemma for P having one single cube K. But then

$$L(f,P) = m_K(f) \cdot vol(K)$$

$$= m_K(f) \cdot \sum_{i=1}^{r} vol(K_i)$$

$$= \sum_{i=1}^{r} m_K(f) \cdot vol(K_i)$$

$$\leq \sum_{i=1}^{r} m_{K_i}(f) \cdot vol(K_i)$$

$$= L(f,Q).$$

The same argument yields $U(f,Q) \leq U(f,P)$. $\square$

Proposition 278 *If P and Q are any two partitions in Part(a.,b.), and if* $f : K(a.,b.) \to \mathbb{R}$ *is a bounded function, then* $L(f,P) \leq U(f,Q)$. *In particular, we have this inequality of real numbers:*

$$\sup\{L(f,P) \mid P \in Part(a.,b.)\} \leq \inf\{U(f,P) \mid P \in Part(a.,b.)\}.$$

Proof Take a common refinement $P \to P', Q \to P'$ of P, Q, then lemma 277 yields $L(f,P) \leq L(f,P') \leq U(f,P') \leq U(f,Q)$. The inequality between suprema and infima is then obvious. $\square$

Definition 196 *Under the condition of proposition 278, if we have the equality*

$$\sup\{L(f,P) \mid P \in Part(a.,b.)\} = \inf\{U(f,P) \mid P \in Part(a.,b.)\},$$

then this number is denoted by $\int_{(a.,b.)} f$ *and is called the* (Riemann) *integral of* f *over* $(a.,b.)$. *The function* f *is then called* integrable. *In more traditional texts, the notation*

$$\int_{(a.,b.)} f(x_1, x_2, \ldots x_n)\, dx_1 \ldots dx_n$$

is also used. For $n = 1$, *where* $(a.,b.) = (a,b)$, *the notations* $\int_a^b f$ *and* $\int_a^b f(x)\, dx$ *are also customary.*

The set of (Riemann) integrable bounded functions $f : K(a.,b.) \to \mathbb{R}$ *is denoted by* $\mathcal{R}(a.,b.)$.

Exercise 154 Show that a constant function $f = c$ is integrable, and that in this case $\int_{(a.,b.)} f = c \cdot vol(K(a.,b.))$.

Exercise 155 The function $f : [0,1] \times [0,1] \to \mathbb{R}$ defined by

$$f(x,y) = \begin{cases} 0 & \text{if } x + y \in \mathbb{Q}, \\ 1 & \text{else.} \end{cases}$$

is not integrable, while the function $f : [0,1] \times [0,1] \to \mathbb{R}$ defined by

$$f(x,y) = \begin{cases} 0 & \text{if } y \leq 1/3 \text{ and } x \leq 2/3, \\ 1 & \text{else.} \end{cases}$$

is integrable. Hint for the second function: Choose adequate partitions to calculate the lower and upper sums.

The following result yields an important method for the construction of integrable functions, and also for the calculation of integrals.

Sorite 279 *Let $(a.,b.)$ be a sequence defining the closed cube $K(a.,b.) \subset \mathbb{R}^n$.*

(i) *The map $\int_{(a.,b.)} : \mathcal{R}(a.,b.) \to \mathbb{R}$ is linear, i.e., for any $f,g \in \mathcal{R}(a.,b.)$ and $\lambda, \mu \in \mathbb{R}$, we have the formulas*

$$\int_{(a.,b.)} \lambda f + \mu g = \lambda \int_{(a.,b.)} f + \mu \int_{(a.,b.)} g.$$

(ii) *Set $f \leq g$ iff $f(x) \leq g(x)$ for all $x \in K(a.,b.)$. Then, if $f \leq g$ for $f,g \in \mathcal{R}(a.,b.)$, we have*

$$\int_{(a.,b.)} f \leq \int_{(a.,b.)} g.$$

(iii) *(Mean Value Theorem of Integral Calculus) If $f : K(a.,b.) \to \mathbb{R}$ is continuous, then there is $x \in K(a.,b.)$ such that*

$$\int_{(a.,b.)} f = f(x) \cdot vol(K(a.,b.)).$$

(iv) *If $f \in \mathcal{R}(a.,b.)$, then $|f| : x \mapsto |f(x)|$ is also integrable, and we have*

$$\left| \int_{(a.,b.)} f \right| \leq \int_{(a.,b.)} |f|.$$

Proof Claim (i) follows from the fact that the supremum or infimum of a linear combination of bounded functions is the linear combination of the suprema and infima of these functions.

Claim (ii) is evident.

As to claim (iii), $Im(f)$ is a closed interval $[a, b]$. And clearly,

$$a \cdot vol(K(a., b.)) \leq \int_{(a.,b.)} f \leq b \cdot vol(K(a., b.)).$$

Therefore,

$$a \leq \frac{\int_{(a.,b.)} f}{vol(K(a., b.))} \leq b.$$

So there is x such that $f(x) = \frac{\int_{(a.,b.)} f}{vol(K(a.,b.))}$, and we are done.

For claim (iv), if $|f|$ is integrable, then, evidently, the claimed inequality holds. So it remains to show that $|f|$ is integrable if f is so. Integrability means that for every $\varepsilon > 0$, there is a partition P such that $0 \leq U(f, P) - L(f, P) < \varepsilon$. Now, if this holds, then for every cube $K \in P$, it is easy to check that we have $M_K(f) - m_K(f) \geq M_K(|f|) - m_K(|f|) \geq 0$, therefore $0 \leq U(|f|, P) - L(|f|, P) \leq U(f, P) - L(f, P) < \varepsilon$, and $|f|$ is integrable.   □

Despite this general result, it is hard to calculate integrals at this stage of the theory. In the following, we shall develop several tools and criteria for calculating integrals. However, it is not true that integration is always explicitly representable by "well-known" functions, on the contrary: integration is a device for generating truly new functions from known ones.

30.3 Measure and Integrability

It can be shown that a function $f : K(a., b.) \rightarrow \mathbb{R}$ is integrable, iff it is continuous except for a "negligible" subset of points. We want to give a precise description of the meaning of the term "negligible". We again need closed cubes $K \subset \mathbb{R}^n$, i.e., by definition, subsets of the form $K = \prod_i [l_i, u_i]$ where $[l_i, u_i] \subset \mathbb{R}$ are closed intervals of $\mathbb{R}$ with $l_i < u_i$. As above, the cube's volume is $vol(K) = \prod_i (u_i - l_i)$.

Definition 197 *A subset $M \subset \mathbb{R}^n$ has measure 0, iff for every $\varepsilon > 0$, there is a covering family $(K_i)_{i \in \mathbb{N}}$ of cubes K_i for M, i.e., $M \subset \bigcup_i K_i$, such that*

$$V = \sum_i vol(K_i) = \lim_{N \to \infty} \sum_{i=0}^{N} vol(K_i)$$

exists, and $V < \varepsilon$.

Exercise 156 Show that, if $M \subset \mathbb{R}^n$ is finite, then it has measure zero.

Example 118 If M is denumerable, i.e., by definition, if there is a bijection $M \overset{\sim}{\to} \mathbb{N}$, then it has measure 0. In fact, we may enumerate the elements of M in a sequence $(m_i)_{i \in \mathbb{N}}$, $M = \{m_i \mid i \in \mathbb{N}\}$. Take the covering family $(K_i)_i$ with the closed cubes $K_i = \overline{K}_{\varepsilon/2^{i+1}}(m_i)$. Then $\sum_i vol(K_i) = \sum_i \frac{\varepsilon}{2^n} \cdot (2^{-n})^i = \frac{\varepsilon}{2^n} \frac{1}{1-2^{-n}} < \varepsilon$.

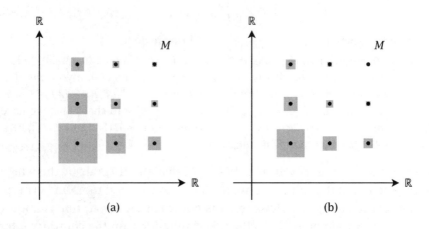

(a)   (b)

Fig. 30.5. (a) A covering of M of measure $\pi^4/36$, (b) a covering of measure $\pi^4/72$.

The set $M = \{(x,y) \in \mathbb{R}^2 \mid x,y \in \mathbb{N} - \{0\}\}$, subset of $\mathbb{R}^2$, is clearly denumerable. For a first covering by squares, each point (i,j) in M is covered by a square of volume (area) $V_{ij} = \frac{1}{(ij)^2}$, i.e., a square of side length $\frac{1}{ij}$ (see figure 30.5 (a)). Then the volume V of the whole covering is

$$V = \sum_{i=1,j=1}^{\infty} V_{ij} = \sum_{i=1,j=1}^{\infty} \frac{1}{(ij)^2} = \frac{\pi^4}{36}.$$

With $\varepsilon = \frac{\pi^4}{18}$, we have $V < \varepsilon$. To proceed from ε to $\varepsilon/2$, simply halve the areas of each square, i.e., $V_{ij} = \frac{1}{2}(ij)^2$. The side length of a point (i,j) is now $\frac{1}{\sqrt{2}ij}$ and $V = \frac{\pi^4}{72}$ which is less than $\varepsilon/2 = \frac{\pi^4}{36}$ (figure 30.5 (b)).

This is quite dramatic, since the example $M = \mathbb{Q}^n \subset \mathbb{R}^n$ yields a set which seems to fill almost all of $\mathbb{R}^n$, but is effectively too small to yield a non-zero measure. In general, the measure of a set $M \subset \mathbb{R}^n$ is defined as the infimum of the sums $\sum_i vol(K_i)$ for all possible covering families $(K_i)_{i \in \mathbb{N}}$ of closed cubes $K_i \subset \mathbb{R}^n$.

A straightforward generalization of example 118 is the following proposition.

Proposition 280 *If $M = \bigcup_{i \in \mathbb{N}} M_i$ is the union of a family $(M_i)_i$ of sets M_i of measure 0, then M has measure 0.*

Proof The idea is to take a covering $(K_i^j)_{j \in \mathbb{N}}$ for each M_i with $\sum_j vol(K_i^j) < \varepsilon \cdot 2^{-(i+1)}$ for every i. Then consider the zigzag covering sequence $K_1^1, K_2^1, K_1^2, K_3^1, K_2^2, K_1^3, \ldots$ of M and we evidently have $\sum_{i,j} vol(K_i^j) < \varepsilon \cdot \sum_i 2^{-(i+1)} < \varepsilon$. $\qquad\square$

For the announced characterization of integrable functions, we need a restatement of continuity.

Definition 198 *For a subset $K \subset \mathbb{R}^n$, and a bounded function $f : K \to \mathbb{R}$, a point $x_0 \in K$, and a positive real number δ, we set*

$$m(x_0, f, \delta) = \inf\{f(x) \mid x \in B_\delta(x_0) \cap K\},$$
$$M(x_0, f, \delta) = \sup\{f(x) \mid x \in B_\delta(x_0) \cap K\}.$$

Then the oscillation $o(f, x_0)$ at x_0 is defined by the (evidently existing) limit

$$o(f, x_0) = \lim_{\delta \to 0} (M(x_0, f, \delta) - m(x_0, f, \delta)).$$

The relation with continuity is as follows:

Proposition 281 *For a subset $K \subset \mathbb{R}^n$, a function $f : K \to \mathbb{R}$, and a point $x_0 \in K$, we have $o(f, x_0) = 0$, iff f is continuous at x_0.*

Proof This is evident, it is left as an exercise for the reader. $\qquad\square$

Lemma 282 *Let $f : K \to \mathbb{R}$ be a bounded function on a closed cube $K \subset \mathbb{R}^n$, $K = K(a., b.)$, and $\varepsilon > 0$. If $o(f, x) < \varepsilon$ for all $x \in K$, then there is a partition $P \in Part(a., b.)$ such that $U(f, P) - L(f, P) < \varepsilon \cdot vol(K)$.*

Proof For each $x \in K$, take a closed cube $\overline{K}(x)$ such that $\varepsilon > M_{K(x)} - m_{K(x)}$. By compactness of K, there is a finite number $\overline{K}(x_1), \ldots \overline{K}(x_r)$ of such cubes, which cover K. Choose a partition of $K \in P(a., b.)$, which is fine enough such that every cube $K_i \in I(P)$ is contained in one of the above cubes $\overline{K}(x_1), \ldots \overline{K}(x_r)$. Then for this partition, $U(f, P) - L(f, P) < \varepsilon \cdot vol(K)$. $\qquad\square$

And this is the announced characterization:

Proposition 283 *Let $K = K(a., b.) \subset \mathbb{R}^n$ be a closed cube and $f : K \to \mathbb{R}$ a bounded function. Then f is integrable, iff the set $NC = \{x \mid x \in K, f$ is not continuous at $x\}$ has measure 0.*

Proof The proof is quite technical and not very inspiring, it uses lemma 282 as well as proposition 280. We therefore omit it here and refer to the beautiful book [39]. □

Exercise 157 Use propositions 280 and 283 to show that if $f, g \in \mathcal{R}(a., b.)$, then $f \cdot g \in \mathcal{R}(a., b.)$.

A special case of the integral is the volume of a general subset $S \subset K$ of a closed cube $K \in \mathbb{R}^n$. We need a couple of topological concepts.

Definition 199 *If $X \subset \mathbb{R}^n$ is given the relative topology as defined in definition 182, then for a subset $Y \subset X$, we define Y^o as the union of all $U \subset Y$ which are open in the relative topology on X, or, equivalently, Y^o is the maximal open subset in Y with respect to the relative topology on X. The set Y^o is called the* interior *of Y.*

The set $\overline{Y}$ is the intersection of all $W \subset X$, containing Y, which are closed in the relative topology on X, or, equivalently, the smallest closed superset of Y in the relative topology of X, is called the closure *of Y.*

The set $\partial Y = \overline{Y} - Y^o$ is called the boundary *of Y (with respect to X).*

Exercise 158 With the preceding notations, show that $\overline{Y} = X - (X - Y)^o$ and $Y^o = X - \overline{(X - Y)}$. Also show that $\overline{\overline{Y}} = \overline{Y}$ and $(Y^o)^o = Y^o$.

Proposition 284 *The characteristic function $\chi_S : K \to \mathbb{R}$ of a subset $S \subset K$ of a closed cube $K = K(a., b.) \subset \mathbb{R}^n$ is integrable, iff its boundary (in K or in $\mathbb{R}^n$, it makes no difference since K is closed) has measure 0.*

Proof The characteristic function χ_S is constant on the relatively open sets S^o and on $K - \overline{S}$. So it is continuous there. In $x \in \partial S$, both values, 0 and 1 are taken in any neighborhood of x. Therefore no neighborhood of x is mapped into a neighborhood $U_{1/2}(f(x))$, and χ_S cannot be continuous on ∂S. The claim now follows from proposition 283. □

Exercise 159 Show that the characteristic function of the subset S of rationals in $K = [0, 1]$ is not integrable, in fact, $\partial S = K$.

Definition 200 *A subset $S \subset K$ of a closed cube $K = K(a., b.) \subset \mathbb{R}^n$ with boundary ∂S having measure 0 is called* Jordan-measurable. *Its* volume $vol(S)$ *is defined as the integral*

$$vol(S) = \int_K \chi_S$$

of its characteristic function. Observe that the omission of the cube K in the volume notation is justified by the fact that the integral to the right does not depend on the surrounding cube K.

Example 119 Definition 200 may be used to calculate the volume or area of a geometric object, for example the shape S bounded by the curve in figure 30.6. To formulate a Riemann integral we define the characteristic function $\chi(A_i)$ of an area element A_i to be equal to 1 if $A_i \cap S \neq \emptyset$, and 0 else.

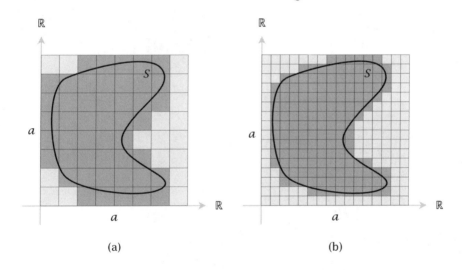

Fig. 30.6. A partition (a) and a refinement (b) of a square that completely contains the shape S bounded by the curve. The squares that intersect S are shown in dark gray.

In figure 30.6 (a), the square K with side length a, which completely contains S, is partitioned into 64 little squares A_i of area $vol(A_i) = \frac{a^2}{64}$. We have the sum

$$\sum_i \chi(A_i) \cdot vol(A_i).$$

This sum counts the number of squares that intersect S. In the concrete example of figure 30.6 (a), this number is 47, thus we have the approximated area $\frac{47}{64} \cdot a^2$.

In figure 30.6 (b) the partition is refined by halving the side length of the squares. Here we have 154 squares, each of side length $\frac{a^2}{256}$, giving a better approximation $\frac{154}{256} \cdot a^2$ for the area of S.

Successive refinements generate squares whose areas tend to 0 and lead to the Riemann integral

$$\lim_{vol(A_i) \to 0} \sum_i \chi(A_i) \cdot vol(A_i) = \int_K \chi_S = vol(S).$$

CHAPTER 31

The Fundamental Theorem of Calculus and Fubini's Theorem

31.1 Introduction

The theory of integration offers several methods for dealing with the a priori quite cryptic concept of integrals. A first group consists of recursive methods. We first deal with one-dimensional integration and then show how higher-dimensional integrals can be reduced to lower-dimensional ones. The former is essentially the fundamental theorem of calculus which interprets integration and differentiation as reciprocal processes. Roughly speaking, in dimension 1, integration of a function f may be settled by finding functions F such that $DF(x) = (f(x))$ (recall that $DF(x)$ is a matrix, whence the parentheses around $f(x)$). The induction step is Fubini's theorem which transforms an integral $\int_{K_1 \times K_2} f$ over a Cartesian product of cubes K_1 and K_2 to a double integral $\int_{K_1} \int_{K_2} f$, where the inner integral $\int_{K_2} f$ is regarded as a function on the cube K_1.

The second group of methods deals with changing variables, a procedure which we know from the chain rule for differentiation. It permits the reinterpretation of seemingly complex functions as simpler ones for a composed function, thereby changing the cube's limit values.

31.2 The Fundamental Theorem of Calculus

This is the one-dimensional "anchorage of induction". To begin with, we need a slight generalization of differentiation concerning the domains where derivatives of functions are defined. Let $f : [a, b] \to \mathbb{R}$ be a function on a closed interval. Then its derivative $f'(x)$ in the interior $[a, b]^o =]a, b[$ of the closed interval $[a, b]$ is defined. We now also need a definition of the derivative at the boundary points a and b of the closed interval. Reviewing all the definitions necessary for the concept of a derivative, one recognizes that the limit process $\lim_{z \to 0} \frac{\|f(a+z) - f(a) - D(a)\|}{\|z\|}$ must be restricted to $z > 0$, and, analogously $\lim_{z \to 0} \frac{\|f(b+z) - f(b) - D(b)\|}{\|z\|}$ must be restricted to $z < 0$. So the one-sided limit of such an expression, if it exists, is the derivative Df_x of f at the boundary points $x = a, b$. It is clearly unique by the arguments known from the discussion of open domains. This is what we shall assume henceforth when saying that $f : [a, b] \to \mathbb{R}$ is differentiable on $[a, b]$.

Proposition 285 (Fundamental Theorem of Calculus) *For a real interval* $[a, b]$, *if* $f \in C^0([a, b])$, *then the integral function*

$$F : [a, b] \to \mathbb{R} : x \mapsto \int_a^x f$$

is in $C^1([a, b])$, *and we have*

$$F'(x) = f(x) \text{ for all } x \in [a, b].$$

In other words, the linear map $D : C^1([a, b]) \to C^0([a, b])$ *is surjective and has a section* $\int_a : f \to \int_a^? f$. *We have* $Ker(D) = Const. \overset{\sim}{\to} \mathbb{R}$. *Any function in the fiber* $D^{-1}f = \mathbb{R} + \int_a^? f$ *is called a* primitive function *of* f. *If* F *is any primitive function of* f, *then we have* $\int_a^b f = F(b) - F(a)$. *The latter is also denoted by* $F|_a^b$. *A primitive function* F *of* f *is also denoted by the so-called* indefinite integral $F = \int f$ *or, more conservatively,* $F = \int f(x) \, dx$, *whereas the Riemann integral* $\int_a^b f$ *is called the* definite integral. *To be clear, the indefinite integral is a function, while the definite integral is a real number.*

Proof Let $y \in [a, b]$ then $F(y + t) - F(y) = \int_y^{y+t} f$. Now, for $\varepsilon > 0$, let $\delta > 0$ be such that $|t| < \delta$ implies $|f(y + t) - f(y)| < \epsilon$. Then we have $|t| \cdot (f(y) - \varepsilon) < \int_y^{y+t} f < |t| \cdot (f(y) + \varepsilon)$, i.e., $|f(y) - \varepsilon| < \frac{|F(y+t) - F(y)|}{|t|} < |f(y) + \varepsilon|$. $\square$

This proposition has a very useful application for calculating one-dimensional integrals. In fact, one often knows primitive functions just by checking their derivative. One may then calculate the definite integral by evaluation of the primitive function at the boundary values. This is the one-dimensional anchorage of the n-dimensional integral operation. We now have to deal with the induction step $n \mapsto n + 1$.

Remark 31 If $a > b$, then the integral $\int_a^b f$ is defined by $-\int_b^a f$, so that we may integrate over the interval between a and b in every case.

An easy consequence of the fundamental theorem of calculus is the following corollary concerning the change of variables. In higher dimensions such a result is also true, but too involved to be exposed in our context, but see [39].

Corollary 286 *If for the real numbers $a \leq b$, the function $g : [a,b] \to \mathbb{R}$ is C^1, and if $f : [c,d] \to \mathbb{R}$ is continuous, and such that $\mathrm{Im}(g) \subset [c,d]$, then*

$$\int_a^b (f \circ g) \cdot g' = \int_{g(a)}^{g(b)} f.$$

Example 120 As an easy example, we are going to integrate $h(x) = \sin(\pi x^2)x$ from 0 to 1. Setting $g(x) = \pi x^2$, then $g'(x) = 2\pi x$ which is "almost" x. By multiplying and dividing $h(x)$ by the constant 2π, we get $h(x) = \frac{1}{2\pi} \sin(\pi x^2) 2\pi x$ which is of the required form, letting $f(x) = \frac{1}{2\pi} \sin(x)$ and $h(x) = f(g(x))g'(x)$.

Thus:

$$\int_0^1 h(x) = \int_0^1 f(g(x))g'(x) = \int_{g(0)}^{g(1)} f(x) = \int_0^\pi \frac{1}{2\pi} \sin(x)$$

$$= -\frac{1}{2\pi} \cos(x) \Big|_0^\pi = -\frac{1}{2\pi} (\cos(\pi) - \cos(0))$$

$$= -\frac{1}{2\pi} (-1 - 1) = \frac{1}{\pi}.$$

Exercise 160 Give a proof of corollary 286 using the calculation of the definite integral by use of primitive functions. Observe that if F is a primitive function for f, then $F \circ g$ is a primitive function for $(f \circ g) \cdot g'$ by the chain rule.

Definition 201 *Let $f, g : X \to \mathbb{R}^n$ be two continuous functions on a subset $X \subset \mathbb{R}^m$. Recall from sorite 265 that we denote by $f \cdot g : X \to \mathbb{R}$ the* *continuous function, which evaluates to the standard scalar product* $(f \cdot g)(x) = (f(x), g(x))$ *of the two vectors* $f(x)$ *and* $g(x)$. *If* $X = K(a., b.)$ *is a closed cube in* $\mathbb{R}^n$, *then the* scalar product *of* f *and* g *is the number*

$$(f, g) = \int_{(a., b.)} f \cdot g.$$

Evidently, $(f, f) \geq 0$, *and we therefore set* $\|f\| = \sqrt{(f, f)}$ *and call this non-negative real number the* norm *of* f.

Sorite 287 *If* $(a., b.)$ *defines a cube* $K(a., b.) \subset \mathbb{R}^m$, *then the scalar product* (f, g) *on functions* f *and* g *in the (infinite-dimensional)* $\mathbb{R}$-*vector space* $C^0(K(a., b.), \mathbb{R}^n)$ *has these properties:*

(i) (f, g) *is a positive definite symmetric bilinear form.*

(ii) *(Schwarz Inequality) We have* $|(f, g)| \leq \|f\| \cdot \|g\|$.

(iii) *(Triangle Inequality) We have* $\|f + g\| \leq \|f\| + \|g\|$.

Proof Claim (i) is evident except for the fact that $(f, f) = \int_{(a., b.)} f \cdot f = 0$ implies $f = 0$. But if $f(x) \neq 0$, with $(f(x), f(x)) = q \neq 0$, say, then by continuity of the scalar product $(f(z), f(z))$ as a function of z, there is a cube neighborhood $K_\delta(x)$ of x such that $(f(z), f(z)) > q/2 > 0$ if $z \in K_\delta(x)$. Then $(f, f) \geq (2\delta)^n q/2$.

The Schwarz inequality was deduced for the standard scalar product in volume 1 by use of Gram's determinant. Here, we use a more general argument: Let $\lambda \in \mathbb{R}$. Then $0 \leq (f + \lambda \cdot g, f + \lambda \cdot g) = \|f\|^2 + \lambda \cdot 2(f, g) + \lambda^2 \cdot \|g\|^2 = Q(\lambda)$. So the quadratic polynomial function $Q(\lambda)$ has only non-negative values, i.e., it cannot have two different zeros and produce negative values in between. This means that $4(f, g)^2 - 4\|f\|^2\|g\|^2 \leq 0$ since the solutions of a quadratic polynomial $c + \lambda b + \lambda^2 a$ are $\lambda = \frac{-a \pm \sqrt{b^2 - 4ac}}{2a}$, and two different real zeros arise iff $b^2 - 4ac > 0$. This yields (ii).

Item (iii) is left as exercise 161. □

Exercise 161 Show that the triangle inequality follows from the Schwarz inequality.

Proposition 288 *For a closed interval* $[a, b] \subset \mathbb{R}$, $a \leq b$, *and two* C^1 *functions* $f, g : [a, b] \rightarrow \mathbb{R}^n$, *we have the equation of* integration by parts

$$\int_a^b Df \cdot g = (f \cdot g)\Big|_a^b - \int_a^b f \cdot Dg.$$

Example 121 To compute the integral

$$\int_a^b x \cos(x)\, dx$$

we use integration by parts, setting $f'(x) = \cos(x)$ and $g(x) = x$. Then $f(x) = \sin(x)$ and $g'(x) = 1$, and

$$\int_a^b f'(x) \cdot g(x)\, dx = (f(x) \cdot g(x))\Big|_a^b - \int_a^b f(x) \cdot g'(x)\, dx$$

$$= (\sin(x) \cdot x)\Big|_a^b - \int_a^b \sin(x)\, dx$$

$$= \sin(b) \cdot b - \sin(a) \cdot a - (-\cos(x))\Big|_a^b$$

$$= \sin(b) \cdot b - \sin(a) \cdot a + \cos(b) - \cos(a).$$

By treating a as a constant and b as a variable, it follows that the primitive function of $x \cos(x)$ is $\cos(x) + x \cdot \sin(x) + C$.

The integral

$$\int_a^b x^2 \sin(x)\, dx$$

looks similar to the previous one, but is slightly more complicated to solve. In fact, it requires applying integration by parts twice. Again, we set $f'(x) = \sin(x)$ and $g(x) = x^2$. Then

$$\int_a^b f'(x) \cdot g(x)\, dx = (f(x) \cdot g(x))\Big|_a^b - \int_a^b f(x) \cdot g'(x)\, dx$$

$$= (-\cos(x) \cdot x^2)\Big|_a^b - \int_a^b -\cos(x) \cdot 2x\, dx$$

$$= (-\cos(x) \cdot x^2)\Big|_a^b + 2\int_a^b \cos(x) \cdot x\, dx.$$

To compute $\int \cos(x) \cdot x\, dx$, we would have to integrate by parts again, but since we have already done so in the first part of the example we simply substitute our solution and get

$$\int_a^b f'(x) \cdot g(x)\, dx = (-\cos(x) \cdot x^2)\Big|_a^b$$

$$+ 2(\sin(b) \cdot b - \sin(a) \cdot a + \cos(b) - \cos(a))$$

$$= -\cos(b) \cdot b^2 + \cos(a) \cdot a^2$$

$$+ 2(\sin(b) \cdot b - \sin(a) \cdot a + \cos(b) - \cos(a))$$

$$= (2 - b^2)\cos(b) + 2b \sin(b)$$

$$+ (a^2 - 2)\cos(a) - 2a \sin(a).$$

The primitive function of $x \cos(x)$ is accordingly

$$(2 - x^2) \cos(x) + 2x \sin(x) + C.$$

Exercise 162 Give a proof of proposition 288 using the fundamental theorem of calculus and the fact $D(f \cdot g) = f^\tau \cdot Dg + g^\tau \cdot Df$ from sorite 265 (vii).

31.3 Fubini's Theorem on Iterated Integration

Proposition 289 (Fubini's Theorem) *Let $(a., b.)$ define a cube $K(a., b.)$ in $\mathbb{R}^n$, and $(c., d.)$ define a cube $K(c., d.)$ in $\mathbb{R}^m$. These data define a sequence $(a.c., b.d.)$ and therefore a cube $K(a.c., b.d.) = K(a., b.) \times K(c., d.)$ in $\mathbb{R}^{n+m}$. Let $f : K(a.c., b.d.) \to \mathbb{R}$ be a continuous function. Then if $x \in K(a., b.)$, we have the restriction function $f_x : K(c., d.) \to \mathbb{R} : y \mapsto f_x(y) = f(x, y)$, which is also continuous. Therefore, we have the integrals $F(x) = \int_{(c.,d.)} f_x$ for each $x \in K(a., b.)$. The function $F : K(a., b.) \to \mathbb{R}$ is continuous, and we have the* iterated integrals

$$\int_{(a.c., b.d.)} f = \int_{(a., b.)} F$$

or, if we denote the variable of $K(a., b.)$ by x and the variable of $K(c., d.)$ by y, the equation

$$\int_{(a.c., b.d.)} f(x, y)\, dx\, dy = \int_{(a., b.)} \left(\int_{(c., d.)} f(x, y)\, dy \right) dx$$

in more traditional terms.

In particular, if a cube is given by $(a_1 a_2 \ldots a_n, b_1 b_2 \ldots b_n)$, and the function $f : K = [a_1, b_1] \times [a_2, b_2] \times \ldots [a_n, b_n] \to \mathbb{R}$ is continuous, then we have the n-fold iterated integral

$$\int_K f(x_1, \ldots x_n)\, dx_1 \ldots dx_n = \int_{a_n}^{b_n} \left(\ldots \left(\int_{a_1}^{b_1} f(x_1, \ldots x_n)\, dx_1 \right) \ldots \right) dx_n.$$

Proof Let P and Q be partitions of $(a., b.)$ and $(c., d.)$, respectively, defining a partition $P \times Q$ of the cube $K(a.c., b.d.)$. Then, using the notation of definition 195, we have

$$L(f, P \times Q) = \sum_{(K,L) \in I(P) \times I(Q)} m_{K \times L}(f) \cdot vol(K \times L)$$

$$= \sum_{K \in I(P)} \left(\sum_{L \in I(Q)} m_{K \times L}(f) \cdot vol(L) \right) \cdot vol(K).$$

But for $x \in K$, $m_{K \times L}(f) \le m_L(f_x)$, hence

$$\sum_{L \in I(Q)} m_{K \times L}(f) \cdot vol(L) \le \sum_{L \in I(Q)} m_L(f_x) \cdot vol(L) \le \int_{K(c.,d.)} f_x = F(x).$$

But this is valid for any $x \in K$, so $\sum_{L \in I(Q)} m_{K \times L}(f) \cdot vol(L) \le m_K(F)$. Therefore

$$L(f, P \times Q) \le \sum_{K \in I(P)} m_K(F) \cdot vol(K) \le L(F, P).$$

A similar argument yields $U(F, P) \le U(f, P \times Q)$, so we have

$$L(f, P \times Q) \le L(F, P) \le U(F, P) \le U(f, P \times Q),$$

but the left and right outer expressions converge to $\int_{(a.c.,b.d.)} f$, so that F turns out to be integrable, too, and the theorem follows.   □

Example 122 Let us calculate the volume of a rhombic pyramid whose base is given by the points $(a, 0, 0), (0, b, 0), (-a, 0, 0), (0, -b, 0)$, see figure 31.1 (a). For reasons of symmetry, it is sufficient to calculate the volume of the tetrahedron over the positive quadrant (see figure 31.1 (b)) and multiply it by 4.

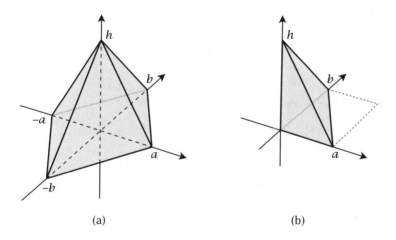

(a)   (b)

Fig. 31.1. The rhombic pyramid (a), and one of the pyramid's four constituting tetrahedra (b). The dashed lines show the extents of the 2-cube over which the integration occurs.

The tetrahedron in question is defined by the plane passing through the points $(a, 0, 0)$, $(0, b, 0)$, and $(0, 0, h)$. Using methods from linear algebra one finds the equation

$$z = f(x, y) = h - \frac{h}{a}x - \frac{h}{b}y$$

for the plane. However, to obtain the volume V_T of the tetrahedron, we must not use f, as it would contribute negative values outside of the tetrahedron's base Δ. The solution is to use $g(x, y) = (f \circ \chi_\Delta)(x, y)$, where χ_Δ is the characteristic function of Δ, returning 1 if $(x, y) \in \Delta$, and 0 otherwise. If we consider a fixed value for x, we have to integrate the function

$$z = f_x(y) = h - \frac{h}{a}x - \frac{h}{b}y.$$

Figure 31.2 shows that, again, we must take care to integrate only where the function f_x is positive, i.e., over the interval $[0, b - \frac{b}{a}x]$.

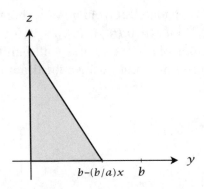

Fig. 31.2. A section of the tetrahedron for a fixed value of x.

So the integral for the tetrahedron's volume is

$$V_T = \int_0^a \int_0^{b - \frac{b}{a}x} \left(h - \frac{h}{a}x - \frac{h}{b}y \right) dy\, dx$$

$$= \int_0^a \left(\frac{h(a-x)y}{a} - \frac{hy^2}{2b} \right) \Bigg|_0^{b - \frac{b}{a}x} dx$$

$$= \int_0^a \left(\frac{h(a-x)(b - \frac{b}{a}x)}{a} - \frac{h(b - \frac{b}{a}x)^2}{2b} \right) dx$$

$$= h \int_0^a \left(b - \frac{2b}{a}x + \frac{b}{a^2}x - \frac{b}{2} + \frac{b}{a}x - \frac{b}{2a^2}x^2 \right) dx$$

$$= h \int_0^a \left(\frac{b}{2} - \frac{b}{a}x + \frac{b}{2a^2}x^2 \right) dx$$

$$= h \left(\frac{b}{2}x - \frac{b}{2a}x^2 + \frac{b}{6a^2}x^3 \right) \Bigg|_0^a$$

$$= h \left(\frac{ab}{2} - \frac{a^2 b}{2a} + \frac{a^3 b}{6a^2} \right)$$

$$= \frac{abh}{6}$$

Because our original pyramid consists of four such tetrahedra, its volume is $V = \frac{2abh}{3}$, which is in accordance to the well-known rule that a pyramid's volume is one third of the product of the area of its base and its height.

Remark 32 The integral may be extended to functions $f : K(a., b.) \to \mathbb{R}^m$ with component functions $f_i = pr_i \circ f, i = 1, 2, \ldots m$, by defining

$$\int_{(a.,b.)} f = \left(\int_{(a.,b.)} f_1, \int_{(a.,b.)} f_2, \ldots \int_{(a.,b.)} f_m \right).$$

In particular, if $m = 2$, we may integrate functions with values in $\mathbb{C}$ by identifying $\mathbb{C}$ with $\mathbb{R}^2$ and integrating separately the real and imaginary parts of such functions.

CHAPTER 32

Vector Fields

32.1 Introduction

In everyday life, when you are presented the weather forecast, a map with the wind velocities may be shown. At every moment, this defines a wind velocity vector at every point of the geographic map (figure 32.1). Or when the water streams are described, one is given a water velocity at every point of the sea surface, at every moment. Or, if we leave the earth and fix a point in the interplanetary space, the sum of all gravitational forces at a given moment yields the gravitational force in that point, a force, which you feel every day by the dominant force field of the earth gravitation, or a force which the earth is subjected to and mainly caused by the sun's gravitational field, in its elliptic orbit. All these are examples of vector fields and, at the same time, illustrate that vector fields are natural and intuitive structures.

The concept of a *vector field* is however a very local one in the sense that one does not know how, for example, a cloud is moved through space by the vector field induced by winds. The calculation of orbits of objects along given vector fields is the subject of the theory of differential equations, see chapter 34. The local character of vector fields makes them a very powerful tool in the definition of complex trajectories, such as planetary orbits or the motion of clouds. They are central concepts in any non-trivial simulation of dynamical systems, a branch of computational science which requires the most powerful computation power and is the effective domain of supercomputing. We should understand that vector

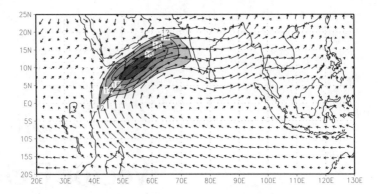

Fig. 32.1. A geographic map showing the wind velocities as a vector field.

fields are, from an information-theoretic point of view, definitions of concepts which describe the behavior only in local contexts, but are targeted at global concepts. A good portion of physics is built upon "local" laws in form of vector fields, but they aim to describe global phenomena such as planetary orbits.

The difficulty to understand the global implication of a vector field leads to an alternate description, the interpretation of a vector field as an operator, which transforms a function into another function through the local calculation of the scalar product of the function's partial derivatives (its differential) with the field's vectors. One shows that vector fields and such so-called "derivation" operators are in one-to-one correspondence. This fact leads to a very powerful method for constructing new vector fields, the Lie product, which is crucial in physics [2], but also in mathematical music theory [28].

32.2 Vector Fields

Let $U \subset \mathbb{R}^n$ be an open set. Recall that $TU = U \times \mathbb{R}^n$. By sorite 265 (iii), we have a bijection

$$p : C^r(U, TU) \xrightarrow{\sim} C^r(U, U) \times C^r(U, \mathbb{R}^n).$$

This applies to the following definition.

Definition 202 *Let $U \subset \mathbb{R}^n$. A C^r-vector field V on U is a C^r-section $V :$ $U \to TU$ of the projection $pr_1 : TU \to U$, or, by the above bijection, a* *section $V : U \to TU$ such that $pr_2 \circ V : U \to \mathbb{R}^n$ is C^r. The set of C^r-vector fields $V : U \to TU$ is denoted by $\mathcal{V}^r(U)$.*

The set $\mathcal{V}^r(U)$ is an $\mathbb{R}$-vector space by the following: If $F, G \in \mathcal{V}^r(U)$, $u \in U$, and $F(u) = (u, F_u)$ and $G(u) = (u, G_u)$ then $(F+G)(u) = (u, F_u + G_u)$. If $\lambda \in \mathbb{R}$, then $(\lambda F)(u) = (u, \lambda F_u)$.

Example 123 Figure 32.2 shows a vector field $F : \mathbb{R}^2 \to T\mathbb{R}^2$ defined by[1]

$$F(x, y) = (x, y; \sin(xy), \cos(xy)).$$

At each point (x, y) the vector $(\sin(xy), \cos(xy))$ anchored at this point is drawn.

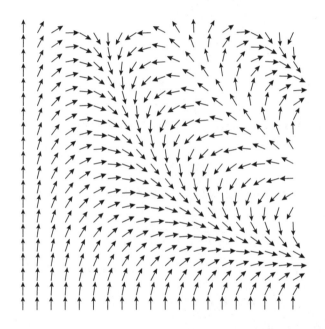

Fig. 32.2. A vector field $F : \mathbb{R}^2 \to T\mathbb{R}^2$. The field is drawn for $0 \le x \le \pi$ and $0 \le y \le \pi$. Note that the length of arrows have been scaled down in order to make the picture clearer.

A vector field $G : \mathbb{R}^3 \to T\mathbb{R}^3$ is shown in figure 32.3. It is defined by

$$G(x, y, z) = (x, y, z; \cos(y), \sin(x), z).$$

---

[1] The correct expression is $F((x, y)) = ((x, y), (\sin(xy), \cos(xy)))$, but for the sake of readability, we use the simplified form above.

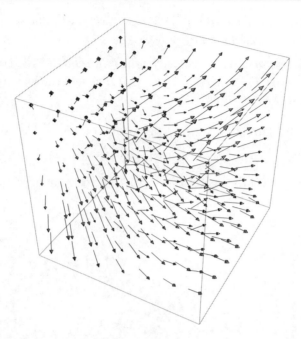

Fig. 32.3. A vector field $G : \mathbb{R}^3 \to T\mathbb{R}^3$. The field is drawn for $-1 \le x \le 1$, $-1 \le y \le 1$ and $-1 \le z \le 1$.

If $f \in C^r(U)$, we have $Tf : TU \to T\mathbb{R} = \mathbb{R}^2$. The *differential of f* is the composition

$$df : pr_2 \circ Tf : TU \to \mathbb{R} : (u,t) \mapsto Df_u(t),$$

a function in $C^{r-1}(TU)$. If $g \in C^r(U)$, one defines

$$(g \cdot df)(u,t) = g(u) \cdot df(u,t).$$

If $g, h : TU \to \mathbb{R}$, we set

$$(g + h)(t,u) = g(t,u) + h(t,u).$$

Example 124 From physics we know the function $f(r) = \frac{1}{r}$, which describes the gravitational potential of a point (x,y) at distance $r = \sqrt{x^2 + y^2}$ from a point mass at 0. Writing f in terms of x and y, we have

$$f(x,y) = \frac{1}{\sqrt{x^2 + y^2}}.$$

Its differential df is

$$df(x, y; u, v) = Df_{(x,y)}(u, v)$$

$$= \left(-\frac{2x}{(x^2 + y^2)^{3/2}}, -\frac{2y}{(x^2 + y^2)^{3/2}} \right) \cdot \begin{pmatrix} u \\ v \end{pmatrix},$$

which describes the gravitational gradient (defined below) at the point (x, y).

Lemma 290 If $f, g \in C^r(U)$, then $d(f \cdot g) = f \cdot dg + g \cdot df$.

Exercise 163 Give a proof of lemma 290 using the fact from sorite 265 that $D(f \cdot g) = f \cdot Dg + g \cdot Df$.

We have an operation of a vector field $F \in \mathcal{V}^r(U)$ on $f \in C^{r+1}(U)$ defined by

$$L_F f = df \circ F : U \to \mathbb{R}.$$

The function $L_F f$ is in $C^r(U)$. It is called the *Lie derivative of f with respect to F*. In particular, if $F \in \mathcal{V}^\infty(U)$ and $f \in C^\infty(U)$, then also $L_F f \in C^\infty(U)$. If for $u \in U$, we have the linear form $Df_u = (\partial_1 f_u, \partial_2 f_u, \dots \partial_n f_u)$, whose transpose $Df_u^\top$ is called the *gradient* of f at u, and if $F(u) = (F_1(u), F_2(u), \dots F_n(u))$, then

$$L_F f(u) = \sum_i \partial_i f_u \cdot F_i(u),$$

the classical standard scalar product of the vectors Df_u and $F(u)$ in $\mathbb{R}^n$. In particular, if $f(u) = pr_j(u)$, the j-th coordinate function, then $\partial_i pr_j = \delta_{ij}$, and therefore, $L_F pr_j(u) = F_j(u)$. In particular, F is uniquely determined by the Lie derivative L_F.

The Lie derivative has these properties:

Lemma 291 If $F \in \mathcal{V}^\infty(U)$ then the Lie derivative $L_F : C^\infty(U) \to C^\infty(U)$ on the real vector space $C^\infty(U)$ is a derivation, i.e.,

(i) L_F is $\mathbb{R}$-linear.

(ii) For $f, g \in C^\infty(U)$, we have $L_F(f \cdot g) = f \cdot L_F g + L_F f \cdot g$.

(iii) If $c \in C^\infty(U)$ is a constant, then $L_F c = 0$.

The set $Der(C^\infty(U))$ of derivations $C^\infty(U) \to C^\infty(U)$, together with the usual pointwise sum $(D + E)(u) = D(u) + E(u)$ and scalar multiplication $(\lambda D)(u) = \lambda \cdot D(u)$ is a real vector space.

Proof Since the derivative of a function $f \in C^{\infty}(U)$ is linear in f, the differential df is also linear in f, but $L_F(f) = df \circ F$ is linear in df, therefore (i) holds.

Claim (ii) follows immediately from lemma 290.

Claim (iii) follows from the fact that $Dc = 0$ for a constant c, whence $dc = 0$, and therefore $L_F(c) = dc \circ F = 0$. The proof that the set $Der(C^{\infty}(U))$ is a real vector space is left as a standard verification to the reader.   □

Example 125 Let f be the potential function from example 124. Consider the constant vector field

$$F(x, y) = (x, y; 1, -1)$$

which can be regarded as a velocity field with constant magnitude $\sqrt{2}$ and direction $(1, -1)$.

The Lie derivative of f with respect to F is

$$
\begin{aligned}
L_F f(x, y) &= (df \circ F)(x, y)\\
&= df(F(x, y))\\
&= df(x, y; 1, -1)\\
&= \left(-\frac{2x}{(x^2 + y^2)^{3/2}}, -\frac{2y}{(x^2 + y^2)^{3/2}} \right) \cdot \begin{pmatrix} 1 \\ -1 \end{pmatrix}\\
&= -\frac{2x}{(x^2 + y^2)^{3/2}} + \frac{2y}{(x^2 + y^2)^{3/2}}
\end{aligned}
$$

If a point (x, y) moves with velocity $\sqrt{2}$ along a straight line with direction $(1, -1)$, $L_F f(x, y)$ is effectively the gravitational force acting on the point in the direction of its movement (figure 32.4).

Proposition 292 *Let $U \subset \mathbb{R}^n$ be an open set. Then the Lie derivative defines a linear isomorphism*

$$L : \mathcal{V}^{\infty}(U) \xrightarrow{\sim} Der(C^{\infty}(U)) : F \mapsto L_F.$$

Proof We have already seen that for the Lie derivative of a vector field $F = (Id_U, F_1, \ldots F_n)$, we have $L_F pr_j = F_j$. Therefore, L is injective, and it is evidently $\mathbb{R}$-linear. The only remaining point is its surjectivity. Now, by injectivity of L, the only candidate for the vector field F yielding a given derivation D is $F_j = D(pr_j), j = 1, \ldots n$. This means that we have to prove the formula

$$D(f) = df \circ (Id_U, D(pr_1), \ldots D(pr_n)).$$

To this end, we make use of an integral form of the mean value theorem. Let $x \in U$ and $y \in B_{\varepsilon(x)}$ for a small ball $B_\varepsilon(x) \subset U$. Then the affine curve

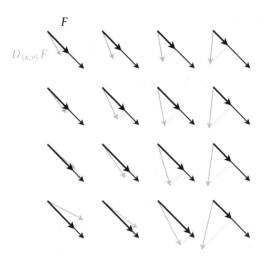

Fig. 32.4. The velocity vector field F and the gradient $Df_{(x,y)}$. The thick vectors have length $L_F f(x, y)$. They measure the gravitational force in the direction of the velocity field.

$$c : [0, 1] \to U : t \mapsto x + t \cdot (y - x)$$

induces a C^∞-function $e = f \circ c : [0, 1] \to \mathbb{R}$, and we have $f(y) - f(x) = e(1) - e(0) = \int_0^1 De\, dt = \sum_j (y_j - x_j) \cdot \int_0^1 D_j e\, dt$. Writing $f_j(y) = \int_0^1 D_j e\, dt$, we have

$$f(y) = f(x) + \sum_j (pr_j(y) - x_j) f_j(y),$$

as a function of y with x being fixed, and where the constituents pr_j and f_j are C^∞-functions with $f_j(x) = D_j f(x)$. Then the properties of D as a derivation yield

$$D(f) = D(f(x)) + \sum_j (D(pr_j) - D(x_j)) f_j + \sum_j (pr_j(y) - x_j) D(f_j)$$

$$= \sum_j D(pr_j) f_j + \sum_j (pr_j(y) - x_j) D(f_j),$$

which, when evaluated at $y = x$, yields

$$D(f)(x) = \sum_j D(pr_j)(x) D_j f(x) = df(x)(D(pr_1)(x), \ldots D(pr_n)(x)),$$

and this proves the above formula. $\qquad \square$

This opens an important technique of vector field construction, the Lie bracket. To this end, consider the following fact:

Proposition 293 *If $D, E \in Der(C^\infty(U))$, then so is the endomorphism*

$$[D, E] = D \circ E - E \circ D.$$

The derivation $[D, E]$ is called the Lie *bracket of D and E. The Lie bracket is $\mathbb{R}$-bilinear and has these properties for all $D, E, F \in Der(C^\infty(U))$:*

   (i) $[D, D] = 0$,

   (ii) (Jacobi Identity) $[D, [E, F]] + [E, [F, D]] + [F, [D, E]] = 0$.

Proof This is a straightforward verification which we omit.   □

Corollary 294 *For two vector fields $F, G \in \mathcal{V}^\infty(U)$ there is a unique vector field $[F, G] \in \mathcal{V}^\infty(U)$ such that $[L_F, L_G] = L_{[F,G]}$. It is called the* Lie *derivative of F and G.*

Proof This follows immediately from proposition 293.   □

An explicit representation of $[F, G]$ in terms of F and G can be given as follows: We know that the i-th component of a vector field F is given by the Lie derivative of the i-th coordinate function $pr_i : U \to \mathbb{R}$, i.e., $F_i = L_F(p_i) = dpr_i \circ F$. Thus we have

$$\begin{aligned}
[F, G]_i &= L_{[F,G]}(pr_i) \\
&= L_F(L_G(pr_i)) - L_G(L_F(pr_i)) \\
&= d(dpr_i \circ G) \circ F - d(dpr_i \circ F) \circ G \\
&= dG_i \circ F - dF_i \circ G \\
&= \sum_j D_j G_i \cdot F_j - D_j F_i \cdot G_j.
\end{aligned}$$

The Lie bracket appears somewhat mysterious here. For an intuitive understanding of its use in control theory, see example 135 in chapter 34.

CHAPTER 33

Fixpoints

33.1 Introduction

Fixpoint theory is a vast field which we shall only discuss for a special case. However, already in this case, it will become clear that this theory is a fundamental tool in many branches of mathematics. In particular, it will turn out that the solution of a differential equation and the definition of a concept by use of the recursion theorem are, both, solutions of a fixpoint problem. This opens the perspective of viewing the solution of a differential equation as a kind of "infinitesimal definition of a concept", while the recursive construction of a concept (such as the arithmetic operations on natural numbers) may be restated as the solution of a "differential equation of a concept".

33.2 Contractions

We encountered distance functions in the discussion of Euclidean spaces, in particular in proposition 213 of volume 1. Here, we shall introduce those spaces which are defined by such distance functions:

Definition 203 *A metric space is a pair* (X, d)*, where* X *is a set and* $d :$ $X \times X \to \mathbb{R}$ *is a* distance function *on* X *with these properties:*

   (i) (Positive Definiteness) *For all* $x \in X$*,* $d(x, y) \geq 0$*, and* $d(x, y) = 0$*, iff* $x = y$*.*

(ii) (Symmetry) *We have $d(x,y) = d(y,x)$ for all $(x,y) \in X \times X$.*

(iii) (Triangle Inequality) *For all $x, y, z \in X$, $d(x,y) + d(y,z) \geq d(x,z)$.*

Example 126 As already indicated, the distance function of a Euclidean space (V, b), which is deduced from the norm, defines a metric, see proposition 213 of volume 1.

Example 127 In the theory of integration, we considered a closed cube $K(a., b.) \subset \mathbb{R}^n$ defined by a sequence $(a., b.)$ of $2n$ real numbers with $a_i \leq b_i$. If we take a function $f \in C^0(K(a., b.))$, then we have the norm $\|f\| = \sqrt{(f, f)} = \sqrt{\int_{(a., b.)} f^2}$ (cf. definition 201). The triangle inequality in sorite 287 guarantees that the function $d(f, g) = \|f - g\|$ on $f, g \in X = C^0(K(a., b.))$ is a distance function.

Exercise 164 For the recursion theorem, we considered the space $X = a^{\mathbb{N}}$ of sequences with values in a set a, see chapter 6.1 in volume 1. If $(x_i)_i, (y_i)_i \in X$, are two different sequences, we set $d((x_i)_i, (y_i)_i) = \frac{1}{2^n}$, where n is the minimal index $i \in \mathbb{N}$ such that $x_i \neq y_i$; if $(x_i)_i = (y_i)_i$ we set $d((x_i)_i, (y_i)_i) = 0$. Clearly, this function is positive definite and symmetric. Prove that the triangle inequality also holds.

For a metric space (X, d) one can also define a topology by the set *Open*(X, d) of *open sets*. To this end, we proceed in complete analogy to the topologies for Euclidean spaces as discussed in section 27.2. If $\varepsilon > 0$ and $x \in X$, we define the ε-*ball around* x by $B_\varepsilon(x) = \{y \mid d(x, y) < \varepsilon\}$. A subset $O \subset X$ is called open, iff it is the union of open balls. The axioms of a topology as listed in sorite 231 are obviously true, just replace $\mathbb{R}^n$ by X. Again, a *closed set* is the complement in X of an open set. For example, the *closed ε-ball around* x is the closed set $\overline{B}_\varepsilon(x) = \{y \mid d(x, y) \leq \varepsilon\}$, complement of the open set $\{y \mid d(x, y) > \varepsilon\}$.

On a metric space (X, d), one may again consider convergent and Cauchy sequences, the definition is in nearly complete analogy with the corresponding definition 177, except that the open cube neighborhoods are not defined here.

Definition 204 *For a metric space (X, d), a sequence $(c_i)_i$ of elements in X is called* convergent *if there is an element $c \in X$ such that for every $\varepsilon > 0$, there is an index N with $c_i \in B_\varepsilon(c)$ for $i > N$. If $(c_i)_i$ converges to c, one writes $\lim_{i \to \infty} c_i = c$. A sequence which does not converge is called* divergent.

A sequence $(c_i)_i$ of elements in X is called a Cauchy sequence, if for every $\varepsilon > 0$, there is an index N with $c_i \in B_\varepsilon(c_j)$ for $i, j > N$. A metric space (X, d), in which every Cauchy sequence converges, is called complete.

Example 128 The Euclidean space (V, b) (see example 126) is a complete metric space, this is the restatement of proposition 232.

The metric function space $C^0(K(a., b.))$ from example 127 is not complete, we state this without a proof. However, the proof idea is simple: One takes a step function $s : K(a., b.) \rightarrow \mathbb{R}$, for example $s(x_1, x_2 \dots x_n) = 0$ if $x_1 \le (a_1 + b_1)/2$, and $s(x_1, x_2 \dots x_n) = 1$ otherwise. This function, which is obviously not in $C^0(K(a., b.))$, is easily seen to be the limit of a sequence $(f_i)_i$ of continuous functions $f_i \in C^0(K(a., b.))$.

Exercise 165 The space $(a^\mathbb{N}, d)$ from exercise 164 is complete. Give a proof thereof.

Exercise 166 If $Y \subset X$ is a closed subset of a complete metric space (X, d), then the metric space $(Y, d|_{Y \times Y})$ is also complete.

The main strategy in the proofs of the recursion theorem and of the main theorem of ordinary differential equations (to be dealt with in the following chapter 34) is the construction of a uniquely determined *fixpoint* $x = k(x)$ *of a map* $k : X \rightarrow X$, where X is a complete metric space. The guarantee that such a fixpoint exists follows from the following central result:

Proposition 295 *Call a function* $k : X \rightarrow Y$ *between metric spaces a* contraction, *iff there is a real number* c *with* $0 < c < 1$ *such that for all* $x, y \in X$, *we have*
$$d(k(x), k(y)) \le c \cdot d(x, y).$$

If $k : X \rightarrow X$ *is a contraction on a complete metric space* (X, d), *then there is a unique fixpoint* $f_k \in X$ *of* k, *i.e.,* $k(f_k) = f_k$, *and if* $k(f) = f$, *then* $f = f_k$. *More precisely, the fixpoint is the limit*

$$f_k = \lim_{i \to \infty} k^i(x)$$

for any point $x \in X$. *We also denote the fixpoint by* Fix(k).

Proof Clearly, a fixpoint is uniquely determined, since if $k(x) = x$ and $k(y) = y$, then $d(x, y) = d(k(x), k(y)) \le c \cdot d(x, y) < d(x, y)$, a contradiction for

$d(x, y) \neq 0$. The existence follows from the completeness of X, if we show that any sequence $(k^i(x))_{i \in \mathbb{N}}$ is Cauchy. But one has $d(k^j(x), k^j(y)) \leq c^j \cdot d(x, y)$. So, suppose $k(x) \neq x$. Then for $N \leq M$,

$$d(k^N(x), k^M(x)) \leq d(k^N(x), k^{N+1}(x)) + \cdots d(k^{M-1}(x), k^M(x))$$

$$= \sum_{i=0}^{M-N-1} d(k^{N+i}(x), k^{N+i}(k(x)))$$

$$\leq \sum_{i=0}^{M-1} d(k^{N+i}(x), k^{N+i}(k(x)))$$

$$\leq \sum_{i=0}^{M-1} c^{N+i} d(x, k(x))$$

$$\leq d(x, k(x)) \cdot c^N \cdot \sum_{i=0}^{M-1} c^i$$

$$\leq d(x, k(x)) \cdot c^N \cdot \frac{1}{1-c},$$

Since $0 < c < 1$, $d(x, k(x)) \cdot c^N \cdot \frac{1}{1-c}$ tends to 0 as $N \to \infty$, thus $d(k^N(x), k^M(x))$ (which is positive) also tends to 0 as $N \to \infty$. Thus $(k^i(x))_i$ is Cauchy and the theorem proved. $\square$

Example 129 The iterative procedure to approximate the fixpoint of $k = \cos : [0, 1] \to [0, 1]$ can be nicely illustrated with a picture. The map k is a contraction, since, by the mean value theorem, for $0 < \xi < 1$, we have $|\cos(x) - \cos(y)| = |x - y| \cdot |\sin(\xi)| \leq |x - y| \cdot \sin(1)$, where $0 < \sin(1) < 1$.

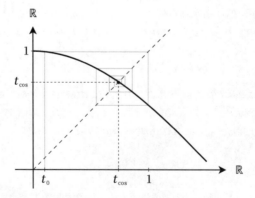

Fig. 33.1. The fixpoint $t_{\cos} \approx 0.739085$. The iteration starts with $t_0 = 0.1$.

We seek the solution to $x = \cos(x)$, i.e., the fixpoint *Fix*(cos). In figure 33.1 we see that curve for $y = \cos(x)$ intersects the straight line $y = x$ somewhere between $x = 0.1$ and $x = 0.9$. The simultaneous solution to both equations is obviously *Fix*(cos).

Starting with $t_0 = 0.01$, draw a line to $t_1 = \cos(t_0)$. To proceed to $\cos(t_1)$, draw a horizontal line to the straight line $y = x$. Hence we get, by projecting downwards to the curve, the value $t_2 = \cos(t_1)$. The procedure is iterated, closing in on the fixed point, until we are satisfied with the approximation. Here we get the approximated fixpoint $t_{cos} \approx 0.739085$.

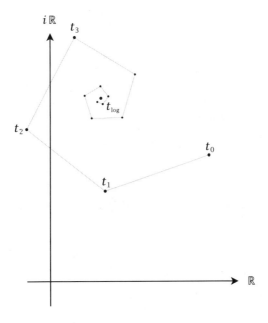

Fig. 33.2. The limit of $t_n = \log^n(1 + i)$ is the fixpoint $t_{\log} \approx 0.318132 + i \cdot 1.337236$.

Example 130 We have seen the logarithm for real values. It can be easily continued to complex values: If $z = |z|e^{i \cdot arg(z)}$, then

$$\log(z) = \log|z| + i \cdot arg(z).$$

For any point $z \in \mathbb{C}$, $z \notin \mathbb{R}$, we have

$$t_{\log} = \lim_{n \to \infty} \log^n(z) \approx 0.318132 + i \cdot 1.337236.$$

In figure 33.2, we start with the point $t_0 = 1 + i$. The result is a logarithmic spiral which approaches $t_{\log}$.

Exercise 167 In recursion theory (see chapter 6.1 in volume 1), we dealt with the unique fixpoint $L_\Phi = \Phi(L_\Phi)$ of the function $\Phi : a^{\mathbb{N}} \to a^{\mathbb{N}}$. Show that we have $d(\Phi(f), \Phi(g)) \leq \frac{1}{2} d(f, g)$. So if g is any sequence, we have

$$L_\Phi = \lim_{i \to \infty} \Phi^i(g).$$

This means that the "concept L_Φ" which we define recursively via Φ, is the limit of a Cauchy sequence.

Corollary 296 *Let $r > 0$ and $x \in X$ for a complete metric space (X, d). If $k : \overline{B}_r(x) \to X$ is a contraction with a constant of contraction c such that $d(k(x), x) \leq (1 - c)r$, then k has a unique fixpoint in $\overline{B}_r(x)$.*

Proof If $y \in \overline{B}_r(x)$, then

$$
\begin{aligned}
d(k(y), x) &\leq d(k(y), k(x)) + d(k(x), x) \\
&\leq c \cdot d(x, y) + (1 - c)r \\
&\leq cr + (1 - c)r \\
&= r.
\end{aligned}
$$

Thus, the contraction maps the complete subspace $\overline{B}_r(x)$ into itself and therefore is a contraction on this ball, hence it has its unique fixpoint in $\overline{B}_r(x)$ according to proposition 295.   □

Example 131 We illustrate corollary 296 with the simple function $f : \mathbb{R} \to \mathbb{R}$, $x \mapsto \frac{1}{3}x + 1$. We have, for $x, y \in \mathbb{R}$,

$$
\begin{aligned}
|f(x) - f(y)| &= |\tfrac{1}{3}x - \tfrac{1}{3}y| \\
&= \tfrac{1}{3}|x - y|.
\end{aligned}
$$

Thus f is a contraction on the whole of $\mathbb{R}$ with contraction constant $c = \frac{1}{3}$. Moreover c is the minimal such constant. Taking $\overline{B}_1(2)$,

$$|f(2) - 2| = \tfrac{1}{3} \leq (1 - c)r = \tfrac{2}{3} \cdot 1 = \tfrac{2}{3}.$$

Thus there is a unique fixpoint in the interval $[1, 3]$. We proceed by splitting the interval into two halves, $[1, 2]$ and $[2, 3]$, and see that the condition from the corollary is fulfilled for $\overline{B}_{1/2}(\frac{3}{2})$, but not for $\overline{B}_{1/2}(\frac{5}{2})$, so there is no need to look further for a fixpoint in $[2, 3]$, since the fixpoint is unique in $[1, 3]$. We continue with this procedure of binary search to find that $\frac{3}{2} = Fix(f)$.

Example 132 (Newton's method) A famous application of fixpoint theory is Newton's method for finding roots of a function. The Taylor series of a function f about a point x is

$$f(x + h) = f(x) + hf'(x) + \tfrac{1}{2}h^2 f''(x) + \ldots.$$

If h is small, we can neglect terms of order greater than 2:

$$f(x + h) \approx f(x) + hf'(x).$$

Given x we want to find a step h such that $f(x + h) = 0$. With this condition we can compute an approximate h that leads us closer to the root:

$$h = -\frac{f(x)}{f'(x)},$$

hence from a given value x_n we get to the next value x_{n+1}:

$$x_{n+1} = x_n - \frac{f(x_n)}{f'(x_n)}.$$

If we set $g(x) = x - \frac{f(x)}{f'(x)}$, then the value we seek is the fixpoint $x = g(x)$, i.e., $x = x - \frac{f(x)}{f'(x)}$, or $0 = -\frac{f(x)}{f'(x)}$, finally $f(x) = 0$.

The theoretical justification is as follows: We are given a C^2 function $f : U \to \mathbb{R}$ in a neighborhood U of a point p. Suppose that $f(p) = 0$ and that $f'(p) \neq 0$. Then there is a closed interval $I = [p - d, p + d]$ such that the function $g(x) = x - \frac{f(x)}{f'(x)}$ is defined on I and is a contraction on I. Since $g(p) = p$, this function fixes p. The point is that according to the fixpoint theorem 295, the sequence $g^i(x_0)$ converges to p for any $x_0 \in I$. So we may find p by successive approximation. The geometric meaning of g is shown in figure 33.3. Let us see why we have a contraction. We have

$$g'(x) = 1 - \frac{f'(x)^2 - f''(x)f(x)}{f'(x)^2},$$

which is defined and continuous in a neighborhood V of p where $f'(x) \neq 0$. Since $g'(p) = 0$, there is a neighborhood $[p - d, p + d]$ of p such that g is defined and $|g'(x)| \leq c < 1$ for $x \in I$. Now, for $x \in I$, by the mean value theorem, we have

$$\begin{aligned}
|g(x) - p| &= |g(x) - g(p)| \\
&= |g'(\xi) \cdot (x - p)| \\
&< c \cdot |(x - p)| \\
&< d
\end{aligned}$$

for a ξ with $p - d < \xi < p + d$, therefore g maps I into I. Further, for any $x, y \in I$, the same argument yields $|g(x) - g(y)| = |g'(\eta) \cdot (x - y)| < c|x - y|$ for a mean value η.

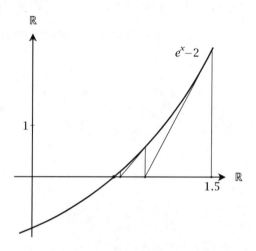

Fig. 33.3. Newton's method for finding the solution of $e^x = 2$. Note that the next point x_{n+1} of the iteration is found as the intersection of the tangent at x_n with the axis $y = 0$.

Figure 33.3 shows a few iterations of the Newton algorithm for finding the root of $f(x) = e^x - 2$, starting with $x_0 = 1.5$. The approximative solution of $e^x = 2$ is 0.693147.

CHAPTER 34

Main Theorem of ODEs

34.1 Introduction

We have learned that the integral $F(x) = F_0 + \int_a^x f$ solves the problem of finding a primitive function F such that $F' = f$ and $F(a) = F_0$. Intuitively, this means that we are given a *velocity field* with value $f(x)$ at time x, and we are looking for a path function F such that its initial position (at time $x = a$) is predefined by F_a and its velocity function F' coincides with the given velocity function f.

Ordinary differential equations generalize this setup and also have an existence theorem for primitive functions, which can be seen as a generalization of the fundamental theorem of calculus. But this generalization is not merely of a technical nature: As already stressed in the introduction to chapter 32, the existence of a global synthesis of the local information provided by a vector field is crucial for the existence of global concepts in natural and computational sciences. But the essence of the local-to-global transition is the existence and uniqueness of global solutions, and this is a kind of convergence towards a fixpoint as described in chapter 33. The desired global concept is the fixpoint of a contraction associated with the vector field. We recall that this parallels the theory of recursively defined entities, which, as we saw in example 167, is also the passage from a "local information flow" to a global concept, guaranteed as a fixpoint of a contraction. Inverting the perspective, the solution of a differential equation defined by a vector field is just another type of concept, which is given by a local information flow and has its existential foundation in a fixpoint theorem for what one could call "informational" contractions. Much like in general recursion theory, where the recursion theorem is the vital basis, the main theorem of ordinary differential equations (ODEs) is the vital basis for all theories which emanate from the globalization or integration of vector field constructions.

34.2 Conservative and Time-Dependent Ordinary Differential Equations: The Local Setup

The local setup is this:

Definition 205 *Let $(\zeta, \eta) \in \mathbb{R} \times \mathbb{R}^n$. Let $I \in \mathbb{R}$ be an open interval containing ζ and $U \subset \mathbb{R}^n$ an open neighborhood of η. Let $F : U \to TU : x \mapsto (x, F_2(x))$ be a continuous vector field on U. A local integral curve of F at (ζ, η) is a C^1 function $y : I \to U$ such that $y(\zeta) = \eta$ and such that the diagram*

$$
\begin{array}{ccc}
I & \xrightarrow{\Delta} & TI \\
y \downarrow & & \downarrow Ty \\
U & \xrightarrow{F} & TU
\end{array}
$$

commutes. Here $\Delta : I \to TI : \xi \mapsto (\xi, 1)$ is the constant vector field with value 1. This means that

$$Ty \circ \Delta = F \circ y,$$

or, setting $y' = pr_2 \circ Ty \circ \Delta$,

$$y'(t) = F_2(y(t)) \quad \text{for all } t \in I,$$

with the initial condition

$$y(\zeta) = \eta.$$

This is the form in which an ordinary differential equation (ODE) is traditionally presented.[1] *The equation is regarded as an equation in the unknown function y of one variable $t \in I$.*

---

[1] In contrast to ODEs, partial differential equations (PDEs) involve partial derivatives of functions of several variables. We shall not deal with PDEs in our modest context and refer the interested reader to [17].

Often, in literature, so-called "time-dependent" vector fields and corresponding differential equations are considered, in contrast to the so-called "conservative" vector field F above. This reads as follows: Instead of a vector field $F : U \rightarrow TU$, one considers a continuous function $F : J \times U \rightarrow TU$, where J is an open interval with $I \subset J$, and then asks for a solution y as above such that

$$
\begin{array}{ccc}
I & \xrightarrow{\ \Delta\ } & TI \\
{\scriptstyle (Id_I, y)}\Big\downarrow & & \Big\downarrow{\scriptstyle Ty} \\
J \times U & \xrightarrow{\ F\ } & TU
\end{array}
\qquad (*)
$$

commutes. This condition can be the written as the equation

$$
y'(t) = F_2(t, y(t))
$$

with the same initial condition as above. But this seemingly more general setup is only a special case of the conservative situation if we apply the following trick. Taking the open set $U^* = J \times U \subset \mathbb{R}^{n+1}$, we extend the time-dependent vector field to a conservative field $F^* : U^* \rightarrow T(U^*) = \mathbb{R}^{n+1} : (t, u) \mapsto (1, F(u))$. The boundary condition now reads $y^*(\zeta) = (\zeta, \eta)$. Then a solution $y^* = (y_1, y) : I \rightarrow U^*$ of the diagram

$$
\begin{array}{ccc}
I & \xrightarrow{\ \Delta\ } & TI \\
{\scriptstyle y^*}\Big\downarrow & & \Big\downarrow{\scriptstyle Ty^*} \\
U^* & \xrightarrow{\ F^*\ } & TU^*
\end{array}
$$

reads as follows: We have $F^* = (\Delta, F)$ and $Ty^* = Ty_1 \times Ty$ for a solution y of equation $(*)$. Therefore, the first coordinate of this diagram is commutative, iff y_1 solves $y_1' = 1$ and $y_1(\zeta) = \zeta$, i.e., $y_1 = Id_I$. So we have this result:

Proposition 297 *With the above notations, a conservative solution y^* for F^* is equivalent to a time-dependent solution y.*

We therefore stick to conservative ODEs and the question of the existence of solutions of such equations.

34.3 The Fundamental Theorem: Local Version

Given the data from definition 205, we denote by $A(F, \zeta, \eta, I, U)$ the set of local integral curves $y : I \rightarrow U$ of $F : U \rightarrow TU$ at (ζ, η). This set can be described by an alternative property: Given F, ζ and η, we have the operator

$$T_{\zeta, \eta, F}(y) = \eta + \int_{\zeta}^{?} F_2 \circ y$$

defined as an endomorphism on the set of curves $y : I \to U$. We then consider the set $B(F, \zeta, \eta, I, U)$ of C^1-curves $y : I \to U$ which are fixpoints of $T_{\zeta, \eta, F}$, i.e.,

$$T_{\zeta, \eta, F}(y) = y.$$

Lemma 298 *With the preceding notations, we have*

$$A(F, \zeta, \eta, I, U) = B(F, \zeta, \eta, I, U).$$

Exercise 168 Give a proof of lemma 298.

This identification of sets A and B induces a beautiful proof and also a method for the construction of solutions of such an ODE: One shows that under a certain condition, a so-called Lipschitz condition, for the vector field F, and by the selection of an adequate subset of functions y, the operator $T_{\zeta, \eta, F}$ is a contraction and therefore has a unique fixpoint which solves the equation.

Definition 206 *A vector field $F : U \to TU$ with second projection $F_2 : U \to \mathbb{R}^n$ is called* locally Lipschitz, *iff for every $u \in U$ there is a neighborhood $N(u) \subset U$ and a positive real number L such that for all $x_1, x_2 \in N(u)$, we have $\|F_2(x_1) - F_2(x_2)\| \leq L \cdot \|x_1 - x_2\|$ for the Euclidean norm.*

Evidently, a locally Lipschitz vector field is continuous, as the condition is in fact something between continuous and differentiable. Here is the local fundamental theorem of ODEs:

Proposition 299 (Local Fundamental Theorem of ODEs) *If for an open set $U \subset \mathbb{R}^n$, the vector field $F : U \to TU$ is locally Lipschitz, then for the initial conditions $\zeta \in \mathbb{R}$ and $\eta \in U$, there is an open interval $I(\zeta)$ containing ζ and an open neighborhood $U(\eta) \subset U$ of η such that $A(F, \zeta, \eta, I(\zeta), U(\eta)) = \{y\}$. This singleton's element y is called the local solution of the differential equation $y' = F \circ y$ at $I(\zeta)$ and $U(\eta)$.*

The proof involves some calculations in Banach spaces (complete metric spaces, defined by norms). But we have a rather explicit procedure which guarantees a contraction as follows.

1. Select a radius $r_1 > 0$ such that $\overline{B}_{r_1}(\eta) \subset U$.

2. Select the radius r_2 such that there is a Lipschitz constant L with $\|F_2(x_1) - F_2(x_2)\| \le L \cdot \|x_1 - x_2\|$ for all $x_1, x_2 \in \overline{B}_{r_2}(\eta)$.

3. Set $r = \min(r_1, r_2)$.

4. Take a number m such that $\|F_2(x)\| \le m$ on all $x \in \overline{B}_r(\eta)$ (which exists since $\overline{B}_r(\eta)$ is compact and F is continuous).

5. Take a positive number $\delta < \frac{r}{2(m+rL)}$.

6. Consider now the open interval $I =]\zeta - \delta, \zeta + \delta[$ around ζ and the set $BC(I, \mathbb{R}^n) = \{f : I \to \mathbb{R}^n \mid f \in C^0, \|f\|_\infty < \infty\}$, where $\|f\|_\infty = \sup_{\xi \in I} f(\xi)$ is the so-called *uniform norm*. We also denote by $d(f, g) = \|f - g\|_\infty$ the derived metric, since no confusion is possible.

7. Next take the constant function $\overline{\eta} : I \to \mathbb{R}^n : \xi \mapsto \eta$, which is evidently in $BC(I, \mathbb{R}^n)$.

8. Finally, consider the closed ball of functions $\overline{B}_r(\overline{\eta}) = \{f \in BC(I, \mathbb{R}^n) \mid d(f, \overline{\eta}) \le r\}$. Then it can be shown (see [23]) that

$$T_{\zeta, \eta, F} : \overline{B}_r(\overline{\eta}) \to BC(I, \mathbb{R}^n),$$

and that this operator is a contraction with a constant c satisfying $d(T_{\zeta, \eta, F}(\overline{\eta}), \overline{\eta}) < (1-c)r$. By corollary 296, and the fact that $BC(I, \mathbb{R}^n)$ is complete with the distance derived from the uniform norm, we infer that there is a unique fixpoint $y \in \overline{B}_r(\overline{\eta})$.

9. This fixpoint may be obtained by starting from $\overline{\eta}$ and then calculating the Cauchy sequence $(T_{\zeta, \eta, F}^i(\overline{\eta}))_i$ which has

$$\lim_{i \to \infty} T_{\zeta, \eta, F}^i \overline{\eta} = y.$$

This is the *Picard-Lindelöf iteration procedure*.

34.4 The Special Case of a Linear ODE

Let us now look at the special case of a linear ODE. It is given by the vector field $T_A : \mathbb{R} \to T\mathbb{R}^n : x \mapsto (x, A(x))$, where $A \in Lin_\mathbb{R}(\mathbb{R}^n, \mathbb{R}^n)$ is a linear endomorphism. This is classically written as

$$y' = A(y)$$

with initial condition $y(\zeta) = \eta$. Evidently, T_A is a continuous vector field. We also have $\|A(x_1) - A(x_2)\| = \|A(x_1 - x_2)\| \leq \|A\| \cdot \|x_1 - x_2\|$ by the Schwarz inequality, and where $\|A\|$ is the Euclidean norm of the matrix of A in the standard basis. Therefore T_A is Lipschitz with a global Lipschitz constant $L = \|A\|$.

Exercise 169 Give a proof of the inequality $\|A(x_1 - x_2)\| \leq \|A\| \cdot \|x_1 - x_2\|$ using the Schwarz inequality in the standard Euclidean space $\mathbb{R}^n$.

Let us now look at the limit process $\lim_{i \to \infty} T^i_{\zeta, \eta, F} \overline{\eta} = y$ when starting from a constant function $\overline{\eta}$ defined by an initial vector $\eta \in \mathbb{R}^n$ and the initial parameter $\zeta = 0$ (this is no essential restriction to the general solution).

Lemma 300 *With the preceding notations, the i-th Picard-Lindelöf iteration $T^i_{0, \eta, T_A} \overline{\eta}$ at the curve parameter t has the shape*

$$T^i_{0, \eta, T_A} \overline{\eta}(t) = \left(\sum_{j=0}^{i} \frac{1}{j!} t^j A^j \right)(\eta). \qquad (*)$$

Exercise 170 Give a proof of the formula $(*)$. Proceed by induction on i.

It follows from the general theory of function limits that the limit function, i.e., the fixpoint of this ODE is the function which at t evaluates to the limit

$$T^\infty_{0, \eta, T_A} \overline{\eta}(t) = \left(\sum_{j=0}^{\infty} \frac{1}{j!} t^j A^j \right)(\eta)$$

of the above series. For obvious reasons, this solution is written as

$$y(t) = e^{tA}(\eta).$$

Example 133 Consider the equation $y' = Ay$ where

$$A = \begin{pmatrix} 0 & 1 \\ -1 & 0 \end{pmatrix},$$

i.e., a clockwise rotation by $\frac{\pi}{2}$.

If we write $y = (y_1, y_2)$, then our equation is equivalent to a system of linear ODEs:

$$y_1' = y_2,$$
$$y_2' = -y_1.$$

To solve this equation, we have to exponentiate the matrix A, so let us look at its powers:

$$A^2 = \begin{pmatrix} -1 & 0 \\ 0 & -1 \end{pmatrix} = -E_2, \quad A^3 = \begin{pmatrix} 0 & -1 \\ 1 & 0 \end{pmatrix} = -A,$$

$$A^4 = \begin{pmatrix} 1 & 0 \\ 0 & 1 \end{pmatrix} = E_2, \quad A^5 = A.$$

Exploiting this periodicity, we can rewrite the exponential and thus solve our equation:

$$\sum_{j=0}^{\infty} \frac{1}{j!} t^j A^j = \sum_{j=0}^{\infty} \frac{1}{(4j)!} t^{4j} E_2 + \sum_{j=0}^{\infty} \frac{1}{(4j+1)!} t^{4j+1} A -$$

$$\sum_{j=0}^{\infty} \frac{1}{(4j+2)!} t^{4j+2} E_2 - \sum_{j=0}^{\infty} \frac{1}{(4j+3)!} t^{4j+3} A$$

$$= \sum_{j=0}^{\infty} \frac{(-1)^j}{(2j)!} t^{2j} E_2 - \sum_{j=0}^{\infty} \frac{(-1)^j}{(2j+1)!} t^{2j+1} A$$

Using the representation of the sine and cosine functions as given in proposition 257, and writing down the matrixes explicitly, we get

$$\sum_{j=0}^{\infty} \frac{1}{j!} t^j A^j = \cos(t) E_2 - \sin(t) A = \begin{pmatrix} \cos(t) & -\sin(t) \\ \sin(t) & \cos(t) \end{pmatrix}.$$

In other words, $y(t) = e^{tA}(\eta)$ describes a clockwise motion on a circle with radius $|\eta|$. Writing the initial condition vector $\eta = (\eta_1, \eta_2)$, the solution to the system of linear equations is:

$$y_1(t) = \cos(t)\eta_1 - \sin(t)\eta_2,$$
$$y_2(t) = \sin(t)\eta_1 + \cos(t)\eta_2.$$

See also example 134.

34.5 The Fundamental Theorem: Global Version

The local fundamental theorem of ODEs is unsatisfactory in that it does only describe solutions around a possibly very small neighborhood of a point of the domain U of the vector field F. However, the uniqueness of solutions offers itself to a process of gluing together local solutions to form a global solution. Here is the setup for the generation of global solutions.

Definition 207 *If $F : U \to TU$ is a Lipschitz continuous vector field and $u, v \in U$, then we say that u is equivalent to v, in signs $u \sim v$, iff there is a local integral curve y for F such that $\{u, v\} \subset Im(y)$.*

Lemma 301 *The relation $\sim$ in definition 207 is an equivalence relation. The equivalence class of $u \in U$ is denoted by $[u]$. The quotient $U/\sim$ is called the* phase portrait *of F, it is denoted by U/F.*

Proof Evidently, $\sim$ is reflexive and symmetric. We have to show that the relation is also transitive. So let $u \sim v \sim w$. Then we have two integral curves $y_1 : I_1 \to U$ and $y_2 : I_2 \to U$ such that $y_1(\xi_1) = v = y_2(\xi_2)$. Let $I_2^* = T^{\xi_1 - \xi_2}(I_2)$, where T^x is the translation by x defined in section 22.3 of volume 1. Then the curve $y_2^* = y_2 \circ T^{\xi_2 - \xi_1} : I_2^* \to U$ is an integral curve such that $y_2^*(\xi_1) = v$, because $(y_2^*)'(t) = y_2'(T^{\xi_2 - \xi_1}(t)) = F_2(y_2(T^{\xi_2 - \xi_1}(t))) = F_2(y_2^*(t))$. By the uniqueness of a solution of the differential equation, both solutions y_2^* and y_1 coincide on the intersection $I_1 \cap I_2^*$. Therefore, we get an integral curve on the union $I_1 \cup I_2^*$, and here $u \sim w$. $\qquad\square$

We now describe the equivalence classes $[u]$ of the phase portrait.

Proposition 302 *Let $u \in U$, then there is a unique local integral curve $y : J \to U$ for F with $y(0) = u$ and such that if $z : I \to U$ is any integral curve for F with $z(0) = u$, then $I \subset J$ and $z = y|_I$. This curve is called the* global integral curve *through u and is denoted by $\int_u F$. We have $Im(\int_u F) = [u]$. In particular, the phase portrait is the collection of images of global integral curves, and U is partitioned into the images of such curves.*

In particular, if $u_1 \sim u_2$, and if $\int_{u_1} F(t_2) = u_2$, then we have $\int_{u_2} F = \int_{u_1} F \circ T^{t_2}$. The domains J_1, J_2 of $\int_{u_1} F, \int_{u_2} F$, respectively, are related by $J_2 = T^{-t_2}(J_1)$.

Proof If we have two integral curves y_1 and y_2 with $y_1(0) = y_2(0) = u$, then by uniqueness, they coincide on the intersection of their domains, so they define an integral curve on the union of their domains. Therefore the union of all integral curves z with $z(0) = u$ is a maximal integral curve y with $y(0) = u$. Clearly for this curve, we have $[u] = Im(y)$. Also, if $\int_{u_1} F$ is maximal, then so is $\int_{u_1} F \circ T^{t_2}$, and therefore it coincides with $\int_{u_2} F$. The statement about the domains is immediate. $\qquad\square$

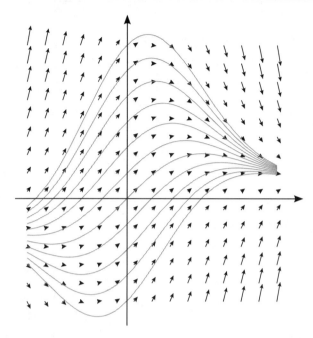

Fig. 34.1. Some integral curves from the phase portrait of the vector field $F(x, y) = (x, y, 1, 1 - 3xy)$, drawn for $-1 \leq x \leq \frac{3}{2}$ and $-1 \leq y \leq \frac{3}{2}$.

Proposition 303 *A global integral curve $\int_w F$, which is not an injective function of the curve parameter, is called a* cycle *of F. If for a curve parameter t_1 of a cycle $\int_w F$, we have $\int_w F(t_1) = \int_w F(t_1 + P)$ for $P \neq 0$, then the cycle domain is $\mathbb{R}$ and $\int_w F(t) = \int_w F(t + P)$ for all t, i.e., the cycle is P-periodic.*

Proof Since $\int_w F(t_1) = \int_w F \circ T^P(t_1)$, and both curves $\int_w F$ and $\int_w F \circ T^P$ are maximal integral curves with common value at t_1, we have $\int_w F = \int_w F \circ T^P$. Therefore, for their common domain J, we have $J = T^P J$, i.e., $J = \mathbb{R}$. □

Example 134 Let $U = \mathbb{R}^2$ and $F(x, y) = (x, y, y, -x)$, the linear vector field defined by a clockwise rotation by $\frac{\pi}{2}$ (see section 34.4). Then the integral curves of this differential equation are all cycles, and their images are concentric circles around the origin.

Example 135 Lie brackets (proposition 293) play an important role in Hamiltonian mechanics, but they are also an essential tool in modern robotics. In non-linear control theory, problems of motion planning are a

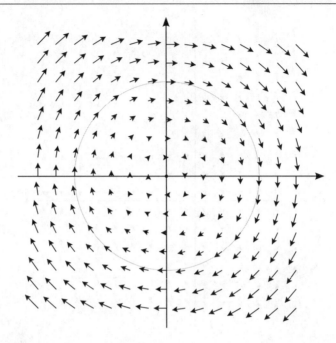

Fig. 34.2. The linear vector field $F(x, y) = (x, y, y, -x)$. The gray circle is one of the local integral curves.

core issue, which is important for multi-legged robot motion, for example. It is here, where the Lie bracket has a very practical interpretation. To this end, one has to give a second interpretation of the Lie bracket $[F, G]$ of two (Lipschitz continuous) vector fields on an open set $U \subset \mathbb{R}^n$. If $u \in U$, then we have integral curves of F and G in a neighborhood of u. Let us focus on this neighborhood. Taking a small $\varepsilon > 0$, we follow a four-part trajectory starting at u on $\int_u F$, see figure 34.3. We interrupt the walk at the point $v = \int_u F(\varepsilon)$. Then we proceed on the integral curve $\int_{\int_u F(\varepsilon)} G$ and stop at point $w = \int_v G(\varepsilon) = \int_{\int_u F(\varepsilon)} G(\varepsilon)$. We now proceed again in the direction of F, but backwards, i.e., we move to $x = \int_w F(-\varepsilon) = \int_{\int_{\int_u F(\varepsilon)} G(\varepsilon)} F(-\varepsilon)$. Finally, having arrived at this point, we move in the direction of G again, but backwards, and we stop at

$$L(\varepsilon) = \int_{\int_{\int_{\int_u F(\varepsilon)} G(\varepsilon)} F(-\varepsilon)} G(-\varepsilon).$$

Finally, we take the limit $\varepsilon \to 0$ and it can be shown that

$$[F, G](x) = \lim_{\varepsilon \to 0} \frac{1}{\varepsilon^2} L(\varepsilon).$$

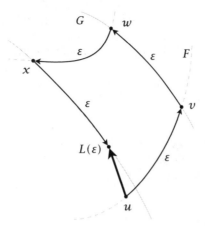

Fig. 34.3. The Lie derivative $[F, G](x)$ can be calculated as the limit of a four-part walk on the integral curves of F and G.

Surprisingly, this interpretation of Lie brackets has a well-known application in common life: when you are parking your car, the combined steering strategy is precisely a Lie bracket action. In fact, we may describe the car's position by the coordinates $Q = (x, y, \theta) \in \mathbb{R}^3$, where x and y are the spatial coordinates of the car's center, while θ is the angle of the car's motion axis. We have two types of movements, which we suppose to be occurring at a constant speed 1: Steering to the right has spatial velocity $(\cos(\theta), \sin(\theta))$, while the angle decreases constantly with rate $-q$, so the three-dimensional velocity field at Q is $F_2(Q) = (\cos(\theta), \sin(\theta), -q)$. Therefore the velocity, i.e., the derivative of the motion coordinates at time t yields

$$F_2(Q) = \frac{dQ}{dt} = (\cos(\theta), \sin(\theta), -q).$$

Similarly, steering to the left yields

$$G_2(Q) = \frac{dQ}{dt} = (\cos(\theta), \sin(\theta), q).$$

Now, interpreting the Lie bracket $[F, G]$ as above, we have the following trajectory for the parking movement: First we drive in forward motion to the right (with angle rate $-q$) moving along F for ε seconds, then we still drive forwards, but steer to the left (now with positive rate q) we follow G for ε seconds. Then we move backwards along F, steering to the right (again with rate $-q$), and again for ε seconds, and conclude with a backwards motion along G (again with rate q), for another ε seconds. The new position approaches the Lie bracket vector field

$$[F,G]_2(Q) = (0, 2q\cos(\theta), 0),$$

which is a vector field in y direction, i.e., a shift movement perpendicular to the street axis, exactly what was desired by the driver.

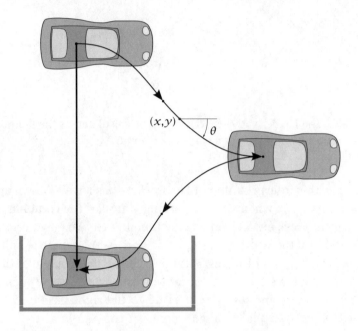

Fig. 34.4. A sideways parking movement of a car can be achieved by going through a sequence of forward and backward movements, illustrating the geometric meaning of Lie brackets.

CHAPTER 35

Third Advanced Topic

35.1 Introduction

Numerical mathematics is a vast field which we can only slightly touch. It is, however, very important to computer science and practice since the way many mathematical theories are presented is far from computational. In fact, some theories are in principle beyond computability. These fields are not as outlandish as one might guess at first sight. We have already encountered the definite imprecision of floating point arithmetic. Another source for problems of computability is the amount of computational effort required to obtain a desired result. The algorithms which would return optimal results are sometimes excessively time-consuming, so one has to look for faster algorithms that deliver less precise, in short, suboptimal, solutions. Numerics is therefore also concerned with the propagation of errors along an algorithmic computation.

35.2 Numerics of ODEs

In this chapter, we want to deal with a problem of numerical solutions of ODEs which is not due to some flaw specific to computers, but mainly to the non-constructive proof of existence of solutions, which can only be made explicit by approximations. In principle, this kind of objects is intimately tied to the topological completeness of $\mathbb{R}$, i.e., their non-constructive nature lies in the non-constructive nature of the real numbers as a such. But it is not only this known gap in constructions, which emerge so dramatically in the case of ODE solutions, it is the fact that such solutions effectively express new kinds of functions, which have properties not shared by already known types of function. This phenomenon already arises in calculus: The integral, when seen as the primitive function of a given function, may generate completely new functions. The main theorem of ODEs is based on a fixpoint theorem, and we know from the recursion theorem, which is a special case of a fixpoint theorem (see exercise 167) that new and basic operations in all fields of mathematics (and in particular in programming) are generated as fixpoints of contractions. So by the very nature of the contraction towards its fixpoint, we have to approach solutions of ODEs by approximations, i.e., by so-called numerical methods. This is nothing more than a computationally effective, but approximative, path towards the ideal fixpoint object.

A priori, the fixpoint theorem for ODE solutions does not suggest a specific formalism for the approximation. However, in the case of linear ODEs, as discussed in section 34.4, the Picard-Lindelöf iteration yields the Taylor approximation by polynomials. It is therefore no great surprise that the numerical methods for ODE solutions refer to Taylor's formula. Whereas Euler's method refers to the approximation by first derivatives, Runge-Kutta's method uses up to the third derivative in Taylor's development and, thus, yields more precise results.

But why do such methods produce sensible results at all? In fact, Taylor's formula involves the derivatives of an unknown function $y(t)$ if the ODE has the shape $\frac{dy}{dt}(t) = F(t, y(t))$ (we only look at one-dimensional ODEs in this chapter). The point is that the very form of an ODE is recursive with regard to the degree of derivation. So, in principle, one may replace the derivative $\frac{d^n y}{dt^n}$ by $\frac{d^{n-1}F}{dt^{n-1}}$. But the trouble begins with the increasingly complicated explicit formulas that arise when one deals with higher derivatives. To manage the growing complexity approximative approaches therefore become mandatory.

Henceforth, we shall consider the standard situation in numerics: a one-dimensional time-dependent ODE, which we state as

$$\frac{dy}{dt}(t) = F(t, y(t)),$$

together with the initial condition $y(t_0) = y_0$. We also assume that general conditions for the uniqueness and existence of a solution y to such an equation are fulfilled. More precisely, we suppose that F is C^∞ in an open neighborhood of the point $(t_0, y_0) \in \mathbb{R}^2$, a condition which is stronger than the Lipschitz condition used in the main theorem 299.

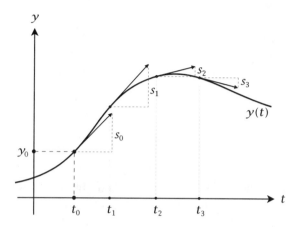

Fig. 35.1. The solution $y(t)$ of an ODE $y'(t) = F(t, y(t))$ with initial condition $y(t_0) = y_0$. At values t_0, t_1, t_2 and t_3, the slopes of the curve are, respectively, $s_0 = F(t_0, y(t_0))$, $s_1 = F(t_1, y(t_1))$, $s_2 = F(t_2, y(t_2))$ and $s_3 = F(t_3, y(t_3))$.

Similarly to the construction of values of a recursively defined object according to the recursion theorem, the solution of an ODE must also start from the initial datum $y(t_0) = y_0$. In contrast to the recursive object, the solution is not a function of a discrete sequence $0, 1, 2, \ldots$ of natural number arguments, but of real numbers t. Since it is, in principle, out of the question to compute the values of the solution at *all* time points, one starts with an increasing sequence of *discrete time points* $(t_k)_{k=0,1,\ldots}$, i.e., $t_0 < t_1 < t_2 < \ldots t_k < t_{k+1} < \ldots$. Using an adequately chosen method, a more or less precise approximation y_k to the exact solution $y(t_k)$ is computed. From the results thus obtained, one proceeds by refining the sequence of time points, and then computing a better approximation.

Such a method $\mathcal{M}$ may be assessed according to its precision for a given sequence $(t_k)_k$ of time points and then by its behavior for a successive refinement of the time sequence. When such approximations are calculated, two types of errors occur:

1. *Round-off errors*: These are errors that are caused by the calculation process itself due to numerical imprecisions, because of the inher-

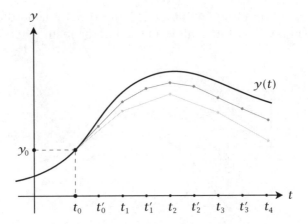

Fig. 35.2. A fictitious method determines a (rather bad) approximation to the solution $y(t)$ of an ODE for the sequence $t_0, t_1, \ldots t_4$ (light gray). After a refinement of the time sequence to $t_0, t_0', \ldots, t_3', t_4$, it yields a better approximation (dark gray).

ently flawed floating point arithmetic of the computer, or because of the choice of the representation of numbers, selected for reasons of improving calculation time, for example.

2. *Truncation errors*: This type of errors is caused by the approximation method $\mathcal{M}$ itself. Truncation errors are two-fold for the following reason. As most common methods $\mathcal{M}$ are based on formulas proceeding inductively with the calculation of y_k, i.e., they calculate the value $y_{k+1} = \mathcal{M}(y_0, y_1, \ldots y_k)$ based on the knowledge of the values $y_0, y_1, \ldots y_k$ (the time arguments are suppressed in our description of the methods).

   • *Local truncation errors*: These errors measure the deviation of the correct value $y(t_{k+1})$ from the value $\mathcal{M}(y(t_0), y(t_1), \ldots y(t_k))$ determined by $\mathcal{M}$.

   • *Cumulative truncation errors*: These errors measure the deviation of the correct value $y(t_N)$ at a final time t_N from the value

   $$y_N = \mathcal{M}(y_0, y_1, \ldots y_{N-1}) = \mathcal{M}(y_0, \mathcal{M}(y_0), \mathcal{M}(y_0, \mathcal{M}(y_0)), \ldots)$$

   calculated by successive application of $\mathcal{M}$, including the round-off errors. Here it is important to note that, *in general, the cumulative truncation error is* not *the sum of the local truncation errors.*

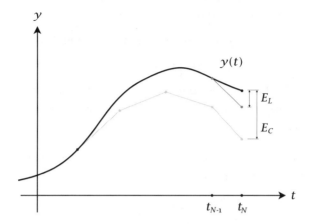

Fig. 35.3. A local truncation error E_L, and the cumulative truncation error E_C at t_N.

Definition 208 *A method $\mathcal{M}$ is called an r-step method iff the calculated value $y_{k+1} = \mathcal{M}(y_0, y_1, \ldots y_k)$ depends only on the last r predecessors $y_{k-r+1}, y_{k-r+2}, \ldots y_k$ of y_{k+1}.*

35.3 The Euler Method

Euler's original method $\mathcal{E}$ is a one-step method, which means that $y_{k+1} = \mathcal{E}(y_k)$. We want to discuss it for the sequence $t_k = t_0 + k \cdot h$, for $k = 0, 1, \ldots N$, of $N + 1$ equidistant time points, with a positive, appropriately small step size h. The idea behind $\mathcal{E}$ is the Taylor formula of degree 1 and remainder term of degree 2:

$$y(t_{k+1}) = y(t_k + h)$$

$$= y(t_k) + h \cdot \frac{dy}{dt}(t_k) + \frac{h^2}{2} \cdot \frac{d^2 y}{dt^2}(t_k + \xi \cdot h)$$

with $0 < \xi < 1$, according to proposition 271. Neglecting the remainder term, we have the approximation

$$y_{k+1} = y(t_k) + h \cdot \frac{dy}{dt}(t_k),$$

where we may replace $y(t_k)$ by y_k and the derivative $\frac{dy}{dt}(t_k)$ by the expression provided by the ODE, i.e., $\frac{dy}{dt}(t_k) = F(t_k, y_k)$, together with the previous value y_k, yielding

$$y_{k+1} = y_k + h \cdot F(t_k, y_k) = \mathcal{E}(y_k), \tag{1}$$

the expression provided by the Euler method.

This method is also called *Cauchy polygon method*, since the piecewise linear approximations can be used to define a polygon connecting the points (t_0, y_0), (t_1, y_1), ... (t_N, y_N).

Example 136 Here is an example of the polygonal shape of the Euler method for the equation

$$\frac{dy}{dt}(t) = \sin(t) - 10 \cdot t \cdot y(t)$$

and the initial condition $t_0 = 0$ and $y_0 = 1$. We apply $\mathcal{E}$ using a step size $h = 0.1$ and a number $N = 10$ of steps starting with 0 on, i.e., we have $t_k = k \cdot 0.1$, with $k = 0, 1, \ldots 10$.

The formula used in Euler's method is then

$$y_{k+1} = y_k + h \cdot (\sin(t_k) - 10 \cdot t_k \cdot y_k)$$
$$= y_k + 0.1 \cdot (\sin(t_k) - k \cdot y_k).$$

k	t_k	y_k	$y(t_k)$	$|y(t_k) - y_k|$
0	0.0	1.0000000000	1.0000000000	0.0000000000
1	0.1	1.0000000000	0.9561023853	0.0438976147
2	0.2	0.9099833417	0.8367953265	0.0731880152
3	0.3	0.7478536064	0.6735741493	0.0742794571
4	0.4	0.5530495452	0.5035696686	0.0494798766
5	0.5	0.3707715613	0.3560822075	0.0146893538
6	0.6	0.2333283345	0.2455946823	0.0122663478
7	0.7	0.1497955811	0.1726399793	0.0228443982
8	0.8	0.1093604431	0.1294062851	0.0200458420
9	0.9	0.0936076977	0.1057920514	0.0121843537
10	1.0	0.0876934607	0.0932629680	0.0055695073

Fig. 35.4. The values determined by the Euler method are in the third column, the values for the exact solution in the fourth column, everything computed with a precision of 10 digits after the decimal point. The fifth column shows their absolute differences.

Table 35.4 shows the values for $y_0, y_1, \ldots y_{10}$ calculated with a precision of 10 digits after the decimal point.

Figure 35.5 compares the exact solution with the values calculated using the Euler method.

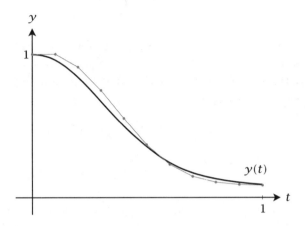

Fig. 35.5. The approximation (in gray) to the solution of the equation from example 136 using Euler's method with step size $h = 0.1$. The exact solution is shown in black.

The main result about the power of approximation power of $\mathcal{E}$ is:

Proposition 304 *If F is C^∞, and if we are given the time sequence $t_k = t_0 + k \cdot h$, with $k = 0, \ldots N$, then the cumulative truncation error $|y(t_N) - y_N|$ defined by $\mathcal{E}$ as specified in formula (1), is not greater than $C \cdot h$ for a constant C which depends only on F, t_0, and t_N, but not on h.*

35.4 Runge-Kutta Methods

The Runge-Kutta method $\mathcal{RK}$ is more precise than the elementary Euler method $\mathcal{E}$, but it is also a one-step method. We want to discuss it again for the sequence $t_k = t_0 + k \cdot h$, with $k = 0, 1, \ldots N$ of $N + 1$ equidistant times for a positive, appropriately small step size h. $\mathcal{RK}$ uses the Taylor formula in the following way: One considers the Taylor development until degree 2, plus the remainder term of degree 3:

$$y(t_{k+1}) = y(t_k + h)$$

$$= y(t_k) + h \cdot \frac{dy}{dt}(t_k) + \frac{h^2}{2} \cdot \frac{d^2y}{dt^2}(t_k) + \frac{h^3}{3!} \cdot \frac{d^3y}{dt^3}(t_k + \xi \cdot h)$$

$$(2)$$

for $0 < \xi < 1$. The first derivative in this formula is again replaced by the right hand side of the ODE $\frac{dy}{dt}(t_k) = F(t_k, y_k)$, additionally the second derivative is the first derivative of the right hand side. Using the chain rule from proposition 264, we get

$$\frac{d^2y}{dt^2}(t) = \frac{\partial F}{\partial t}(t, y) + \frac{\partial F}{\partial y}(t, y) \cdot \frac{dy}{dt}(t)$$

$$= \frac{\partial F}{\partial t}(t, y) + \frac{\partial F}{\partial y}(t, y) \cdot F(t, y(t)).$$

Again, we neglect the cubic term and obtain the approximation

$$y_{k+1} = y_k + h \cdot F(t_k, y_k) + \frac{h^2}{2} \cdot \left[\frac{\partial F}{\partial t}(t_k, y_k) + \frac{\partial F}{\partial y}(t_k, y_k) \cdot F(t_k, y_k) \right] \quad (3)$$

with a local truncation error

$$\frac{h^3}{3!} \cdot \frac{d^3y}{dt^3}(t_k + \xi \cdot h).$$

If controllable information were available about the partial derivative expressions for $\frac{\partial F}{\partial t}$ and $\frac{\partial F}{\partial y}$ that could be applied for the arguments t_k and y_k, everything would work out nicely, but this is often not the case. The $\mathcal{RK}$ method circumvents this problem through a formula which eventually guarantees a local truncation error not greater than $C \cdot h^5$, but does only use the function F, not its partial derivatives. We shall omit a discussion of this result, since its proof is quite involved.

There are in fact several Runge-Kutta method variants, however, here we shall only present the so-called *classical fourth-order Runge-Kutta method*. The method is best described by the successive determination of four variables. Here is the formula:

$$y_{k+1} = \mathcal{RK}(y_k)$$

$$= y_k + \frac{h}{6} (p_1 + 2 \cdot p_2 + 2 \cdot p_3 + p_4),$$

where

$$p_1 = F(t_k, y_k),$$

$$p_2 = F(t_k + \tfrac{h}{2}, y_k + \tfrac{h}{2} \cdot p_1),$$

$$p_3 = F(t_k + \tfrac{h}{2}, y_k + \tfrac{h}{2} \cdot p_2),$$

$$p_4 = F(t_k + h, y_k + h \cdot p_3).$$

This formula needs a few comments: The expression $\mathcal{RK}(y_k)$ looks like an Euler method, except that the slope is not taken at one, but at four different points. The first slope p_1 is the slope at the Euler point (t_k, y_k), just as before. The second slope p_2 is the slope taken at the midpoint $\tfrac{h}{2}$ for the Euler approximation, i.e., at the argument $(t_k + \tfrac{h}{2}, y_k + \tfrac{h}{2} \cdot p_1)$. The third value p_3 is again the slope at a midpoint, but this time the slope is taken at the y-value $y_k + \tfrac{h}{2} \cdot p_2$ instead of $y_k + \tfrac{h}{2} \cdot p_1$ The fourth and last value p_4 is the slope at the endpoint $t_k + h = t_{k+1}$, with y-value $y_k + h \cdot p_3$. The final slope used to compute y_{k+1} is a weighted average of the slopes p_1 (weight $\tfrac{1}{6}$), p_2 (weight $\tfrac{1}{3}$), p_3 (weight $\tfrac{1}{3}$) and p_4 (weight $\tfrac{1}{6}$). For an illustration of this process, see figure 35.6.

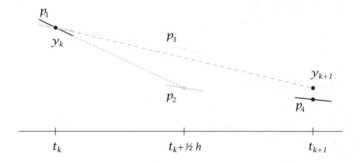

Fig. 35.6. The classical Runge-Kutta method uses four derivatives to calculate the value y_{k+1}. The first derivate p_1 is the slope at the start point (t_k, y_k) (black), the second p_2 at the midpoint $t_k + \tfrac{1}{2}h$, where the y-value is determined by following the slope p_1 from y_k (gray). The third derivative is calculated similarly, but this time using the slope p_2 (light gray). The last derivative is calculated at the endpoint t_{k+1} following the slope p_3. The resulting value y_{k+1} is calculated at t_k, following a slope that is a weighted average of p_1, p_2, p_3 and p_4.

Example 137 We take up again the ODE from example 136 and apply the fourth-order Runge-Kutta method using the same step size $h = 0.1$. The results are shown in table 35.7.

k	t_k	y_k	$y(t_k)$	$|y(t_k) - y_k|$
0	0.0	1.0000000000	1.0000000000	0.0000000000
1	0.1	0.9561021361	0.9561023853	0.0000002492
2	0.2	0.8367935726	0.8367953265	0.0000017539
3	0.3	0.6735735888	0.6735741493	0.0000005605
4	0.4	0.5035853290	0.5035696686	0.0000156604
5	0.5	0.3561408033	0.3560822075	0.0000585958
6	0.6	0.2457182467	0.2455946823	0.0001235645
7	0.7	0.1728260648	0.1726399793	0.0001860854
8	0.8	0.1296241703	0.1294062851	0.0002178852
9	0.9	0.1059991483	0.1057920514	0.0002070969
10	1.0	0.0934268053	0.0932629680	0.0001638374

Fig. 35.7. The values determined by the classical fourth-order Runge-Kutta method are in the third column, the values for the exact solution in the fourth column, everything computed with a precision of 10 digits after the decimal point. The fifth column shows their absolute differences.

As can been seen, the deviations are much smaller than the deviations for Euler's method. We omit a graphical illustration of the results, since, at this resolution, no difference between the calculated values and the exact value can be discerned.

We shall not give a proof of the classical $\mathcal{RK}$ method, but rather show, how a Runge-Kutta type approximation can be produced with respect to the degree 3 Taylor approximation (2). Our solution will yield a Runge-Kutta formula with $p_3 = p_4 = 0$. Let us start with a formula

$$y_{k+1} = y_k + h\left[uF(t_k, y_k) + vF(t_k + vh, y_k + \mu hF(t_k, y_k))\right] \qquad (4)$$

with unknown constants u, v, v and μ. We now use Taylor's formula for the function F of two variables:[1]

---

[1] The Taylor formula for functions in two variables is

$$F(x+d, y+e) = F(x, y) + d \cdot \frac{\partial F}{\partial x}(x, y) + e \cdot \frac{\partial F}{\partial y}(x, y) + \text{ higher terms in } d \text{ and } e,$$

for details, see [15].

$$F(t_k + vh, y_k + \mu h F(t_k, y_k)) =$$

$$F(t_k, y_k) + vh\frac{\partial F}{\partial t}(t_k, y_k) + \mu h F(t_k, y_k)\frac{\partial F}{\partial y}(t_k, y_k) + Rh^2,$$

which, when inserted in (4), yields

$$y_{k+1} = y_k + h(u + v)F(t_k, y_k)$$

$$+ h^2\left[vv\frac{\partial F}{\partial t}(t_k, y_k) + \mu v F(t_k, y_k)\frac{\partial F}{\partial y}(t_k, y_k)\right] + Rvh^3.$$

Then, choosing $u + v = 1$, $vv = \frac{1}{2}$ and $\mu v = \frac{1}{2}$ yields

$$y_{k+1} = y_k + hF(t_k, y_k)$$

$$+ \frac{h^2}{2}\left[\frac{\partial F}{\partial t}(t_k, y_k) + F(t_k, y_k)\frac{\partial F}{\partial y}(t_k, y_k)\right] + Rvh^3,$$

which is an approximated Taylor expansion as in (3), up to a term in h^3.

Example 138 We take the ODE $\frac{dy}{dt} = 2y$ with $y_0 = 3$ and $t_0 = 1$ and calculate an approximation of the solution by the third-order Runge-Kutta method $\mathcal{RK}$, that we have just developed, for $h = 0.2$ and $N = 3$.

Then we have $\frac{dy}{dt} = F(t, y)$ where $F(t, y) = 2y$. Therefore $\frac{\partial F}{\partial t}(t, y) = 2\frac{dy}{dt} = F(t, y)$ and $\frac{\partial F}{\partial y}(t, y) = 2\frac{dy}{dy} = 2$.

By substituting these values, we can specialize our formula, neglecting the cubic term Rvh^3,

$$y_{k+1} = y_k + hF(t_k, y_k) + \frac{h^2}{2}(F(t_k, y_k) + F(t_k, y_k) \cdot 2)$$

$$= y_k + hF(t_k, y_k) + \frac{3h^2}{2}F(t_k, y_k)$$

$$= y_k + 2hy_k + 2\frac{3h^2}{2}y_k$$

$$= y_k(1 + 2h + 3h^2)$$

The values calculated according to our conditions are shown in table 35.8.

k	t_k	y_k	$y(t_k)$	$\|y(t_k) - y_k\|$	$\|y(t_k) - y_{Ek}\|$
0	1.0	3.0000000000	3.0000000000	0.0000000000	0.0000000000
1	1.2	4.2720000000	4.4754740929	0.2034740929	0.2754740929
2	1.4	6.0833280000	6.6766227855	0.5932947855	0.7966227855
3	1.6	8.6626590720	9.9603507682	1.2976916962	1.7283507682

Fig. 35.8. The values for the third-order approximation of example 138. To facilitate a comparison between this method and the Euler method, the sixth column shows the differences for the Euler method.

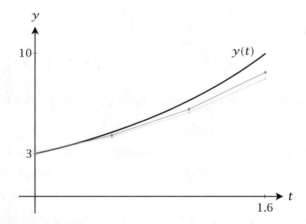

Fig. 35.9. The approximation (in gray) to the solution of the equation from example 138 by a third-order Runge-Kutta method with step size $h = 0.2$. The exact solution is shown in black. As comparison the Euler approximation of the same solution is drawn in light gray.

PART IV

Selected Higher Subjects

CHAPTER 36

Categories

36.1 Introduction

We have encountered a considerable number of types of mathematical structure, such as sets and functions, digraphs and digraph morphisms, groups and group homomorphisms, R-modules and R-linear maps, automata and morphisms of automata, or metric spaces and continuous maps. All these structures have some common features which give rise to a unifying concept: that of a category. A category is a collection of objects and relations between such objects, called morphisms, together with a small number of rules for combining and comparing morphisms. Categories were introduced to mathematics in 1945 by Samuel Eilenberg and Saunders Mac Lane in their paper *General Theory of Natural Equivalences* to provide a systematic account for the relations between algebraic structures (such as groups) and topological spaces. In the sixties the mathematician Alexander Grothendieck made use of special categories, called toposes, to solve important problems of algebraic geometry (the general theory of solutions of polynomial equations). Simultaneously, the mathematician William Lawvere used toposes to create a synthesis between geometry and logic, a theory which has profound implications also in theoretical computer science, see [18] and [37]. Moreover, toposes are also used for new directions in mathematical music theory. So categories are a universal and profound theoretical approach with which every computer scientist should have some acquaintance.

The theory of categories has two basic subjects: (1) functors and natural transformations between categories, (2) universal constructions of new objects. Functors describe the relations between different categories, while natural transformations describe relations between different functors. We have encountered functors, for example, in the theory of vector spaces, where each vector space V is sent to the identity matrix $E^{\dim(M)}$, whereas the linear maps $f : V \to W$ are sent to transformation matrixes of size $\dim(W) \times \dim(V)$ (see section 22.1 in volume 1) in such a way that the composition of maps commutes with the composition of matrixes.

Universal constructions deal with the creation of new objects from a specific diagram system of given objects and morphisms, in such a way that these new objects are the best possible solution with respect to a specific property. For example, the Cartesian product $X \times Y$ of two sets, together with its projections $pr_X : X \times Y \to X$ and $pr_Y : X \times Y \to Y$, has the universal property described in proposition 57 in section 6.2 of volume 1. This prototype of a property is in fact also encountered in other categories, for example the category of digraphs or the category of groups. Category theory deals with the unified description and analysis of such constructions.

If one should give category theory a unifying stamp, one would say that while algebra deals with the solution of *algebraic equations*, category theory deals with the solution of *diagram equations*.

36.2 What Categories Are

The prototype of a category is the set $\mathbb{M}(R) = \coprod_{m,n} \mathbb{M}_{m,n}(R)$ of all matrixes over a ring R, which we introduced in definition 152, section 20.1, of volume 1. What are the characteristic features of this structure? A matrix $M \in \mathbb{M}_{m,n}(R)$ represents a linear map $R^n \to R^m$ between modules. Such a representation needs two types of concepts: map and module, or in set-theoretic terms, function and object. But the concept of object is in fact superfluous. One may replace the object R^n by the identity matrix E_n, since there is a 1-to-1 correspondence between the two. We have already used this identification in the notation $M : E_n \to E_m$ in section 20.1 of volume 1. This means that we replace module objects by special maps, the identities on those module objects. But what about the elements of modules? Are they lost in our new conceptual space? No, in fact, we know that the linear maps $x : R^1 \to R^m$ are the elements of $\mathbb{M}_{1,m}(R)$, i.e., the elements of R^m. Therefore, every element of a module is recovered by a matrix. So we may really work in the environment of linear maps $R^n \to R^m$ between modules R^n and R^m in the exclusive language of matrixes in $\mathbb{M}(R)$.

Another example for the obsolete role of objects as compared to maps or morphisms is the situation for sets. We know that two sets x and y are equal iff their identities Id_x and Id_y are so. We may therefore write $f : Id_x \to Id_y$ instead of the traditional notation $f : x \to y$ for a set function f. Again, we have to check what happens with set elements for this conceptual transition. But there is a bijection $\epsilon : x \overset{\sim}{\to} Set(1, x)$ via $\epsilon(\xi)(0) = \xi$ for $\xi \in x$.

What is the characteristic property of an identity map? For set maps, an identity Id has this property: For every set function f, if the composition $f \circ Id$ is defined, then it equals f, and similarly from the other side: for every matrix or set function g, if the composition $Id \circ g$ is defined, then it equals g. Same property, mutatis mutandis, for matrix identities E_n. This motivates the following definition:

Definition 209 *A category* **C** *is a collection of entities* $f, g, h, \ldots$ *which are called* morphisms, *together with a partial composition, i.e., for some morphisms f and g, a new morphism $f \circ g$ of* **C** *is defined. An* identity *or* object *in* **C** *is by definition a morphism e such that, whenever defined, we have $e \circ f = f$ and $g \circ e = g$. We have these axioms:*

1. *If $f \circ g$ and $g \circ h$ are both defined, $(f \circ g) \circ h$ is defined.*

2. *Whenever one of the two compositions $(f \circ g) \circ h$ or $f \circ (g \circ h)$ is defined, both are defined and they are equal; we denote the resulting morphism by $f \circ g \circ h$.*

3. *For every morphism f there are two identities, a "left" identity e_L and a "right" identity e_R, such that $e_L \circ f$ and $f \circ e_R$ are defined (and necessarily equal to f).*

Lemma 305 *Two identities e and e' of* **C** *can be composed iff they are equal, and then $e \circ e = e$ (identities are idempotent.).*

Proof If e_L is a left identity for an identity e, then, by the property of identities, we have $e_L \circ e = e$ and $e_L \circ e = e_L$. So an identity is equal to its left identity, and the same for the right identity. In particular, $e \circ e = e$ is defined. If $e \ne e'$ for two identities, then the existence of $e \circ e'$ means also that $e \circ e' = e$ and $e \circ e' = e'$ by the defining properties of identities, a contradiction. Therefore $e \circ e'$ is not defined.   □

Lemma 306 *The left, resp. right, identities of a morphism f are uniquely determined by f.*

Proof Let $e \circ f = e' \circ f = f$ be two left identities. Then we have $f = e' \circ f = e' \circ (e \circ f)$, so the composition $(e' \circ e) \circ f$ is also defined, and therefore also $e' \circ e$, but then by lemma 305, $e = e'$; the same argumentation works for right identities.   □

The right and left identities are called $e_R = dom(f)$, the *domain of f*, and $e_L = codom(f)$, the *codomain of f*. We write $f : e_R \to e_L$ to fix this information. For given objects r and l in $\mathbf{C}$, the collection of morphisms $f : r \to l$ with $dom(f) = r$ and $codom(f) = l$ is denoted by $\mathbf{C}(r, l)$, or also $Hom_C(r, l)$, or $Hom(r, l)$ if $\mathbf{C}$ is clear. Evidently, every morphism is a member of a unique collection $\mathbf{C}(r, l)$, viz. that with $r = dom(f)$ and $l = codom(f)$. In other words, the collections $\mathbf{C}(r, l)$ define a partition of $\mathbf{C}$ in a non-set-theoretic common sense. In order to denote an identity e in an evident way, one usually writes $e = Id_e$.

Lemma 307 *If $f : r \to l$ and $g : r' \to l'$ are morphisms in a category $\mathbf{C}$, then their composition $g \circ f$ is defined iff $codom(f) = l = r' = dom(g)$.*

Proof If e is a right identity for g, then if $g \circ f$ is defined, we have $g \circ f = (g \circ e) \circ f = g \circ (e \circ f)$, therefore $e \circ f$ is defined and necessarily equals f, so $dom(g) = codom(f)$. Conversely, if $dom(g) = codom(f)$, then for the right identity e of g which equals the left identity of f, both, $g \circ e$ and $e \circ f$ are defined, hence also $g \circ f$.   □

Before we go over to a discussion of prominent examples of categories, we will introduce a number of concepts which we will recognize as generalizations of well-known concepts from previous theories.

Definition 210 *Let f, g, f', g', h be morphisms of a category $\mathbf{C}$.*
  (i) *A morphism f is mono, or a monomorphism, iff for any two compositions $f \circ g$ and $f \circ g'$, the equality $f \circ g = f \circ g'$ implies $g = g'$.*
  (ii) *A morphism f is epi, or an epimorphism, iff for any two compositions $g \circ f$ and $g' \circ f$, the equality $g \circ f = g' \circ f$ implies $g = g'$.*
  (iii) *A morphism f is called a section if there is a left inverse g, i.e., $g \circ f = dom(f)$.*

(iv) *A morphism f is called a* retraction *if it has a* right inverse *h, i.e.,* $f \circ h = dom(h)$.

(v) *A morphism f that is a section and a retraction is* iso, *or an* isomorphism. *Its right and left inverse coincide, it is uniquely determined, called* inverse *and denoted by* f^{-1}.

(vi) *If* $dom(f) = codom(f) = c$, *then f is called* endo, *or an* endomorphism *of c. The collection of endomorphisms of c is denoted by* $End(c)$.

(vii) *An endomorphism which is an isomorphism is called* auto, *or an* automorphism. *The collection of automorphisms of c is denoted by* $Aut(c)$.

If $End(c)$ and $Aut(c)$ are sets, they define monoids and groups, respectively, with the identity Id_c as unit, called *endomorphism monoids of the object c* and *automorphism groups of the object c*, respectively.

36.3 Examples

Here is a list of important examples of categories, also including those discussed in previous and some later chapters. If adequate, we shall add some specific comments on these categories. The reader should verify the axioms of a category reviewing the chapters where the specific objects have been introduced.

Example 139 Given a ring R, the category $\mathbb{M}(R)$ is an ordinary set, the morphisms are the matrixes, the compositions $M \circ N$ are the matrix products $M \cdot N$, the identities are the identity matrixes $E_n, n \in \mathbb{N}$. The *elements in the objects* E_n are by definition the morphisms $M : E_1 \to E_n$, they identify with R^n, as we have already seen above.

Exercise 171 Given a field R, show that a matrix $M \in \mathbb{M}_{m,n}(R)$ in the category $\mathbb{M}(R)$ is a section iff $rk(M) = n$. It is a retraction iff $rk(M) = m$, and it is an isomorphism iff $det(M) \neq 0$. See definition 167, section 22.1 in volume 1, for the definition of the rank of a matrix.

Example 140 The morphisms in the category **Sets** are the set functions $f : x \to y$. The set of morphisms $f : x \to y$ is denoted by $Set(x, y)$. The identities are the identities Id_x, one for each set x.

To satisfy the axioms of definition 209, only one representative of a set is allowed, i.e., we should find a way to identify x and Id_x, for each set x. There are two ways to do so: Either we replace sets by identities, or vice versa. The first alternative leads to an infinite regress: Because then we would have replaced each function f, which in fact is a set, by its identity Id_f, and then each Id_f, which is a set, by its identity Id_{Id_f}, and so on *ad infinitum*. Therefore we are left with the second alternative: to replace every identity Id_x by the set x. Assuming the axiom of foundation, i.e., all sets are founded, this replacement will terminate.

An *element of a set object* x is by definition a morphism $f : 1 \to x$, this identification having already been discussed above.

Exercise 172 Show that in the category of sets **Sets**, a morphism is mono iff it is injective. Show that it is epi iff it is surjective. Show that it is iso iff it is a bijection, in other words, it is iso iff it is mono and epi. Show that two morphisms $f, g : x \to y$ are equal iff they coincide on the elements of x, i.e., iff for each element $\xi : 1 \to x$, we have $f \circ \xi = g \circ \xi$. In a category, an object which has this property, shared by the set 1 in **Sets**, is said to *generate the category*.

In category theory, (di)graphs may have infinite sets of arrows and vertexes. The definition of possibly infinite (di)graphs and morphisms between them is literally the same as for finite graphs. One just omits the restriction of finiteness. We admit (di)graphs without this restriction in what follows.

Example 141 The category of (now possibly infinite) digraphs **Digraph** has as morphisms the elements of the sets $Digraph(\Gamma, \Delta)$ of morphisms between digraphs Γ and Δ, while the identities are the identities of digraphs, the identification of digraphs and morphisms is similar to the situation for sets discussed before.

Example 142 Given a digraph $\Gamma : A \to V^2$, the *path category of* Γ is the set of paths $Path(\Gamma)$ in Γ, as defined in section 12.2 of volume 1. The composition $p \circ q$ of two paths p and q is the product of paths pq. The identities are the lazy paths v at vertex $v \in V$. Since the length of paths is added under composition, and since lazy paths have length 0, there are no isomorphisms except the lazy paths. However, every morphism is epi and mono since the paths determine the arrows which they are built of. This is an example of a category, where, in contrast to **Sets**, "epi" and "mono" is not equivalent to "iso".

Example 143 For (now possibly infinite) graphs, we have the category **Graph** composed of sets $Graph(\Gamma, \Delta)$ of morphisms between graphs Γ and Δ, while the identities are the identities of graphs, the identification of graphs and morphisms is similar to the situation for sets discussed before.

Example 144 The morphisms of the category **Mon** of monoids are the elements of the sets $\textbf{Mon}(M, N) = Monoid(M, N)$ of monoid homomorphisms $f : M \to N$. Observe that here the set-theoretic elements $m \in M$ of a monoid M are no longer automatically accessible from the trivial, one-element monoid $1 = \{0\}$. In fact, the image of the neutral element $0 \in 1$ under a monoid homomorphism must be the neutral element $e \in M$. How then can we access the other elements? The solution is the word monoid $Word(x)$ generated by a single letter x, by the universal property (proposition 111 in section 15.1 of volume 1) of the word monoid, since then we have the set bijection $Monoid(Word(x), M) \xrightarrow{\sim} M$.

Example 145 A monoid M is a category, whose morphisms are the elements of M, the composition is the composition in the monoid; the unique object is the neutral element $e \in M$.

Example 146 The category **Gr** of groups works in complete analogy with the category **Mon**. In fact this is an example of a so-called *subcategory*, i.e., the members of the collection **Gr** are monoid homomorphisms, i.e., members of the category **Mon**. The objects are just special monoids, i.e., those M which have the property $M^* = M$. Such a subcategory is also a *full* subcategory, which means that for every pair of objects M and N in **Gr**, we have $\textbf{Gr}(M, N) = Group(M, N) = Monoid(M, N)$, i.e., the morphisms between objects of the subcategory are *all* the morphisms between these objects in the larger category. It is easily seen that we have a bijection $Group(\mathbb{Z}, G) \xrightarrow{\sim} G$ for any group G.

Example 147 The category **Rings** of rings is composed of all sets of morphisms $\textbf{Rings}(R, S) = Ring(R, S)$ between rings R and S. Again, here the set-theoretic elements $r \in R$ of a ring R are described by the universal property of the monoid algebra $\mathbb{Z}[X]$ over the integers and one indeterminate X since we have exactly one ring homomorphism $\mathbb{Z} \to R$ and then we have the bijection of sets $Ring(\mathbb{Z}[x], S) \xrightarrow{\sim} S$.

Example 148 The category $\textbf{Lin}_R$ of modules over a ring R consists of the sets $\textbf{Lin}_R(R,V) = Lin_R(V,W)$ of R-linear module homomorphisms $f : V \to W$.

Example 149 The category $\textbf{Aff}_R$ of R-modules is composed of the sets $\textbf{Aff}_R(V,W) = Aff_R(V,W)$ of R-affine module homomorphisms $f : V \to W$. The category $\textbf{Lin}_R$ is a proper subcategory of $\textbf{Aff}_R$: For a pair V,W of modules, the linear homomorphisms form a proper subset of the set of affine homomorphisms.

Example 150 Recall that in definition 141, section 19.3 of volume 1, automata morphisms $(\sigma,\alpha) : (i,M) \to (j,N)$ are pairs $(\sigma : S \to T, \alpha : \mathcal{A} \to \mathcal{B})$ between state spaces S,T and alphabets $\mathcal{A},\mathcal{B}$ with conditions (i), (ii) (see definition 141). The set of morphisms $(i,M) \to (j,N)$ is denoted by $Automata((i,M),(j,N))$. The category $\textbf{Automata}$ of automata comprises all sets $Automata((i,M),(j,N))$. See sorite 166 in section 19.3, volume 1, to check the standard category properties of this collection of morphisms defining the category of automata $\textbf{Automata}$.

Example 151 The category of acceptors $\textbf{Acceptors}$ is composed of the sets $Acceptors((i,M,F),(j,N,G))$ of morphisms of acceptors $(\sigma,\alpha) :$ $(i,M,F) \to (j,N,G)$. The category of acceptors is not a subcategory of the category of automata since many acceptors may give rise to the same automaton. Nevertheless, every morphism of acceptors $(\sigma,\alpha) : (i,M,F) \to$ (j,N,G) gives rise to a morphism of automata $(\sigma,\alpha) : (i,M) \to (j,N)$ and the composition of two morphisms of acceptors induces the composition of the corresponding morphisms of automata. This is a typical situation of a functor, i.e., a transfer of a category to another, to be discussed later in section 36.4.

Example 152 Take the subcategory of $\textbf{Sets}$ composed of all open sets $U \subset \mathbb{R}^n$ as objects, the morphisms $f : U \to V$ being the continuous maps. Then this category, denoted by C^0, is composed of the sets of morphisms $C^0(U,V)$ of continuous maps $f : U \to V$ between objects U and V.

Example 153 (See chapter 40) For metric spaces (X,d) and (Y,e), one has the category $\textbf{Metr}$ whose sets of morphisms $\textbf{Metr}((X,d),(Y,e))$ consists of the continuous set maps $f : X \to Y$ for the topologies on the metric spaces X and Y.

Example 154 The subcategory of C^0 consisting of differentiable maps between open subsets of $\mathbb{R}^n$ is denoted by **Diff** and comprises the sets of morphisms $\mathbf{Diff}(U,V) = Diff(U,V)$ for open sets $U \subset \mathbb{R}^n$ and $V \subset \mathbb{R}^m$ consisting of differentiable maps $f : U \to V$.

Example 155 The subcategory of C^r consisting of r times continuously differentiable maps between open subsets of $\mathbb{R}^n$ is denoted by C^r and comprises the sets of morphisms $C^r(U,V)$ for open sets $U \subset \mathbb{R}^n$ and $V \subset \mathbb{R}^m$ consisting of r times continuously differentiable maps $f : U \to V$.

Example 156 (See chapter 41) The neural category $C\mathcal{N}$ is a special case of a path category associated with the digraph $\mathcal{E} \to \mathcal{N}^2$. Observe that in this case, the graph is not built on finite sets, but the definition of a (di)graph is valid for infinite sets of arrows and vertexes, too, without any change.

Example 157 For every category **C** there is the *opposite category* $\mathbf{C}^{opp}$. Its morphisms are the same, but composition is defined as $f \circ^{opp} g = g \circ f$, i.e., it is defined iff the composition with opposite factors is defined in **C**. This opposite construction exchanges the domains and codomains of morphisms. Intuitively, an arrow $f : x \to y$ in **C** corresponds to a arrow $f : y \to x$ in $\mathbf{C}^{opp}$.

Example 158 For every couple **C** and **D** of categories, whose morphisms are all sets (all the categories we have dealt with so far have this property), we have the *Cartesian product category* $\mathbf{C} \times \mathbf{D}$. Its morphisms are the pairs (f,g) of morphisms f in **C** and g in **D**. The composition $(f_1,g_1) \circ (f_2,g_2) = (f_1 \circ f_2, g_1 \circ g_2)$ of two morphisms (f_1,g_1) and (f_2,g_2) is defined iff both $f_1 \circ f_2$ and $g_1 \circ g_2$ are defined. In particular, $dom((f,g)) = (dom(f), dom(g))$ and $codom((f,g)) = (codom(f), codom(g))$.

36.4 Functors and Natural Transformations

As already mentioned in the introduction, the principal subject of the original paper by Eilenberg and Mac Lane were not categories, but functors and natural transformations, i.e., specific structure-conserving relations between given categories. In other words, functors are the "morphisms" between categories:

Definition 211 *If* **C** *and* **D** *are categories, a* functor $F : \mathbf{C} \to \mathbf{D}$ *is an assignment which attributes to every morphism c in* **C** *a morphism $F(c)$ in* **D** *such that:*

(i) $F(c)$ *is an identity if c is so.*

(ii) *If $c \circ c'$ is defined in* **C***, then $F(c) \circ F(c')$ is defined and $F(c \circ c') = F(c) \circ F(c')$.*

In particular, functors carry isomorphisms to isomorphisms. Moreover, the composition $F \circ G : \mathbf{C} \to \mathbf{E}$ of two functors $F : \mathbf{C} \to \mathbf{D}$ and $G : \mathbf{D} \to \mathbf{E}$ is a functor, and composition of functors is associative if defined.

Definition 212 *Some notations and properties pertaining to functors:*

(i) *Two categories* **C** *and* **D** *are called* isomorphic *if there exists an isomorphism of functors, i.e., there exist two functors $F : \mathbf{C} \xrightarrow{\sim} \mathbf{D}$ and $F^{-1} : \mathbf{D} \xrightarrow{\sim} \mathbf{C}$ such that $F^{-1} \circ F = Id_\mathbf{C}$ and $F \circ F^{-1} = Id_\mathbf{D}$. Here $Id_\mathbf{C}$ and $Id_\mathbf{D}$ denote the identity functors on* **C** *and* **D***.*

(ii) *A functor F is called* full *if $F(Hom_\mathbf{C}(x, y)) = Hom_\mathbf{D}(F(x), F(y))$ for all object pairs x and y.*

(iii) *A functor F is called* faithful *if $F : Hom_\mathbf{C}(x, y) \to Hom_\mathbf{D}(F(x), F(y))$ is injective for all pairs x and y.*

(iv) *A functor F is called* fully faithful *if it is full and faithful, i.e., the map $F : Hom_\mathbf{C}(x, y) \to Hom_\mathbf{D}(F(x), F(y))$ is a bijection.*

(v) *Functors F are also called "covariant" since they are opposed to functors $F : \mathbf{C}^{opp} \to \mathbf{D}$, which are called "contravariant" but still denoted by $F : \mathbf{C} \to \mathbf{D}$.*

The properties (i) and (ii) in definition 211 show that there is a *category of all categories*, denoted by **Cat**, whose objects are the categories, while the functors $F : \mathbf{C} \to \mathbf{D}$ between two categories **C** and **D** are the morphisms of **Cat**.

Example 159 If **C** and **D** are any two categories, pick one object d in **D**. Then we have the constant functor $[d] : \mathbf{C} \to \mathbf{D}$ which to every morphism in **C** assigns the object (i.e., the identity) d.

Example 160 Recalling example 145, which shows that every monoid is a category, the monoid homomorphisms $F : M \to N$ are precisely the functors between these monoids *qua* categories.

Example 161 The assignment $\Gamma \mapsto |\Gamma|$ and $f = (u, v) : (\Gamma \to \Delta) \mapsto f : |\Gamma| \to |\Delta|$ is a functor $|?| : \mathbf{Digraph} \to \mathbf{Graph}$, this was in fact shown in exercise 46 of section 10.2 in volume 1.

Example 162 Let $f = (u, v) : \Gamma \to \Delta$ be a morphism of digraphs. We write $f(a) = u(a)$ and $f(x) = v(x)$ for arrows a and vertexes x, respectively, in Γ. Then the function $Path(f) : Path(\Gamma) \to Path(\Delta)$ which sends a path $p = a_n a_{n-1} \ldots a_0$ to $Path(f)(p) = f(a_n) f(a_{n-1}) \ldots f(a_0)$ and a lazy path x to $f(x)$, is a functor on the path categories. Observe that the latter are just sets with a composition operation. The assignment $Path : \Gamma \mapsto Path(\Gamma)$ is also a functor $\mathbf{Digraph} \to \mathbf{Cat}$.

Example 163 Recall from proposition 167, section 19.3 in volume 1, that we have a map $(i, M, F) \mapsto (i : M : F)$ from the acceptor (i, M, F) to the language $(i : M : F)$, which we regard as an object of the category **Sets**. This defines the *language* functor $Acceptors \to \mathbf{Sets}$.

Example 164 Given the category $\mathbf{Lin}_{\mathbb{R}}$ of real vector spaces with linear homomorphisms, one has the assignment $?^* : V \mapsto V^* = \mathbf{Lin}_{\mathbb{R}}(V, \mathbb{R})$ of the dual space. Moreover, if $f : V \to W$ is in $\mathbf{Lin}_{\mathbb{R}}(V, \mathbb{R})$, we have a linear map $f^* : W^* \to V^* : h \mapsto h \circ f$. This assignment defines a contravariant functor $?^* : \mathbf{Lin}_{\mathbb{R}} \to \mathbf{Lin}_{\mathbb{R}}$, i.e., a functor $?^* : \mathbf{Lin}_{\mathbb{R}} \to \mathbf{Lin}_{\mathbb{R}}^{opp}$.

This idea generalizes to any space X instead of the one-dimensional $\mathbb{R}$, i.e., one defines $\mathbf{Lin}_{\mathbb{R}}(?, X) : V \mapsto \mathbf{Lin}_{\mathbb{R}}(V, X)$, and analogously $\mathbf{Lin}_{\mathbb{R}}(f, X) : \mathbf{Lin}_{\mathbb{R}}(W, X) \mapsto \mathbf{Lin}_{\mathbb{R}}(V, X) : h \mapsto h \circ f$ for a linear map $f : V \to W$. This yields the contravariant functor $\mathbf{Lin}_{\mathbb{R}}(?, X) : \mathbf{Lin}_{\mathbb{R}} \to \mathbf{Lin}_{\mathbb{R}}$.

Example 165 More generally, for any category $\mathbf{C}$, such that all collections of morphisms $\mathbf{C}(x, y)$ are sets, if any object y of $\mathbf{C}$ is fixed, one has this contravariant functor $\mathbf{C}(?, y) : \mathbf{C} \to \mathbf{Sets}$ given by $\mathbf{C}(?, y)(x) = \mathbf{C}(x, y)$, while for a morphism $f : x_1 \to x_2$ in $\mathbf{C}$, one has the set map $\mathbf{C}(f, y) : \mathbf{C}(x_2, y) \to \mathbf{C}(x_1, y) : h \mapsto h \circ f$.

Example 166 If we have a Cartesian product category $\mathbf{C} \times \mathbf{D}$ (see example 158) the assignments $pr_1 : (f, g) \mapsto f$ and $pr_2 : (f, g) \mapsto g$ define two functors $pr_1 : \mathbf{C} \times \mathbf{D} \to \mathbf{C}$ and $pr_2 : \mathbf{C} \times \mathbf{D} \to \mathbf{D}$.

Example 167 On the category **Diff**, one has the tangent assignment $T : U \mapsto TU$, which on a differentiable map $f : U \to V$ yields the tangent map

$Tf : TU \to TV$ on the tangent bundles. By proposition 264, this defines the *tangent functor* $T :$ **Diff** $\to$ **Sets**.

Example 168 For this example, we refer to chapter 40. On the category **Metr**, one has the assignment $\mathcal{H} : (X, d) \mapsto (\mathcal{H}(X, d), h(d))$ of the Hausdorff-metric space, which on a continuous map $f : X \to Y$ of metric spaces yields the continuous map $\mathcal{H}(f) : \mathcal{H}(X) \to \mathcal{H}(Y)$. By proposition 347, this defines a functor $\mathcal{H} :$ **Metr** $\to$ **Metr**.

The most important functors are diagrams in a category, which are defined in a graphical way as follows. Observe that this general setup comprises the informal terminology concerning diagrams and commutative diagrams, which has been used throughout the entire course of this book.

Definition 213 *A* diagram in a category **C** *is a functor* $\mathcal{D} : Path(\Gamma) \to$ **C** *on the path category of a digraph* Γ. *The digraph* Γ *is called the* diagram scheme *of* $\mathcal{D}$. *A diagram* $\mathcal{D}$ *is called* commutative, *if for any two paths* $p, q : v \to w$ *in* Γ *with common domain* v *and codomain* w, *the images* $\mathcal{D}(p)$ *and* $\mathcal{D}(q)$ *are equal (see figure 36.1).*

Here is the operational restatement of what basic data a diagram in a category really requires.

Lemma 308 *Given a digraph* Γ *and a category* **C**, *an assignment* $F : p \mapsto F(p)$ *from paths in* Γ *to morphisms in* **C** *is a functor if the assignment* $F : a \mapsto F(a), F : v \mapsto F(v)$ *on the arrows* a *and vertexes* v *of* Γ *has the property: Whenever* $head(a) = tail(b)$, *we have* $codom(F(a)) = dom(F(b))$, *and whenever* $v = head(a)$ *or* $v = tail(a)$, *then* $F(v) = codom(F(a))$ *or* $F(v) = dom(F(a))$, *respectively. This means that defining such a diagram means defining its values on arrows and vertexes plus the condition on the tails and heads. Therefore, one also denotes a diagram by its restriction to arrows and vertexes, i.e., by a morphism on "digraphs", where* **C** *is understood as digraph whose arrows are the morphisms and whose vertexes are the objects, the heads corresponding to the codomains, and the tails to the domains:*

$$\mathcal{D} : \Gamma \to \mathbf{C}$$

Proof Clearly the condition $head(a) = tail(b)$ is necessary by the definition of a functor. If conversely, the assignment $F : v \mapsto F(v)$ on the arrows and vertexes of Γ has the named property, then the functor F^* on the path category of Γ is

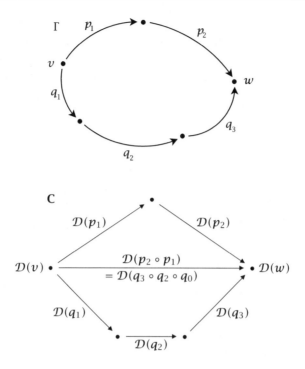

Fig. 36.1. Part of the path category of a digraph Γ and its image in **C** under the functor $\mathcal{D}$.

defined as follows: Let $p = a_n a_{n-1} \ldots a_0$ be a path in Γ. This is also the composition of the "morphisms" a_i in the path category. So we have to define $F^*(p) = F(a_n) \circ F(a_{n-1}) \circ \ldots F(a_0)$. This is well defined if the domains $\mathrm{dom}(F(a_i))$ and codomains $\mathrm{codom}(F(a_{i-1}))$ of successive morphisms coincide. But this is guaranteed by the property stated in the lemma. So the definition is legitimate. And then, if we are given a second path $q = b_m b_{m-1} \ldots b_0$ whose tail is the head of p, then we have $F^*(qp) = F(b_m) \circ F(b_{m-1}) \circ \ldots F(b_0) \circ F(a_n) \circ F(a_{n-1}) \circ \ldots F(a_0) = (F(b_m) \circ F(b_{m-1}) \circ \ldots F(b_0)) \circ (F(a_n) \circ F(a_{n-1}) \circ \ldots F(a_0)) = F^*(q) \circ F^*(p)$. On the lazy path at vertex x, we necessarily have $F^*(x) = \mathrm{dom}(v)$ if $tail(f) = x$ or $F^*(x) = \mathrm{codom}(v)$ if $head(f) = x$. If there is no arrow starting or ending in x, no condition on the image $F^*(x)$ is required. $\qquad\square$

The original Eilenberg-Mac Lane paper cited above had something more involved in mind, using categories and functors, but also structured relations between different functors $F, G : \mathbf{C} \to \mathbf{D}$ on the same categories. Let us give an example of such a relation before we state the general definition.

For every set function $f : A \to B$ there are two set functions $i_A : A \to Word(A)$ and $i_B : B \to Word(B)$. We know from proposition 111 in section 15.1 of volume 1, that there is a unique monoid homomorphism $Word(f) : Word(A) \to Word(B)$, which generates the following commutative diagram

$$
\begin{array}{ccc}
A & \xrightarrow{\ i_A\ } & Word(A) \\
{\scriptstyle f}\downarrow & & \downarrow{\scriptstyle Word(f)} \\
B & \xrightarrow{\ i_B\ } & Word(B)
\end{array}
$$

of sets. This means that we have two functors: $Id_{\mathbf{Sets}} : \mathbf{Sets} \to \mathbf{Sets}$ and $Word : \mathbf{Sets} \to \mathbf{Sets}$, which are related, in each object, by the embedding i_X of a set X in its word monoid. This relation was called "natural" by Eilenberg and Mac Lane.

This is the idea of a natural transformation:

Definition 214 *If $F, G : \mathbf{C} \to \mathbf{D}$ are two functors, a* natural transformation *$t : F \to G$ is a system of morphisms $t(c) : F(c) \to G(c)$ in $\mathbf{D}$, for each object c in $\mathbf{C}$, such that for every morphism $f : x \to y$ in $\mathbf{C}$, we have the following commutative diagram in $\mathbf{D}$:*

$$
\begin{array}{ccc}
F(x) & \xrightarrow{\ t(x)\ } & G(x) \\
{\scriptstyle F(f)}\downarrow & & \downarrow{\scriptstyle G(f)} \\
F(y) & \xrightarrow{\ t(y)\ } & G(y)
\end{array}
$$

So the above example is a natural transformation

$$i : Id_{\mathbf{Sets}} \to Word.$$

Natural transformations can be composed in an evident way, and the composition is associative. For every functor F we have the natural identity Id_F. We therefore have the category $Func(\mathbf{C}, \mathbf{D})$ of functors $F : \mathbf{C} \to \mathbf{D}$ and natural transformations $Nat(F, G)$ between two functors $F, G : \mathbf{C} \to \mathbf{D}$. Properties between such functors are said to be *natural* if they relate to the category $Func(\mathbf{C}, \mathbf{D})$, for example, $F \overset{\sim}{\to} G$ is a natural isomorphism iff it is an isomorphism among the natural transformations from F to G.

Exercise 173 Take any two categories $\mathbf{C}$ and $\mathbf{D}$. Then for any two objects d and e in $\mathbf{D}$, $Nat([d], [e])$, the collection of natural transformations between two constant functors, is in bijection with the collection $\mathbf{D}(d, e)$ of morphisms from d to e. In other words, the natural transformations between constant functors $[d] : \mathbf{C} \to \mathbf{D}$ recover the target category $\mathbf{D}$.

36.5 Limits and Colimits

The most prominent use of natural transformations occurs in the construction of universal objects, which we shall discuss now. Let us do one example, which we have used all over the book: the Cartesian product. We have encountered Cartesian products in different contexts, but the typical one is the set-theoretic environment. Given two sets a and b, the Cartesian product was described by its universal property in proposition 57 of section 6.2 in volume 1. We need two maps $pr_a : a \times b \to a$ and $pr_b : a \times b \to b$. The universal property tells us that for any couple $v : c \to a$ and $w : c \to b$, there is exists a unique map $(v, w) : c \to a \times b$ such that $v = pr_a \circ (v, w)$ and $w = pr_b \circ (v, w)$. Intuitively, we could rephrase this as follows: Take the category whose objects are the diagrams $D = (v : c \to a, w : c \to b)$, while, given two such "objects", $D_1 = (v_1 : c_1 \to a, w_1 : c_1 \to b)$ and $D_2 = (v_2 : c_2 \to a, w_2 : c_2 \to b)$ the morphisms $f : D_1 \to D_2$ are the maps $f : c_1 \to c_2$ such that $v_1 = v_2 \circ f$ and $w_1 = w_2 \circ f$, i.e., the following diagram commutes:

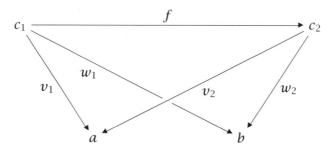

In this category, the Cartesian product is a special object: there is a unique arrow $(a, b) : c \to a \times b$ into the Cartesian product $a \times b$ for every object $D = (v : c \to a, w : c \to b)$. We have already encountered such objects in set theory: the set $1 = \{0\}$ has this property, we had written $! : c \to 1$ for that unique arrow.

Definition 215 *An object e in a category $\mathbf{C}$ such that there is a unique morphism $! : c \to e$ for every object c in $\mathbf{C}$, is called a* final *object, it is usually denoted by 1. A final object i in the opposite category $\mathbf{C}^{opp}$ is called an* initial *object in $\mathbf{C}$; it is usually denoted by 0. This means that for every object c in $\mathbf{C}$, there is a unique morphism $! : 0 \to c$. The notations $!$, 1, and 0 are ambiguous, but very useful and are applied whenever possible.*

Example 169 In the category **Sets**, the set $1 = \{0\}$ is final and the empty set $0 = \varnothing$ is initial. A set e is final iff $card(e) = 1$, the only initial set is the empty set.

Example 170 In the category **Digraph**, the digraph $1_{\text{Digraph}} : 1 \rightarrow 1^2 (\overset{\sim}{\rightarrow} 1)$ is final while the empty digraph $0_{\text{Digraph}} : 0 \rightarrow 0^2 = 0$ is initial.

Exercise 174 Show that in a category any two initial (final) objects are isomorphic.

So Cartesian products are a special case of final elements in some category built from diagrams. Let us make this precise.

Definition 216 *Suppose we are given a category* **C**, *a digraph (the diagram scheme)* Δ *and a diagram* $\mathcal{D} : \Delta \rightarrow$ **C**. *For an object* c *in* **C**, *we consider the constant diagram* $[c] : \Delta \rightarrow$ **C**. *Then a natural transformation* $s : [c] \rightarrow \mathcal{D}$ *is called a* cone *on* Δ. *A natural transformation* $t : \mathcal{D} \rightarrow [c]$ *is called a* cocone *over* Δ.

This looks more complicated than it is. In fact, a cone is this: For every vertex v of Δ we are given a morphism $s(v) : c \rightarrow \mathcal{D}(v)$ such that for any arrow $a : v \rightarrow w$ of Δ, one has a commutative triangle of morphisms in **C**:

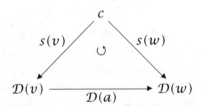

Same for cocones: For every vertex v of the digraph Δ, we are given a morphism $t(v) : \mathcal{D}(v) \rightarrow c$ such that for any arrow $a : v \rightarrow w$ of Δ, one has a commutative triangle of morphisms in **C**:

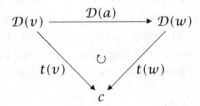

Example 171 The special case of the situation of the Cartesian product is that of a discrete diagram scheme $\Delta = \{1, 2\}$ consisting of two points 1 and 2, which are mapped to the objects a and b, i.e., $\mathcal{D}(1) = a$ and $\mathcal{D}(2) = b$. The projections onto the factors define a cone, where the commutativity condition is empty. The cone is the pair of projections for the object $c = a \times b$, i.e., $s(1) = pr_a$ and $s(2) = pr_b$:

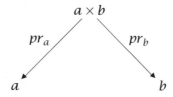

Example 172 More generally, a fiber product $a \times_d b$, introduced in section 6.3 of volume 1, deals with the situation of a diagram scheme δ consisting of three points $1, 2, 3$ and two arrows $1 \to 3, 2 \to 3$, the cone then is the commutative diagram of sets

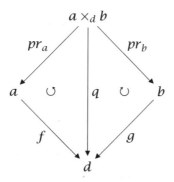

with f and g the maps given for the definition of the fiber product, with $q = f \circ pr_a = g \circ pr_b$ and with $\mathcal{D}(1) = a$, $\mathcal{D}(2) = b$ and $\mathcal{D}(3) = d$.

Example 173 Refer to the coproduct construction in section 6.2 of volume 1 for this example. Dually to example 171, the situation of the coproduct is again that of a discrete diagram scheme $\Delta = \{1, 2\}$ consisting of two points 1 and 2, which are mapped to the objects a and b, i.e., $\mathcal{D}(1) = a$ and $\mathcal{D}(2) = b$. The injections into the cofactors define a cocone, where the commutativity condition is empty. The cocone is the pair of injections for the object $c = a \sqcup b$, i.e., $t(1) = in_a$ and $t(2) = in_b$:

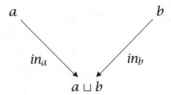

Definition 217 *Given a diagram* $\mathcal{D} : \Delta \to \mathbf{C}$, *the cone category* $Cone(\mathcal{D})$ *has as objects all cones* $t : [c] \to \mathcal{D}$, *c objects in* $\mathbf{C}$, *whereas the morphisms* $f : t_1 \to t_2$ *between cones* $t_1 : [c_1] \to \mathcal{D}, t_2 : [c_2] \to \mathcal{D}$ *are the morphisms* $f : c_1 \to c_2$ *in* $\mathbf{C}$ *such that* $t_1 = t_2 \circ f$. *The same data define the cocone category* $Cocone(\mathcal{D})$. *Here the objects are the cocones* $t : \mathcal{D} \to [c]$, *c objects in* $\mathbf{C}$, *whereas the morphisms* $f : t_1 \to t_2$ *between cocones* $t_1 : \mathcal{D} \to [c_1]$ *and* $t_2 : \mathcal{D} \to [c_2]$ *are the morphisms* $f : c_1 \to c_2$ *in* $\mathbf{C}$ *such that* $f \circ t_1 = t_2$.

Here is the general role of a Cartesian product, a fiber product or a co-product played in this category-theoretic setting:

Definition 218 *For a diagram* $\mathcal{D} : \Delta \to \mathbf{C}$, *a limit of* $\mathcal{D}$ *is a final object in the cone category* $Cone(\mathcal{D})$. *This object (which is determined up to isomorphism as we know from exercise 174) is denoted by* $\lim(\mathcal{D})$. *Thus, the limit is not an object of* $\mathbf{C}$, *but a cone, including the constant "top object" of the cone. But often, if the rest is clear, one also writes* $\lim(\mathcal{D})$ *to denote that top object.*

Dually, a colimit of $\mathcal{D}$ *is an initial object in the cocone category* $Cocone(\mathcal{D})$. *This object (which is determined up to isomorphism as we know from exercise 174) is denoted by* $colim(\mathcal{D})$. *Thus, the colimit is not an object of* $\mathbf{C}$, *but a cocone, including the constant "bottom object" of the cocone. But often, if the rest is clear, one also writes* $colim(\mathcal{D})$ *to denote that bottom object.*

A category $\mathbf{C}$ *for which all limits (of finite diagram schemes) exist is called* (finitely) complete; *a category* $\mathbf{C}$ *for which all colimits (of finite diagram schemes) exist is called* (finitely) cocomplete.

Exercise 175 Here is an intuitive interpretation of limits in classical terms of limits in analysis. Let $\mathbf{C}$ be the category, where objects are the real numbers $x \in \mathbb{R}$, and whose morphisms are all pairs $f = (x, y)$ of real numbers, such that $x \leq y$. Take a subset $S \subset R$ and look at the diagram $\mathcal{D}$ defined by the pairs $x \to y$, i.e., $x \leq y$ in S. Show that $colim(\mathcal{D})$ exists and equals $\sup(S)$ iff $\sup(S) < \infty$. Also show that $\lim(\mathcal{D})$ exists and equals $\inf(S)$ iff $\inf(S) > -\infty$.

The first observation is that for the category of sets, the limit and colimit of the discrete diagram with two points is precisely the product and the coproduct, this is clear from the discussion preceding the general definitions. But there is more, we have arbitrary limits and colimits in **Sets**. Since this is a fundamental fact which is often used in the construction of limit and colimit objects, we also include an explicit description of such constructions in the following proposition.

Proposition 309 *The category* **Sets** *of sets is complete and cocomplete. More precisely: Let $\Delta : A \to V^2$ be a diagram scheme and $\mathcal{D} : \Delta \to$ **Sets** a diagram of sets and set maps.*

A limit can be constructed by the following method: Consider the Cartesian product $D = \prod_{v \in V} \mathcal{D}(v)$, and take the subset $L \subset D$ consisting of all families $(x_v)_{v \in V}$ such that for every arrow $a : v \to w$ in A, we have $\mathcal{D}(a)(x_v) = x_w$. Then $\lim(\mathcal{D})$ is represented by the cone $l : [L] \to \mathcal{D}$ such that for all $v \in V$, $l(v) : L \to \mathcal{D}(v)$ is the restriction of the projection $pr_l : D \to \mathcal{D}(v)$ to L.

A colimit can be constructed in the following way: Consider the family $F = (\mathcal{D}'(v))_{v \in V}$, with $\mathcal{D}'(v) = \{v\} \times \mathcal{D}(v)$, of pairwise disjoint sets $\mathcal{D}'(v)$, each of which is evidently equipollent to $\mathcal{D}(v)$ by the second projection $\mathcal{D}'(v) \to \mathcal{D}(v)$. We replace the transition maps $\mathcal{D}(a) : \mathcal{D}(v) \to \mathcal{D}(w)$ for arrows $a : v \to w$ by the evident maps $\mathcal{D}'(a) : \mathcal{D}'(v) \to \mathcal{D}'(w)$. Take the (now disjoint) union $U = \bigcup_{v \in V} \mathcal{D}'(v)$. On this set, we have a relation $x \sim y$ iff there is an arrow $a : v \to w$ in A with $x \in \mathcal{D}'(v), y \in \mathcal{D}'(w)$ and $y = \mathcal{D}'(a)(x)$. The equivalence relation $\approx$ generated by $\sim$ defines the set $C = U/\approx$, together with the maps $c(v) : \mathcal{D}(v) \to C$, which are induced by the bijections $\mathcal{D}(v) \overset{\sim}{\to} \mathcal{D}'(v)$, the injections $\mathcal{D}'(v) \to U$, and the canonical map $U \to C$. The colimit $colim(\mathcal{D})$ is represented by the cocone $c : \mathcal{D} \to [C]$.

Proof For the cone $t : [c] \to \mathcal{D}$, we define the following set map $f : c \to L = \lim(\mathcal{D})$. If $x \in c$, then for every vertex v of the diagram scheme Γ, we have a map $t(v) : c \to \mathcal{D}(v)$. Take $f(x) = (t(v)(x))_{v \in V}$. Then the functorial naturality relation for any arrow $v \xrightarrow{a} w \in A$ of Γ means $\mathcal{D}(a) \circ t(v) = t(w)$. Therefore $f(x) \in L$. Clearly, by construction, f is a morphism of cones $f : t \to \lim(\mathcal{D})$. But it is also the only candidate, since the commutation relation defining cone morphisms means $pr_v(f(x)) = t(v)(x)$.

For the cocone $t : \mathcal{D} \to [c]$, we take the following map $f : C \to c$. Let $x \in C$. Then $x = c(v)(\xi)$ for a vertex v of Γ and an element $\xi \in \mathcal{D}(v)$. The we define $f(x) = t(v)(\xi)$. We have to show that this is a well-defined map. If $\xi \approx$

η, where $\eta \in \mathcal{D}(w)$, then, by the definition of $\approx$, this means that there is a sequence of vertexes $v = v_0, v_1, \ldots v_n = w$ and maps $\mathcal{D}(a_i) : \mathcal{D}(v_i) \rightarrow \mathcal{D}(v_{i+1})$ or $\mathcal{D}(a_i) : \mathcal{D}(v_{i+1}) \rightarrow \mathcal{D}(v_i)$ and a sequence of elements $\xi = \xi_0, \xi_1, \ldots \xi_n = \eta$ with $\xi_i \in \mathcal{D}(v_i)$ such that for $\mathcal{D}(a_i) : \mathcal{D}(v_i) \rightarrow \mathcal{D}(v_{i+1})$, we have $\xi_{i+1} = \mathcal{D}(a_i)(\xi_i)$ or for $\mathcal{D}(a_i) : \mathcal{D}(v_{i+1}) \rightarrow \mathcal{D}(v_i)$, we have $\xi_i = \mathcal{D}(a_i)(\xi_{i+1})$. But then, by the naturality of the cocone t, this means that $t(v_i)(\xi_i) = t(v_{i+1})(\xi_{i+1})$, all i, so we have $t(v)(\xi) = t(w)(\eta)$, and f is well defined. Since by construction, every element of C is an image of an element of one of the sets $\mathcal{D}(v)$, the function f is unique. $\qquad \square$

Again, this construction yields the following, not really complicated restatement of what it means to have a limit: Given any cone $t : [c] \rightarrow \mathcal{D}$, there is a unique set map $f : c \rightarrow \lim(\mathcal{D})$ such that for all vertexes v of the diagram scheme Δ, we have a commutative diagram

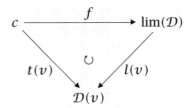

of set maps.

And dually: Given any cocone $s : \mathcal{D} \rightarrow [d]$, there is a unique set map $f : \mathrm{colim}(\mathcal{D}) \rightarrow c$ such that for all vertexes v of the diagram scheme Δ, we have a commutative diagram

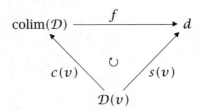

of set maps.

Exercise 176 Compare this to the constructions of the Cartesian product (proposition 57, section 6.2 in volume 1), the more general fiber product (proposition 65, section 6.3 in volume 1), and the coproduct (proposition 58, section 6.2 in volume 1), and verify that those objects are constructed following the construction rules of limits and colimits in proposition 309.

36.6 Adjunction

We conclude this short introduction to categories by a construction which has been used in the theory of Turing machines under the title of "Turing adjunction" (see section 19.3.2 in volume 1). The concept of adjunction is in fact a most fruitful one in category theory, it has far-reaching applications in all mathematical branches, and, most prominently in formal logic and topos theory, i.e., the foundations of computer science. So this very short section should not mislead the reader: the subject is as central as it is short, but it is too advanced to be treated in length in such an introductory text. In this section, we suppose that all collections $\mathbf{C}(x, y)$ of morphisms with given domain and codomain are sets.

Definition 219 *Given two functors $F : \mathbf{C} \to \mathbf{D}$ and $G : \mathbf{D} \to \mathbf{C}$, we say that F is* left adjoint *to G or that G is* right adjoint *to F, iff the functors*

$$\mathbf{D}(F(?), ?) : \mathbf{C}^{opp} \times \mathbf{D} \to \mathbf{Sets} : (x, y) \mapsto \mathbf{D}(F(x), y)$$

and

$$\mathbf{C}(?, G(?)) : \mathbf{C}^{opp} \times \mathbf{D} \to \mathbf{Sets}(x, y) \mapsto \mathbf{C}(x, G(y))$$

are (naturally) isomorphic. One also writes this fact in these symbols:

$$\frac{x \to G(y)}{F(x) \to y}$$

meaning that morphisms in the numerator correspond one-to-one to morphisms in the denominator.

Example 174 We have encountered one very important instance of an adjoint pair of functors in proposition 59 of section 6.2 in volume 1. We take $\mathbf{C} = \mathbf{D} = \mathbf{Sets}$ and fix a set a. The first functor is $F = a \times ? : x \mapsto a \times x$. The second functor is $G = ?^a : x \mapsto x^a$. From proposition 59 of section 6.2 in volume 1, we know that for each pair (x, y) of sets, there is a canonical isomorphism

$$\delta_{x,y} : Set(a \times x, y) \overset{\sim}{\to} Set(x, y^a),$$

defined by $\delta_{x,y}(f)(\xi)(\alpha) = f(\xi, \alpha)$ for $f : a \times x \to y, \xi \in x, \alpha \in a$. It is easy to check that this isomorphism defines a natural transformation and therefore we have the adjunction

$$\frac{x \to a^y}{a \times x \to y}$$

in **Sets**.

CHAPTER 37

Splines

37.1 Introduction

Splines are a generic term for a widespread technique for constructing continuous curves and surfaces which have to pass through given points and, locally, realize a specific type of functions, such as polynomials, differentiable functions, or other functions satisfying similar qualifications. The name "spline" comes from ship construction where a spline is a thin narrow wooden or metallic strip fitted into a groove in the edge of a board. The special case of Bézier splines is of great importance in industrial shape design. It was first developed by the aviation and automobile industry, in the late fifties by James Ferguson (Boeing), Pierre Bézier (Renault), and Paul de Faget de Casteljau (Citroën). In the following we shall mainly sketch the theory of Bézier splines, keeping however in mind the generic approach underlying all these splining methods.

37.2 Preliminaries on Simplexes

Before delving into the subject of splines proper, a small parenthesis on affine maps on polyhedra is required.

Definition 220 *Given a natural number d, the* standard simplex Δ_d *of dimension d is the set of points $(\xi_0, \xi_1, \ldots \xi_d) \in \mathbb{R}^{d+1}$ with $\sum_i \xi_i = 1$ and $0 \le \xi_i \le 1$, for all $i \in \{0, \ldots d\}$.*

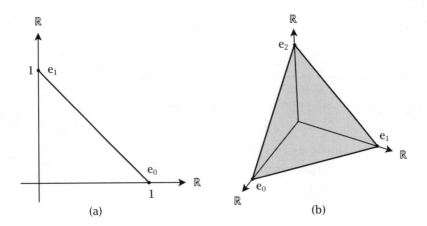

Fig. 37.1. (a) The standard simplex $\Delta_1 \subset \mathbb{R}^2$, (b) the standard simplex $\Delta_2 \subset \mathbb{R}^3$.

Definition 221 *Given natural numbers n and d, and a sequence* $(P_i)_{i=0,...d}$ *of points* $P_i \in \mathbb{R}^n$, *the* affine simplex $\Delta(P_0,...P_d)$ *is the unique map* $\Delta(P_0,...P_d) : \Delta_d \to \mathbb{R}^n$ *induced by an affine map on* $\mathbb{R}^{d+1}$ *such that* $\Delta(P_0,...P_d)(e_i) = P_i$, *for all* $i = 0, 1,...d$, *for the standard basis vectors* $e_i = (0,...,0,1,0,...0) \in \mathbb{R}^{d+1}$, *the number 1 being at position* $i + 1$. *We write* $\Delta[P_0,...P_d] = Im(\Delta(P_0,...P_d))$ *for the image set of the affine simplex.*

Exercise 177 Show the existence and uniqueness of the map $\Delta(P_0,...P_d)$, which extends to an affine map on $\mathbb{R}^{d+1}$. To this end, show that the restriction of an affine map to Δ_d is uniquely determined by its values on the basis vectors e_i.

Exercise 178 Show that the following formula holds:

$$\Delta[P_0,...P_d] = \left\{ \sum_{i=0}^{d} \lambda_i P_i \,\middle|\, \sum_{i=0}^{d} \lambda_i = 1, 0 \le \lambda_i \le 1 \right\}.$$

Lemma 310 *If* $\Delta(P_0,...P_d) : \Delta_d \to \mathbb{R}^n$ *is an affine simplex, and if* $f : \mathbb{R}^n \to \mathbb{R}^m$ *is an affine map, then the composed map* $f \circ \Delta(P_0,...P_d) : \Delta_d \to \mathbb{R}^m$ *is the affine simplex defined by the* f-*images of the points* P_i, *i.e.,*

$$f \circ \Delta(P_0,...P_d) = \Delta(f(P_0),...f(P_d)).$$

Proof Clearly, the composition $f \circ \Delta(P_0, \ldots P_d)$ is an affine simplex, since the composition of restrictions of affine maps is a restriction of an affine map. The formula is then evident. □

Definition 222 *A subset $X \subset \mathbb{R}^n$ is called* convex, *iff for all $x, y \in X$ the closed line $\Delta[x, y]$ is a subset of X. Since for any non-empty family $(C_i)_{i \in I}$ of convex sets $C_i \subset \mathbb{R}^n$ containing a given set D, the intersection $\bigcap_i C_i$ is convex and contains D, there is a minimal convex set containing D, namely the intersection of all convex sets in $\mathbb{R}^n$ containing D. This set is called* the convex hull of D, *we denote it by $Conv(D)$.*

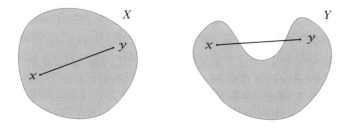

Fig. 37.2. X is convex, while Y is obviously not.

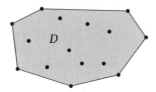

Fig. 37.3. In gray, the convex hull $Conv(D)$ of the set of points D.

Example 175 Let $u, v \in \mathbb{R}^n$ be two vectors, $u \neq 0$. Then the set $H = \{x \mid (u, x - v) = 0\}$ is called a *hyperplane in* $\mathbb{R}^n$. To one side of H we have the set consisting of those x such that $(u, x - v) \geq 0$, we call it the closed positive halfspace defined by H and denote it by H^+. Then H^+ is convex.

Lemma 311 *Let $(P_i)_{i=0\ldots d}$ be a finite sequence of points $P_i \in \mathbb{R}^n$. Then the convex hull of the set $\{P_i \mid i = 0, \ldots d\}$ is the image $\Delta[P_0, \ldots P_d]$ of the affine simplex $\Delta(P_0, \ldots P_d)$.*

Proof Clearly, for any two points $a, b \in \Delta[P_0, \ldots P_d]$, the straight line $\Delta[a, b]$ is in $\Delta[P_0, \ldots P_d]$, so $\Delta[P_0, \ldots P_d]$ is convex. Conversely, every point in $\Delta[P_0, \ldots P_d]$ is in its convex hull. In fact, this is true by the definition of convexity for $d = 1$. Then, any point $x \in \Delta[P_0, \ldots P_d]$ is the combination $\xi_0 \cdot P_0 + \sum_{i=1}^{d} \xi_i \cdot P_i$. If $\xi_0 = 0$ we are done by induction. Else, we have

$$\xi_0 \cdot P_0 + \sum_{i=1}^{d} \xi_i \cdot P_i = \xi_0 \cdot P_0 + (1 - \xi_0) \cdot \sum_{i=1}^{d} \frac{\xi_i}{1 - \xi_0} \cdot P_i,$$

where the second sum $u = \sum_{i=1}^{d} \frac{\xi_i}{1-\xi_0} \cdot P_i$ is a point in $\Delta[P_1, \ldots P_d]$, so it is in the convex hull of $P_1, \ldots P_d$, and therefore, by the definition of convexity, the combination $\xi \cdot P_0 + (1 - \xi_0) \cdot u$ is in the convex hull of $\{P_i \mid i = 0, \ldots d\}$. $\square$

Lemma 312 *The convex hull of a set $X \subset \mathbb{R}^n$ is the union of all affine simplex images $\Delta[P_0, \ldots P_d]$ of finite sequences $(P_0, \ldots P_d)$ of points in X.*

Proof Since $X \subset Y$ implies $Conv(X) \subset Conv(Y)$, each $\Delta[P_0, \ldots P_d]$ is a subset of $Conv(X)$. Conversely, this union is convex. In fact, if $a \in \Delta[P_0, \ldots P_d]$ and $b \in \Delta[Q_0, \ldots Q_e]$, then also $a, b \in \Delta[P_0, \ldots P_d, Q_0, \ldots Q_e]$, and we are done, since the latter set is a convex subset of $Conv(X)$. $\square$

Proposition 313 *Let $D \subset \mathbb{R}^n$ and $f : \mathbb{R}^n \to \mathbb{R}^m$ be an affine map (see definition 168 in section 22.3 of volume 1). We have*

$$f(Conv(D)) = Conv(f(D)).$$

In particular, the f-image of a convex set $D \subset \mathbb{R}^n$ is convex.

Proof According to lemma 312, we have $Conv(D) = \bigcup_{P_0, \ldots P_d \in D} \Delta[P_0, \ldots P_d]$. So

$$\begin{aligned}
f(Conv(D)) &= \bigcup_{P_0, \ldots P_d \in D} f(\Delta[P_0, \ldots P_d]) \\
&= \bigcup_{P_0, \ldots P_d \in D} \Delta[f(P_0), \ldots f(P_d)] \\
&= \bigcup_{Q_0, \ldots Q_d \in f(D)} \Delta[Q_0, \ldots Q_d] = Conv(f(D)).
\end{aligned}$$

$\square$

37.3 What are Splines?

In order to define a spline, recall that in integration theory (section 30.2), we defined sets $Part(a., b.)$ of partitions of n-dimensional cubes $K(a., b.)$

in $\mathbb{R}^n$, yielding the cubes $K \in P$, where $P = (P_i)_i \in Part(a.,b.)$ is a partition. Besides these data, one needs a type $\mathcal{T}_m(K)$ of functions $f : K \to \mathbb{R}^m$ defined on cubes K. Usually, types like polynomial functions, differentiable functions, or polynomial functions of degree at most d, etc., are available. The only common feature required by the different types is that a type forms a real vector space. Thus a type $\mathcal{T}_m(K)$ is a real vector space of functions with values in $\mathbb{R}^m$. One prominent example is the type $Pol_m^d(K)$ consisting of all functions $f : K \to \mathbb{R}^m$ where the coordinate functions $f_i : K \to \mathbb{R}$ are polynomials of total degree at most d in the n coordinate variables $x_1, \ldots x_n$ of K. For $d = 1$ one speaks of *linear splines*, for $d = 2$ *quadratic splines*, and for $d = 3$ one has the famous *cubic splines*. Here is the definition of a spline:

Definition 223 *Given a type $\mathcal{T}_m(K)$ of functions on cubes $K \subset \mathbb{R}^n$ with values in $\mathbb{R}^m$, an n-dimensional spline function on a partition $P \in Part(a.,b.)$ is a function $f : K(a.,b.) \to \mathbb{R}^m$ such that on each cube $K \in I(P)$, the restriction $f|_K$ to K is in $\mathcal{T}_m(K)$. If, moreover, a condition C on the behavior of the functions $f|_K$ on the cubes K of the partition is required, one calls f a* spline of type $\mathcal{T}$ with condition C. *In particular, for $n = 1$ one speaks of* spline curves, *for $n = 2$, of* spline surfaces.

The simplest case involves one-dimensional splines, defined for a partition $a = x_0 < x_1 < \ldots x_k = b$ of the interval $[a,b]$, taking the type Pol_1^1 of linear polynomials. It is obvious that there is exactly one linear spline on the interval $[a,b]$ satisfying these conditions.

Often, the conditions C meet the etymology of "spline" in the sense that, for two adjacent cubes K_1 and K_2 of the partition the derivatives of $f_1 = f|_{K_1}$ and $f_2 = f|_{K_2}$ on the intersection $\partial K_1 \cap \partial K_2$ must coincide. For example, for one-dimensional cubes and the type of differentiable functions, one is given a partition $a = x_0 < x_1 < \ldots x_k = b$ of the interval $[a, b]$ and then requires that, for any two adjacent intervals $K_1 = [x_{i-1}, x_i]$ and $K_2 = [x_i, x_{i+1}]$, one has $\frac{df_1}{dx}(x_i) = \frac{df_2}{dx}(x_i)$. The next proposition gives the reason why cubic splines are so popular.

Proposition 314 *For each one-dimensional partition $P = (a = x_0 < x_1 < \ldots x_k = b)$, and two sequences $(y_0, y_1, \ldots y_k)$, $(t_0, t_1, \ldots t_k)$, with $y_i, t_i \in \mathbb{R}^m$, with the intervals $K_i = [x_i, x_{i+1}]$, $i = 0, \ldots k - 1$, there is a uniquely determined spline function $f : [a,b] \to \mathbb{R}^m$ of type Pol_m^3 with the condition $f(x_i) = y_i$ for $i = 0, 1, \ldots k$ and such that $Df_{i-1}(x_i) = Df_i(x_i) = t_i$, for $i = 1, \ldots k - 1$, and $Df_0(x_0) = t_0, Df_{k-1}(x_k) = t_k$, where $f_j = f|_{K_j}$.*

Exercise 179 Give a proof of proposition 314, which is based on the following core fact. One considers the single unit interval $[0,1]$ or real numbers. Two arbitrary values $y_0, y_1 \in \mathbb{R}$, as well as two arbitrary slope numbers $t_0, t_1 \in \mathbb{R}$ are given. Then there is a unique cubic polynomial $f(X) = a + bX + cX^2 + dX^3$ such that $f(0) = y_0$, $f(1) = y_1$, $f'(0) = t_0$, $f'(1) = t_1$. In fact, the conditions mean that we have this linear system of four equations for the unknown coefficients a, b, c, d:

$$y_0 = a,$$
$$y_1 = a + b + c + d,$$
$$t_0 = b,$$
$$t_1 = b + 2c + 3d,$$

which yields

$$a = y_0,$$
$$b = t_0,$$
$$c = 3(y_1 - y_0) - 2t_0 - t_1,$$
$$d = 2y_0 - 2y_1 + t_0 + t_1.$$

Example 176 Required is a function $f : [1,5] \to \mathbb{R}$ satisfying the following conditions:

$$f(1) = 1, \qquad f'(1) = 2,$$
$$f(3) = 4, \qquad f'(3) = 0,$$
$$f(4) = 2, \qquad f'(4) = -3,$$
$$f(5) = 2, \qquad f'(5) = 2.$$

These conditions induce a partition of $[1,5]$ into the three intervals $K_0 = [1,3]$, $K_1 = [3,4]$, $K_2 = [4,5]$. The restriction of f to K_i is modeled by a polynomial of degree 3:

$$f|_{K_0}(X) = a_0 + b_0 X + c_0 X^2 + d_0 X^3$$
$$f|_{K_1}(X) = a_1 + b_1 X + c_1 X^2 + d_1 X^3$$
$$f|_{K_2}(X) = a_2 + b_2 X + c_2 X^2 + d_2 X^3$$

The coefficients a_i, b_i, c_i and d_i are determined by substituting the conditions on K_i in $f|_{K_i}$ and $f'|_{K_i} = b_i + 2c_i X + 3d_i X^2$. The result is the following system of equations:

For $f|_{[1,3]}$:

$$a_0 + b_0 1 + c_0 1^2 + d_0 1^3 = 1$$
$$b_0 + 2c_0 1 + 3d_0 1^2 = 2$$
$$a_0 + b_0 3 + c_0 3^2 + d_0 3^3 = 4$$
$$b_0 + 2c_0 3 + 3d_0 3^2 = 0$$

For $f|_{[3,4]}$:

$$a_1 + b_1 3 + c_1 3^2 + d_1 3^3 = 4$$
$$b_1 + 2c_1 3 + 3d_1 3^2 = 0$$
$$a_1 + b_1 4 + c_1 4^2 + d_1 4^3 = 2$$
$$b_1 + 2c_1 4 + 3d_1 4^2 = -3$$

For $f|_{[4,5]}$:

$$a_2 + b_2 4 + c_2 4^2 + d_2 4^3 = 2$$
$$b_2 + 2c_2 4 + 3d_2 4^2 = -3$$
$$a_2 + b_2 5 + c_2 5^2 + d_2 5^3 = 2$$
$$b_2 + 2c_2 5 + 3d_2 5^2 = 2$$

Solving for the unknowns a_i, b_i, c_i, d_i the resulting polynomials are

$$f|_{[1,3]}(X) = -\tfrac{1}{2} + \tfrac{3}{4}X + X^2 - \tfrac{1}{4}X^3,$$
$$f|_{[3,4]}(X) = 50 + 45X - 12X^2 + X^3,$$
$$f|_{[4,5]}(X) = 142 - 83X + 16X^2 - X^3.$$

and, piecing together, the required function f is:

$$f(x) = \begin{cases} -\tfrac{1}{2} + \tfrac{3}{4}x + x^2 - \tfrac{1}{4}x^3 & \text{if } 1 \le x \le 3, \\ 50 + 45x - 12x^2 + x^3 & \text{if } 3 \le x \le 4, \\ 142 - 83x + 16x^2 - x^3 & \text{if } 4 \le x \le 5. \end{cases}$$

Exercise 180 Let $K = \prod_i [a_i, b_i]$ be an n-dimensional cube with a set $P = \{P_1, \ldots P_k\} \subset K$ of k points. Then there is a polynomial spline $f : K \to \mathbb{R}^m$

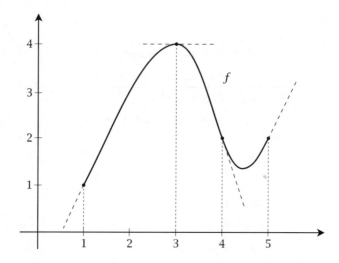

Fig. 37.4. The function $f : [1, 5] \to \mathbb{R}$ from example 176.

which is a C^1 function, satisfies $Df_{P_i} = 0$, for all $i = 1, 2, \ldots k$, and is of degree ≤ 3 in each coordinate of the cubes K_j of the partition of the spline.

To solve the problem, first reduce it to $m = 1$, and proceed by induction on n as follows. Define the partition $P \in Part(a., b.)$ through the n projections $pr_j(P) \subset \mathbb{R}$ into the j-th coordinate space, and take as condition that $Df_{P_i} = 0$ on the points P_i, which are now special points of the cubes of the partition. Now, let $x = (w, z) \in \mathbb{R}^{n-1} \times \mathbb{R}$ be a point of the cube K. If $z = b_n$, the value of the recursively given function $g(w)$ on $K' = \prod_{i=1,\ldots n-1}[a_i, b_i]$ with the set $P' = p(P)$, p being the projection into the first $n - 1$ coordinates, is taken. Else, let $u, v \in pr_n(P)$ be the unique pair such that $u \leq z < v$. Take the recursively given functions $g_u, g_v : K' \to \mathbb{R}$ relative to the fibers $K_u = pr_n^{-1}(u) \cap K$ and $K_v = pr_n^{-1}(v) \cap K$, and apply $F(X)$ to $X = z$, where $F(X)$ is the unique cubic polynomial such that $F(u) = g_u(w)$, $F(v) = g_v(w)$ and $F'(u) = F'(v) = 0$ guaranteed by exercise 179.

37.4 Lagrange Interpolation

The spline approach is well illustrated by the Lagrange interpolation technique. The situation is that of one-dimensional splines. One is interested in a type $\mathcal{T}$ of functions which are determined by a number of so-called *control points*, i.e., values $P_0, P_1, \ldots P_d$ in the spline codomain $\mathbb{R}^m$, and then looks for polynomial functions $f : [a, b] \to \mathbb{R}^m$ such that they not only take the values $P_0 = f(a)$ and $P_d = f(b)$ at the boundary of the interval, but also take the control values $P_i = f(x_i), i = 1, \ldots d - 1$ for an increasing sequence $a = x_0 < x_1 < x_2 \ldots x_{d-1} < x_d = b$ of arguments x_i, called *knots*.

The simplest situation involves real-valued functions, i.e., $m = 1$. We know from proposition 143, section 16.2, in volume 1, that there is exactly one polynomial $f \in \mathbb{R}[X]$ of degree $\leq d$ such that $f(x_i) = P_i$, for all $i \in \{0, \ldots d\}$. The Newton interpolation formula allows us to calculate the coefficients of f. But this formula does not explicitly refer to the control points. Lagrange's formula meets this requirement. For the following development we consider the real vector space $\mathbb{R}^d[X] \subset \mathbb{R}[X]$ of those polynomials with $deg(f) \leq d$. Evidently, the sequence $(1, X, \ldots X^d)$ is a basis of $\mathbb{R}^d[X]$, i.e., $dim(\mathbb{R}^d[X]) = d + 1$.

Proposition 315 *Given a sequence $a = x_0 < x_1 < x_2, \ldots x_{d-1} < b = x_d$ of real numbers, the sequence $(L_i)_{i=0,1,\ldots d}$ of Lagrange polynomials*

$$L_i = \prod_{j=0, j \neq i}^{d} \frac{X - x_j}{x_i - x_j}$$

in X is a basis of $\mathbb{R}^d[X]$. More precisely, and more generally, if $P_0, P_1, \ldots P_d$ is any sequence of control points in $\mathbb{R}^m$, then the unique polynomial function $f : \mathbb{R} \to \mathbb{R}^m$ of degree $\leq d$ in each coordinate, such that $f(x_i) = P_i$, for all i, is given by

$$f = \sum_i P_i L_i.$$

In particular, we have the identity $1 = \sum_i L_i$.

Proof Obviously the Lagrange polynomials L_i have degree $\leq d$ and make possible the condition $f(x_i) = P_i$, since $L_i(x_j) = \delta_{ij}$. If in every coordinate of a polynomial function f the polynomials are of degree $\leq d$, then they must coincide with the Lagrange polynomial expressions by proposition 140, section 16.2, volume 1 (this is not the fundamental theorem of algebra, but an easy preliminary result concerning the number of roots of polynomials). □

Now, in a certain sense this is the best we can expect if we want to traverse all control points. Let us restate the formula in more geometric terms. We know that the standard simplex Δ_d is the convex hull of the basis vectors $e_0, e_1, \ldots e_d \in \mathbb{R}^{d+1}$, its elements are described as linear combinations $x = \sum_i \xi_i e_i$, where $1 = \sum_i \xi_i$ and $0 \leq \xi_i \leq 1$. So $\Delta_d \subset \Lambda_d$, the subset of $\mathbb{R}^{d+1}$ defined by the solutions of the linear equation $1 = \sum_i \xi_i$. In other words, we have the *Lagrange curve*

$$L : [a, b] \to \Lambda_d : x \mapsto (L_0(x), L_1(x), \ldots L_d(x)),$$

and then, taking the unique map $\delta : \Lambda_d \to \mathbb{R}^m$, with $\delta(e_i) = P_i$, which extends to an affine map on $\mathbb{R}^{d+1}$, the above interpolation function f has this shape:

$$f = \delta \circ L.$$

Now, the Lagrange curve L hits all basis points e_i of the standard simplex Δ_d, starts at e_0, and ends at e_d. But it is *not* contained in Δ_d. This is quite obvious, since it is a C^1-curve which cannot stay within Δ_d at the transient vertexes e_i for $i \neq 0, d$, if its derivative does not vanish, see figure 37.5. The reason that we want a curve to be contained in a simplex is that this guarantees a certain degree of predictable behavior. In practice, a curve satisfying this condition will be contained in the convex hull of its control points and will not wildly run around as may do Lagrange polynomials of higher degrees. This, then, is the problem which Bézier interpolation and its splines solve.

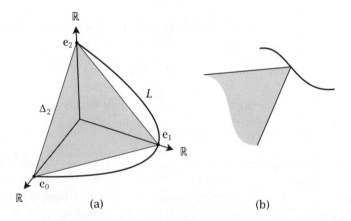

Fig. 37.5. (a) The Lagrange curve L on the simplex Δ_2 in $\mathbb{R}^3$. (b) A curve with non-vanishing differential cannot lie completely within the simplex if it passes through a corner point.

37.5 Bézier Curves

The Bézier approach is best understood in the "generic" environment of the standard simplex Δ_d. In fact, much as the Lagrange interpolation, which is the affine image of the Lagrange curve in Λ_d, a Bézier curve is the affine image of one single standard Bézier curve B on Δ_d. What is the role of the basis points e_i for the Bézier method? Recall that the Lagrange curve L traverses all basis points, with the additional condition that the order in which the curve passes through the points must be $e_0, e_1, \ldots e_d$. In contrast, Bézier curves do not satisfy the condition that intermediate basis points lie on the curve. Instead a geometrically very useful condition is required: the Bézier curve B starts at e_0, ends at e_d and remains within Δ_d for all curve parameters. This goal is met by an ingenious interpretation of the following observation. Let us concentrate on the elementary curve parameter interval $I = [0, 1]$. The one-dimensional curve $b(x) = (x, 1 - x)$ in Δ_1 has the coordinate sum 1, which, by the binomial theorem 255 in section 27.4.1, is rewritten as

$$1 = (x + (1 - x))^d = \sum_{i=0}^{d} \binom{d}{i} x^i (1 - x)^{d-i}.$$

This formula describes $d + 1$ functions, the *Bernstein polynomials of degree d*:

$$B_i^d(X) = \binom{d}{i} X^i (1 - X)^{d-i}.$$

Their sum is 1 for all $x \in \mathbb{R}$:

$$1 = \sum_{i=0}^{d} B_i^d(x),$$

and for $x \in [0, 1]$, we have

$$0 \le B_i^d(x) \le 1.$$

This means that the *Bernstein curve*

$$B^d : [0, 1] \to \Delta_d : x \mapsto (B_0^d(x), \ldots B_d^d(x))$$

of degree d has values in the standard d-dimensional simplex, and such that $B^d(0) = e_0$ and $B^d(1) = e_d$. See figure 37.6 and compare it to figure 37.5 for Lagrange polynomials.

Here is the analogue of the property for Lagrange polynomials:

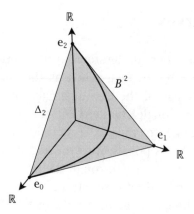

Fig. 37.6. The Bernstein curve B^2 on the simplex Δ_2 in $\mathbb{R}^3$. The curve is completely contained in the simplex.

Proposition 316 *The sequence $(B_i^d)_{i=0,\ldots d}$ of Bernstein polynomials of degree d is a basis of $\mathbb{R}^d[X]$.*

Proof Since the number of Bernstein polynomials $(B_i^d)_{i=0,1,\ldots d}$ is $d + 1$, it suffices to show that these polynomials are linearly independent. Let $0 = f(x) = \sum_i c_i \cdot B_i^d(x)$ be the zero polynomial function of $x \in \mathbb{R}$. Then $f(1) = c_d = 0$. Its derivative f' is also the zero function, so $f'(1) = c_{d-1}(B_{d-1}^d)'(1) = 0$, whence $c_{d-1} = 0$, etc., for higher derivatives, and we conclude that all c_i vanish. $\square$

Definition 224 *Given an affine simplex $\Delta(P_0, P_1, \ldots P_d) : \Delta_d \to \mathbb{R}^n$, the Bézier curve defined by the control points $P_0, P_1, \ldots P_d$ is the composed map*

$$B(P_0, P_1, \ldots P_d) = \Delta(P_0, P_1, \ldots P_d) \circ B^d$$

of the Bernstein curve B^d and the affine simplex $\Delta(P_0, P_1, \ldots P_d)$. Denote by $B[P_0, P_1, \ldots P_d]$ the image of $B(P_0, P_1, \ldots P_d)$ in $\mathbb{R}^n$.

Corollary 317 *Let $B(P_0, P_1, \ldots P_d)$ be a Bézier curve of degree d, then its image $B[P_0, P_1, \ldots P_d]$ is contained in the convex hull of the control points. If $f : \mathbb{R}^n \to \mathbb{R}^m$ is any affine map, then we have*

$$f \circ B(P_0, P_1, \ldots P_d) = B(f(P_0), f(P_1), \ldots f(P_d)).$$

Therefore the affine image of a Bézier curve is the Bézier curve of the affine image of its control points. This is the so-called covariance *of the Bézier construction.*

Proof The convex hull of the control points is the image of the standard simplex, and this contains the Bernstein curve B^d, whence the claim concerning the convex hull of control points. Moreover, we have

$$f \circ B(P_0, P_1, \ldots P_d) = f \circ \Delta(P_0, P_1, \ldots P_d) \circ B^d$$
$$= \Delta(f(P_0), f(P_1), \ldots f(P_d)) \circ B^d$$
$$= B(f(P_0), f(P_1), \ldots f(P_d)),$$

by lemma 310 and definition 224. $\qquad\square$

For the shaping of Bézier splines, the most important fact is

Proposition 318 *If $B(P.) = B(P_0, P_1, \ldots P_d)$ is a Bézier curve of degree d, then we have*

$$\frac{dB(P.)}{dx}(P_0) = d \cdot (P_1 - P_0),$$
$$\frac{dB(P.)}{dx}(P_d) = d \cdot (P_d - P_{d-1}),$$

i.e., the tangent to the Bézier curve at the beginning and ending points is just the d-fold difference of the first and last consecutive control points, respectively.

Proof Since the derivative of an affine function F inducing the map $e_i \mapsto P_i$ is its linear part, we have the derivative matrix $DF = (P_0, P_1, \ldots P_d)$. This applies to the derivative $DB^d(0) = (DB_0^d(0), DB_1^d(0), \ldots DB_d^d(0)) = (-d, d, 0, \ldots 0)$, whence $\frac{dB(P.)}{dx}(P_0) = d \cdot (P_1 - P_0)$. The second formula results from $DB^d(1) = (0, \ldots 0, -d, d)$. $\qquad\square$

Note that this last property is extremely useful for controlling the initial and final tangent of a Bézier curve, and thus enables this splining technique: In order to spline two Bézier curves $B(P.)$ and $B(P'.)$ of degrees d and d', respectively, at a common control point $P_d = P'_0$, one must satisfy the equation

$$P_d - P_{d-1} = P'_1 - P'_0.$$

Example 177 Let $B : [0, 1] \to \mathbb{R}^2$ be the Bézier curve with control points

$$P_0 = (\tfrac{1}{2}, 2), P_1 = (1, 3), P_2 = (2, \tfrac{1}{2}), P_3 = (3, 1), P_4 = (\tfrac{5}{2}, \tfrac{5}{2}).$$

Then

$$B(x) = P_0 B_0^4(x) + P_1 B_1^4(x) + P_2 B_2^4(x) + P_3 B_3^4(x) + P_4 B_4^4(x)$$

$$= (\tfrac{1}{2}, 2)(1-x)^4 + (1,3)4(1-x)^3 x + (2, \tfrac{1}{2})6(1-x)^2 x^2$$

$$+ (3,1)4(1-x)x^3 + (\tfrac{5}{2}, \tfrac{5}{2})x^4$$

$$= (\tfrac{1}{2} + 2x + 3x^2 - 2x^3 - x^4, 2 + 4x - 21x^2 + 26x^3 - \tfrac{17}{2}x^4).$$

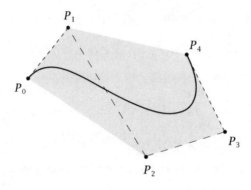

Fig. 37.7. The Bézier curve with control points P_0, P_1, P_2, P_3 and P_4, in that order.

The curve $B(x)$, for $x \in [0,1]$, including the control points, is shown in figure 37.7. Note that the curve lies completely within the convex hull (in light gray) of $\{P_0, P_1, P_2, P_3, P_4\}$.

37.5.1 Bernstein and de Casteljau Recursion

For the practical calculations related to Bézier curves, two recursive procedures are important: the calculation of Bernstein polynomials and the calculation of curve parameters.

Concerning Bernstein polynomials, we have these formulas:

Proposition 319 *For degree $d > 0$ and $i = 1, 2, \ldots d$, one has*

$$B_i^d = X B_{i-1}^{d-1} + (1 - X) B_i^{d-1}.$$

Proof Recall from lemma 254 the recursive formula $\binom{d-1}{i-1} + \binom{d-1}{i} = \binom{d}{i}$ for binomial coefficients. Then we have

$$XB_{i-1}^{d-1} + (1 - X)B_i^{d-1} = \binom{d-1}{i-1}X^i(1-X)^{d-i} + \binom{d-1}{i}X^i(1-X)^{d-i}$$

$$= \binom{d}{i}X^i(1-X)^{d-i}$$

$$= B_i^d.$$

□

The procedure for calculating curve points at parameter value x uses the following result:

Proposition 320 (de Casteljau Algorithm) *Given the control points P_0, $P_1, \ldots P_d$ of a Bézier curve of degree d, the value $B(P_0, P_1, \ldots P_d)(x)$ is calculated by the following recursive formula, defining a sequence of points P_i^j for $j = 1, 2, \ldots d$ and $i = 0, 1, \ldots d - j$, via*

$$P_i^j = (1 - x)P_i^{j-1} + xP_{i+1}^{j-1}.$$

where $P_i^0 = P_i$. Then we have

$$B(P_0, P_1, \ldots P_d)(x) = P_0^d(x).$$

Proof The algorithm follows from the recursive formulas for Bernstein polynomials in proposition 319, we omit it and refer to [36]. □

This is best visualized by a schema, which we show for $d = 3$. In this schema a value at the tail of an arrow is multiplied by the factor above the arrow. The value at the head of the arrow is the sum of two of those scaled values:

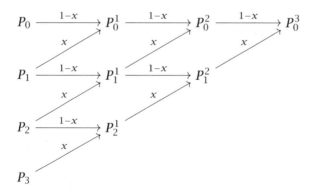

Figure 37.8 illustrates two realizations of de Casteljau's algorithm.

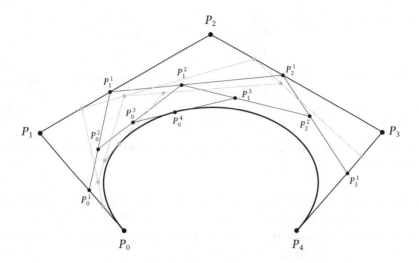

Fig. 37.8. De Casteljau's algorithm for the control points P_0, P_1, P_2, P_3 and P_4. The P_i^j are the result of an execution of the algorithm for $x = \frac{7}{12}$. Also shown in gray is a second execution for $x = \frac{3}{4}$.

Putting together Bézier curves is now straightforward. We are given a partition $a = x_0 < x_1 < \ldots x_k = b$ of the interval $[a, b]$. For each interval $K_i = [x_i, x_{i+1}]$, $i = 0, 1, \ldots k - 1$, we have an affine bijection $t_i : K_i \xrightarrow{\sim} [0, 1]$ with $t_i(x) = \frac{x - x_i}{x_{i+1} - x_i}$. Then we take the stretched Bézier curves $B_i = B(\ldots) \circ t_i$ connecting successive last and first control points for the i-th Bézier curve, and finally glue them together to a Bézier spline $f : [a, b] \to \mathbb{R}^n$, with $f|_{K_i} = B_i$.

37.6 Tensor Product Splines

Bézier and other curves that follow similar construction principles can be used to generate surfaces and higher-dimensional shapes. We present here an elegant method for combining a number of curves to such shapes of higher complexity. The method is part of a general theory of mathematical structures, called *tensor products*. We shall, however, not develop this theory in its full-fledged generality, but only use some elementary notations indicating that behind our ad-hoc approach a far-reaching method is hidden.

The idea is this: Often, the real vector space $\mathbb{R}^n$ is not given without further information, but having its dimension decomposed as a product $n = n_1 \cdot n_2 \cdot \ldots n_k$. Assuming such a factorization, the members of the canonical basis $e_0, e_1, \ldots e_{n-1}$ of $\mathbb{R}^n$ are reindexed by a sequence of indexes $(i_1, i_2, \ldots i_k)$ with $0 \leq i_1 \leq n_1 - 1, 0 \leq i_2 \leq n_2 - 1, \ldots 0 \leq i_k \leq n_k - 1$. In this reindexing system, we write $e_{i_1 \otimes i_2 \otimes \ldots i_k}$ for $e_{s(i_1, i_2, \ldots i_k)}$ under a fixed bijection of sets[1] $s : n_1 \times n_2 \times \ldots n_k \overset{\sim}{\rightarrow} n$. The specific choice of s is irrelevant, so we omit it in the following discussion. The symbol $\otimes$ is the *tensor product sign*, which we only use as an abstract symbol without any further meaning here. In this representation, the vector space $\mathbb{R}^n$ is denoted by $\mathbb{R}^{n_1 \otimes n_2 \otimes \ldots n_k}$. For example, in $\mathbb{R}^{2 \otimes 2}$, we have the basis $(e_{0 \otimes 0}, e_{0 \otimes 1}, e_{1 \otimes 0}, e_{1 \otimes 1})$, corresponding to, say, the canonical basis (e_0, e_1, e_2, e_3). We shall see that this is in fact a very practical notation for higher-dimensional spline theory.

The tensor product is used as follows: For a given factorization $n = n_1 \cdot n_2 \cdot \ldots n_k$, there is a map

$$t(n_1, n_2, \ldots n_r) : \mathbb{R}^{n_1} \times \mathbb{R}^{n_2} \times \ldots \mathbb{R}^{n_k} \rightarrow \mathbb{R}^{n_1 \otimes n_2 \otimes \ldots n_k}$$

which is defined as follows: If $x_i = \sum_{j=0,1,\ldots n_i - 1} \xi_j^i \cdot e_j \in \mathbb{R}^{n_i}$ for $i = 1, 2, \ldots k$, then

$$t(n_1, n_2, \ldots n_k)(x_1, x_2, \ldots x_k) = \sum \xi_{i_1}^1 \xi_{i_2}^2 \ldots \xi_{i_k}^k \cdot e_{i_1 \otimes i_2 \otimes \ldots i_k},$$

where the sum is over all indexes $0 \leq i_1 \leq n_1 - 1, 0 \leq i_2 \leq n_2 - 1, \ldots 0 \leq i_k \leq n_k - 1$. This map is linear in each argument x_i.

If we work with standard simplexes Δ_d, then we also adapt the dimensional notation, i.e., we write $\Delta_{d_1 \otimes d_2 \otimes \ldots d_k}$ for the Δ_d, where $d + 1 = \prod_i (d_i + 1)$. In fact, if the vertexes of Δ_{d_i} are the e_{j_i}, then the vertexes of $\Delta_{d_1 \otimes d_2 \otimes \ldots d_k}$ are the basis vectors $e_{j_1 \otimes j_2 \otimes \ldots j_k}$.

Proposition 321 *Let $d_1, d_2 \ldots d_k$ be a sequence of natural numbers, then the map $t(d_1 + 1, d_2 + 1, \ldots d_k + 1)$ maps $\Delta_{d_1} \times \Delta_{d_2} \times \ldots \Delta_{d_k}$ into $\Delta_{d_1 \otimes d_2 \otimes \ldots d_k}$.*

Exercise 181 Give a proof of proposition 321. Observe that the coefficients $\xi_{i_1}^1 \xi_{i_2}^2 \ldots \xi_{i_k}^k$ in the definition of $t(d_1 + 1, d_2 + 1, \ldots d_k + 1)$ also arise in the development of the product $\prod_i \sum_{j=0,\ldots d_i} \xi_j^i$, where all factors are sums yielding 1.

---

[1] Recall here that natural numbers are ordinal sets.

These facts now make it possible to construct higher-dimensional shapes. Suppose that we are given a curve $R^i : [0,1] \to \Delta_{d_i}$ of a certain type—a Bernstein curve B^{d_i}, for example—for each degree $d_i, i = 1, 2, \ldots k$. Then the direct product of these data yields this application:

$$\bigotimes_i R^i : [0,1]^k \xrightarrow{R^1 \times \cdots R^k} \Delta_{d_1} \times \cdots \Delta_{d_k} \xrightarrow{t(d_1+1,\ldots d_k+1)} \Delta_{d_1 \otimes \cdots d_k}$$

Once we have understood this map, the rest is straightforward. We only need to define the unique map $\Delta(P_{0\otimes\cdots 0}, \ldots P_{d_1\otimes\cdots d_k}) : \Delta_{d_1 \otimes d_2 \otimes \cdots d_k} \to \mathbb{R}^n$ with control points $P_{j_1\otimes\cdots j_k} \in \mathbb{R}^n$ and to compose it with $\bigotimes_i R^i$. This yields the k-dimensional shape map

$$\Delta(P_{0\otimes\cdots 0}, \ldots P_{d_1\otimes\cdots d_k}) \circ \bigotimes_i R^i : [0,1]^k \to \mathbb{R}^n.$$

Here, the indexing of the control points by the product of k indexes suggests that the control points are arranged in a k-dimensional grid. For example, if $k = 3$, we view the control points as distributed in a three-dimensional grid, although their positions may be completely messed up. Let us look at the frequent example of $k = 2$ and $n = 3$. This means we are considering a surface, defined by two independent parameters $x, y \in [0,1]$, with values in $\mathbb{R}^3$. If R^1 and R^2 are Bernstein curves, then we have a *Bézier surface* with values in the three-dimensional real space. Such surfaces are used as functions on the rectangular units of partitions for two-dimensional splines. For example, in automobile industry and similar applications, this construction is essential.

Example 178 Consider the construction of a spline surface based on Bernstein polynomials B^3 and B^2 on the simplexes Δ_3 and Δ_2. The construction described above is the composition of 3 functions:

First

$$f = R^1 \times R^2 : [0,1]^2 \to \Delta_3 \times \Delta_2,$$

where $R^1 = B^3$, and $R^2 = B^2$ are the Bernstein polynomials. Next

$$g = t(4,3) : \Delta_3 \times \Delta_2 \to \Delta_{3\otimes 2},$$

and finally

$$h = \Delta(P_{0\otimes 0}, \ldots P_{3\otimes 2}) : \Delta_{3\otimes 2} \to \mathbb{R}^3,$$

which requires 12 control points $P_{0\otimes 0}, \ldots P_{3\otimes 2}$.

Now, a pair $(x, y) \in [0, 1]^2$ is mapped by f to the pair $(B^3(x), B^2(y))$, a pair $(u, v) \in \Delta_3 \times \Delta_2$ is mapped to $\sum_{i,j} u_i v_j e_{i \otimes j}$, and this result is projected into $\mathbb{R}^3$ by "substituting" each basis vector $e_{i \otimes j}$ by the corresponding vector $P_{i \otimes j}$.

Putting everything together yields

$(h \circ g \circ f)(x, y) =$

$$
\begin{array}{lll}
(1-x)^3(1-y)^2 P_{0 \otimes 0} & + & 2y(1-x)^3(1-y)P_{0 \otimes 1} & + & (1-x)^3 y^2 P_{0 \otimes 2} & + \\
3x(1-x)^2(1-y)^2 P_{1 \otimes 0} & + & 6xy(1-x)^2(1-y)P_{1 \otimes 1} & + & 3xy^2(1-x)^2 P_{1 \otimes 2} & + \\
3x^2(1-x)(1-y)^2 P_{2 \otimes 0} & + & 6x^2 y(1-x)(1-y)P_{2 \otimes 1} & + & 3x^2 y^2(1-x)P_{2 \otimes 2} & + \\
x^3(1-y)^2 P_{3 \otimes 0} & + & 2x^3 y(1-y)P_{3 \otimes 1} & + & x^3 y^2 P_{3 \otimes 2} &
\end{array}
$$

An alternative approach, which leads to the same result, considers what it means to add a further parameter y to a single parameter Bézier curve. Looking back at figure 37.7, imagine that the control points are not fixed, but are themselves described by Bézier splines parameterized by the second variable y. Having $m + 1$ control points P_i, $0 \le i \le m$, each a Bézier curve with $n + 1$ control points P_{ij}, $0 \le j \le n$, which makes $(m+1)(n+1)$ control points in total, we have the following:

$$
P_i(y) = \sum_{j=0}^{n} P_{ij} B_j^n(y), \quad \text{for } 0 \le i \le m.
$$

Putting the pieces together, we get the surface function

$$
B(x, y) = \sum_{i=0}^{m} P_i(y) B_i^m(x)
$$

$$
= \sum_{i=0}^{m} \sum_{j=0}^{n} P_{ij} B_j^n(y) B_i^m(x).
$$

Putting $m = 3$ and $n = 2$, we get the same function as above, with $P_{ij} = P_{i \otimes j}$.

Figure 37.9 shows a Bézier surface for sixteen control points, i.e., the case where $m = n = 3$.

37.7 B-Splines

There are still other types of curves, which may be better suited for the construction of specific curve shapes, such as B-splines or their gener-

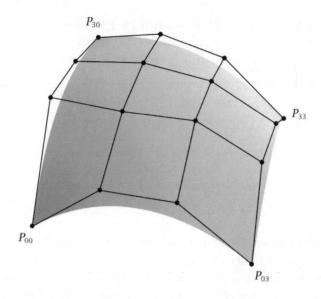

Fig. 37.9. Bézier surface with 4×4 control points.

alization to NURBS, i.e., Non Uniform Rational B-Splines. Like the Bernstein curve, the B-spline curve is of the type discussed above, i.e., having its image within the standard simplex Δ_d, and not only on the surrounding "linear space" Λ_d, such as is the case for the Lagrange curve L. And they are also defined by recursion for a given degree d, which counts the number $d + 1$ of control points $P_0, \ldots P_d$ in $\mathbb{R}^n$. They take up the idea, known from Lagrange interpolation, of an increasing sequence $0 = x_0 \le x_1 \le \cdots x_m = 1$ of $m + 1$ parameter "knots" and to prescribe function values defined according to these knots. The functions f_i^p to be defined are given for every order $p \in \mathbb{N}$ with $p \le m - d - 1$ and for $i = 0, 1, \ldots d$, and constructed recursively:

1. The functions of order 0 are the characteristic functions of the closed-open knot intervals, i.e.,

$$f_i^0 = \chi_{[x_i, x_{i+1}[}$$,

2. The functions of order $p > 0$ are built from functions of order $p - 1$, i.e.,

$$f_i^p(x) = \frac{x - x_i}{x_{i+p} - x_i} f_i^{p-1}(x) + \frac{x_{i+p+1} - x}{x_{i+p+1} - x_{i+1}} f_{i+1}^{p-1}(x).$$

Since consecutive x_i and x_{i+1} may be equal, the denominator of a factor can be 0, in which case we set the factor to 0.

For every order p, there is a B-spline curve $BS^p : [0,1] \to \Delta_d$, yielding the affine simplex

$$BS^p(P_0, \ldots P_d) = \Delta(P_0, \ldots P_d) \circ BS^p.$$

The condition that the resulting curve be contained in the standard simplex Δ_d requires that $x_i = 0$, for $i < p$, and $x_i = 1$, for $i > m - p$. If, in addition, it is desired that the curve starts at P_0 and ends at P_d, we set $x_p = 0$ and $x_{m-p} = 1$. If the knots x_i, $p \leq i \leq m - p$ are evenly spaced, the B-spline is said to be *uniform*.

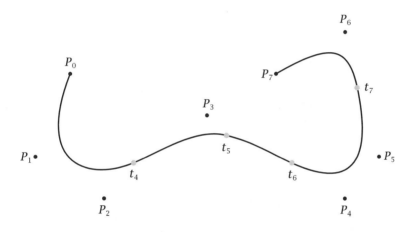

Fig. 37.10. A cubic B-spline with eight control points (in black), and twelve knots (in gray), the first four lie at P_0, the last four at P_7.

Example 179 A cubic B-spline has $p = 3$, i.e, if there are $(d + 1)$ control points, then at least $(d+1)+4$ parameter knots are required. For example, a non-uniform cubic B-spline with $d = 7$ (figure 37.10) is given by the following eight control points

$$P_0 = (0,0), \quad P_1 = (-\tfrac{1}{2}, -1), \quad P_2 = (\tfrac{1}{2}, -\tfrac{3}{2}), \quad P_3 = (2, -\tfrac{1}{2}),$$
$$P_4 = (4, -\tfrac{3}{2}), \quad P_5 = (\tfrac{9}{10}, -1), \quad P_6 = (4, \tfrac{1}{2}), \quad P_7 = (3, 0)$$

and twelve knots

$$x_0 = x_1 = x_2 = x_3 = 0,$$
$$x_4 = \tfrac{1}{3}, x_5 = \tfrac{1}{2}, x_6 = \tfrac{3}{5}, x_7 = \tfrac{9}{10},$$
$$x_8 = x_9 = x_{10} = x_{11} = 1.$$

NURBS are derived from B-splines by putting numerical weights w_i at the control points, i.e., P_i is replaced by $w_i \cdot P_i$. We omit the details here and refer to [19].

A lot of information on curves and surfaces can be found in a standard text on computer graphics, see [11].

CHAPTER 38

Fourier Theory

38.1 Introduction

Fourier theory goes back to the work of the mathematician and politician Jean Baptiste Joseph Fourier (1768-1830). It was initiated in his work from 1807 on solutions of the partial differential heat equation $D_{xx}^2 T(x,t) = c \cdot D_t T(x,t)$ that describes the infinitesimal behavior of the temperature T of a metallic pole as a function of position x and time t. Fourier's method introduces infinite sums of sine and cosine functions, the famous structures which were later coined Fourier series. This paper "Théorie de la propagation de la chaleur dans les solides" was strongly criticized by his fellow mathematicians because of the lack of mathematical rigor. Nonetheless, it received the mathematics prize in 1811 and thus became a starting point of a big mathematical theory, which has an incredible omnipresence in modern natural science. Fourier was not only a mathematical revolutionary, he was also politically active and even was put to jail as a terrorist during the French revolution.

Mathematically, Fourier's achievement provides us with a generalization of the idea of representing vectors as finite linear combinations to vector spaces of not too pathological functions $f(x)$ of a real variable x, having a given period p, i.e., $f(x + p) = f(x)$. This setup goes beyond algebra in that the sum representation $f = \sum_i \lambda_i \cdot e_i$ need not be finite, and therefore, the summation means convergence for some metric topology on the given vector space of functions. One is also given a scalar product on that space, such that orthogonality and length of functions are defined. Fourier found an orthonormal basis, namely $(e_n)_{n \in \mathbb{Z}}$, with $e_n(x) = e^{inx}$ for $p = 2\pi$, a basis which by the Euler equation $e^{inx} = \cos(nx) + i\sin(nx)$ yields the Fourier series in terms of sine and cosine functions.

However, mathematically speaking, this basis is far from trivial. We know from chapter 257 that Euler's equation and all the properties of the related functions are far from elementary. The very definition of the exponential function or, equivalently, of the sine and cosine function requires convergent power series, for example. The Fourier basis is at the same time extraordinary since it is of a striking simplicity: just the integer powers $e_n(x) = e^{inx} = (e^{ix})^n$ of a single germinal function e^{ix}. But the latter is a complex mathematical construction. There are many justifications for such a choice of basis functions. One of them is their appearance as solutions of basic differential equations in mechanics, such as Hooke's law $m \cdot D_t^2 x = -k \cdot x$ of a mass m at position x, which is tied to a spring with the spring constant k. The solution of this equation is in fact a sine function, since $D^2 \sin = -\sin$ is a differential equation of the above type. But this justification is a physical one relating to an elementary mechanical situation. Mathematically speaking, there are many other orthonormal bases besides Fourier's basis.

The prominence of Fourier's basis is that it has a strong relation to important differential equations in physics and that therefore, highly developed technologies, including Fast Fourier Transform (FFT) algorithms, are available. Modern audio technology (including compression methods such as MP3 and its successors) is based on Fourier's theory, and image processing in computer graphics is unthinkable without Fourier's theory. Moreover, the recent development of wavelets is based on Fourier transforms. The latter is a limit situation, where the period p tends to infinity. We append a short exposition of Fourier transforms, because this theory is needed to deal with the theory of wavelets, which we present in chapter 39. But a transcendental justification of Fourier theory, as it is preconized by the old theory of Fourier analysis of the human ear, or by the belief that musical harmony is best understood by Fourier analysis, has not been confirmed by modern research, see [28], and pertains rather to the mysticism of overtones than to science.

38.2 Spaces of Periodic Functions

Since Fourier theory is built upon integrability of periodic functions, we first need to define what a periodic function is as well as what it means to call a periodic function integrable.

Definition 225 *Let $p > 0$ be a real number. A function $f : \mathbb{R} \to \mathbb{C}$ is called* periodic with period p *or p-periodic iff $f = f \circ T^p$, i.e., for all $x \in \mathbb{R}$,* $f(x) = f(x + p)$.

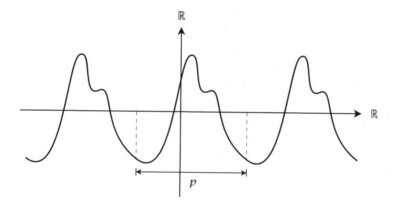

Fig. 38.1. A periodic function with period p.

For the following proposition, recall from remark 32 that a function $f : X \to \mathbb{R}^n$ is integrable iff each of its components $f_i = pr_i \circ f$, for $i = 1, 2, \ldots n$ is so, and that integration for $f : X \to \mathbb{C}$ is regarded as the special case of $n = 2$.

Proposition 322 *For a p-periodic function $f : \mathbb{R} \to \mathbb{C}$, the following statements are equivalent:*

(i) *f is integrable on every closed interval $[a, b]$.*

(ii) *f is integrable on the interval $[0, p]$.*

Proof We use the criterion in proposition 283 of integrability of a function. Observe that our function has complex values, and therefore the set of non-continuity is just the union of the sets of non-continuity for the real part and the imaginary part of the function. Now, (ii) is a special case of (i). Conversely, if f is integrable on $[0, p]$, then it has a set $N \subset [0, p]$ of non-continuity of measure zero. Since f is p-periodic, the set N_n of non-continuity of f on a shifted interval $T^{np}[0,p]$ is also of measure zero. But the interval $[a,b]$ is a subset of a finite union of such shifted intervals $T^{np}[0,p]$, and therefore the finite union of the N_n has measure zero by proposition 280, and the same is true for the intersection of this set with $[a,b]$.   □

Definition 226 *A p-periodic function $f : \mathbb{R} \to \mathbb{C}$ is called integrable iff it satisfies the equivalent properties of proposition 225. The set of p-periodic integrable functions is denoted by $\int_p(\mathbb{R},\mathbb{C})$.*

Lemma 323 *If $f \in \int_p(\mathbb{R},\mathbb{C})$, then for all intervals $I = [a, a+p]$, the value of $\int_I f$ is one and the same.*

Proof Suppose that f is continuous. By proposition 286, setting $\xi(x) = x + p$, we have $\int_0^p f(x)dx = \int_0^p f(x+p)dx = \int_0^p f(\xi(x))\xi'(x)dx = \int_a^{a+p} f(\xi)\,d\xi$. The theorem of change of variable is also valid for integrable functions, see [14], chapter 3.9, for a proof, so the same argument works in the general case.   □

Proposition 324 *Given a period p, let $f, g \in \int_p(\mathbb{R},\mathbb{C})$. Then:*

(i) *$f + g \in \int_p(\mathbb{R},\mathbb{C})$, where $(f+g)(x) = f(x) + g(x)$, the sum being in $\mathbb{C}$.*

(ii) *$f \cdot g \in \int_p(\mathbb{R},\mathbb{C})$, where $(f \cdot g)(x) = f(x) \cdot g(x)$, the product being in $\mathbb{C}$.*

(iii) *With the previously defined sum and product, $\int_p(\mathbb{R},\mathbb{C})$ is a commutative ring containing the subring of constant functions, which is identified with $\mathbb{C}$. Therefore, $\int_p(\mathbb{R},\mathbb{C})$ is called the $\mathbb{C}$-algebra of integrable p-periodic functions.*

(iv) *If $\overline{f}$ denotes the conjugate function defined by $\overline{f}(x) = \overline{f(x)}$, then $\overline{f} \in \int_p(\mathbb{R},\mathbb{C})$ and $\overline{\overline{f}} = f$. Moreover, $\overline{f+g} = \overline{f} + \overline{g}$ and $\overline{f \cdot g} = \overline{f} \cdot \overline{g}$, i.e., conjugation is a ring automorphism, and it leaves the subring of constant functions $\mathbb{C}$ invariant.*

(v) *The absolute value function $|f|(x) = |f(x)|$ is also an element of $\int_p(\mathbb{R},\mathbb{C})$, and so are the functions $f^+ = (f + |f|)/2$ and $f^- = (f - |f|)/2$.*

Proof Observe that if $f, g \in \int_p(\mathbb{R},\mathbb{C})$ are integrable, then so are continuous functions applied to the pair map $(f,g) : \mathbb{R} \to \mathbb{C}^2$, for example $+ : \mathbb{C}^2 \to \mathbb{C}, \cdot : \mathbb{C}^2 \to \mathbb{C}$, or the special case $|f| = \sqrt{Re(f)^2 + Im(f)^2}$. Up to straightforward verifications, this yields all the statements of the proposition.   □

Fourier's theorem is concerned with a special subalgebra of $\int_p(\mathbb{R}, \mathbb{C})$, which is defined by functions which are a patchwork of C^1-functions, more precisely:

Definition 227 *A p-periodic function is called* piecewise smooth *iff there is a partition $P \in Part(0, p)$ of the period-interval $[0, p]$ such that for every interval $J \in I(P)$, there is a function $g_J \in C^1(J, \mathbb{C})$ such that $g_J|_{J^o} = f|_{J^o}$. The set of p-periodic, piecewise smooth functions is denoted by $PC_p^1(\mathbb{R}, \mathbb{C})$, it is a subset of $\int_p(\mathbb{R}, \mathbb{C})$ by proposition 283. It is closed under sums and products of functions and contains $\mathbb{C}$, therefore it is called the $\mathbb{C}$-algebra of* piecewise smooth p-periodic functions.

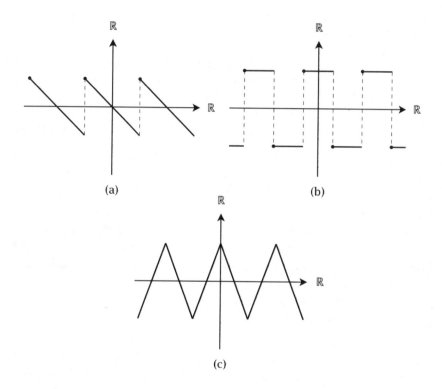

Fig. 38.2. Three curves in $PC_p^1(\mathbb{R}, \mathbb{R})$: (a) the sawtooth curve, (b) the rectangle curve, (c) the triangle or zigzag curve.

The Fourier theory deals with the description of the algebra $PC_p^1(\mathbb{R}, \mathbb{C})$ in terms of special orthonormal $\mathbb{C}$-vector space bases.

38.3 Orthogonality

The special functions mentioned before are the trigonometric functions cos and sin, or their complex counterpart: the exponential function for imaginary arguments, i.e., $e^{ix} = \cos(x) + i\sin(x)$, where $x \in \mathbb{R}$. In this basic section, we shall focus our attention to the period $p = 2\pi$. Later we shall present more general, but mathematically straightforward, statements for arbitrary periods.

Notation 1 *For $n \in \mathbb{Z}$, we denote by e_n the function in $PC_{2\pi}^1(\mathbb{R}, \mathbb{C})$ with $e_n(x) = e^{inx}$.*

Lemma 325 *For any non-empty finite set $B \subset \mathbb{Z}$, the sequence $(e_b)_{b \in B}$ is linearly independent over $\mathbb{C}$.*

Proof Suppose that $\sum_b \lambda_b e_b = 0$. This is equivalent to $e_k \cdot \sum_b \lambda_b e_b = \sum_b \lambda_b e_{b+k} = 0$ for any $k \in \mathbb{Z}$. So we may suppose that the equation has the shape $\sum_{j=0,\ldots m} \mu_m \cdot (e^{it})^j = 0$ for all $t \in \mathbb{R}$. This means that the polynomial $\sum_{j=0,\ldots m} \mu_j \cdot (e^{it})^j = 0$ of degree $\leq m$ vanishes for an infinity of different values e^{it}, which by proposition 140, section 16.2, in volume 1, is only possible if it is the zero polynomial. □

In particular, for $N \in \mathbb{N}$, the sequence $(e_{-N}, e_{-N+1}, \ldots e_{-1}, e_0, e_1, \ldots e_N)$ is the basis of a $(2N + 1)$-dimensional $\mathbb{C}$-vector space U_N, whose elements $\sum_{n=-N}^N c_n \cdot e_n$ are called *trigonometric polynomials of degree $\leq N$*. If $N_1 < N_2$, then we have the canonical embedding $i_{N_1} : U_{N_1} \to U_{N_2}$ as well as the canonical retraction $p_{N_1} : U_{N_2} \to U_{N_1}$ of i_{N_1}, given by $p_{N_1}(e_n) = 0$ for $N_2 \geq n > N_1$. We denote by U_∞ the limit $U_\infty = \lim_N U_N$ with respect to the retractions p_N, recall chapter 36 for a discussion of limits in categories. What are the elements of U_∞? By definition of a limit, they are the sequences $(S_N)_{N \in \mathbb{N}}$, where $S_N = \sum_{-N}^N c_{N,n} \cdot e_n$, and where $p_{N_1}(S_{N_2}) = S_{N_1}$ for all $N_1 < N_2$. So the coefficients $c_{N,n}$ are in fact independent of the index N, and we may therefore denote the limit element $(S_N)_{N \in \mathbb{N}}$ by a symbolic infinite sum $\sum_{\lim} c_n \cdot e_n$, which represents the sequence of complex numbers $(c_n)_{n \in \mathbb{Z}}$, but also takes into account that the finite partial sums $\sum_{-N}^N c_n \cdot e_n$ are the elements of the vector spaces U_N. Of course, this notation is introduced in view of the fact that in many interesting situations, the infinite sum is a function in $PC_{2\pi}^1(\mathbb{R}, \mathbb{C})$ and not just a formal construction in a limit set.

Next, we introduce a form on $PC_{2\pi}^1(\mathbb{R}, \mathbb{C})$, which is slightly more general than the bilinear forms encountered on real vector spaces in chapter 24 of volume 1 and in definition 201 and sorite 287. For $f, g \in PC^1_{2\pi}(\mathbb{R}, \mathbb{C})$, we define the complex number[1]

$$\langle f, g \rangle = \frac{1}{2\pi} \int_0^{2\pi} f \cdot \overline{g},$$

where the product $f \cdot \overline{g}$ of f and the conjugate of g is in the algebra $PC^1_{2\pi}(\mathbb{R}, \mathbb{C})$. In particular, $\langle f, f \rangle$ is a non-negative real number, and one defines the *norm of a function f* with respect to this form by

$$\|f\|_2 = \sqrt{\langle f, f \rangle}.$$

Here are the important properties of this generalized bilinear form:

Sorite 326 *The form $\langle ?, ? \rangle : PC^1_{2\pi}(\mathbb{R}, \mathbb{C})^2 \to \mathbb{C}$ has the following properties for $f, g, h \in PC^1_{2\pi}(\mathbb{R}, \mathbb{C})$ and $\lambda \in \mathbb{C}$:*

(i) $\langle g, f \rangle = \overline{\langle f, g \rangle}$,

(ii) $\langle f + g, h \rangle = \langle f, h \rangle + \langle g, h \rangle$,

(iii) $\langle \lambda \cdot f, h \rangle = \lambda \cdot \langle f, h \rangle$,

(iv) $\langle f, g + h \rangle = \langle f, g \rangle + \langle f, h \rangle$,

(v) $\langle f, \lambda \cdot g \rangle = \overline{\lambda} \cdot \langle f, g \rangle$,

(vi) (Schwarz Inequality) $|\langle f, g \rangle| \leq \|f\|_2 \cdot \|g\|_2$,

(vii) $\|f\|_2 \geq 0$, *and* $\|f\|_2 = 0$ *iff the set* $\{x \mid 0 \leq x \leq 2\pi, f(x) \neq 0\}$ *is finite*,

(viii) (Triangle Inequality) $\|f + g\|_2 \leq \|f\|_2 + \|g\|_2$.

Due to properties (i) to (v), the form $\langle ?, ? \rangle$ is called Hermitian.

Proof Except for statement (vii), this sorite is straightforward, in particular, the Schwarz inequality is proved by the same method as the Schwarz inequality in sorite 287. As to (vii), the finiteness condition is clearly sufficient for the vanishing of $\|f\|_2$, conversely, if the set $\{x \mid 0 \leq x \leq 2\pi, f(x) \neq 0\}$ is infinite, its intersection with the points in $[0, 2\pi]$ where f is continuous is non-empty. From this fact, one immediately deduces the non-vanishing of $\|f\|_2$ as we already did in the proof of sorite 287. □

With respect to this Hermitian form, one writes $f \perp g$ for the equation $\langle f, g \rangle = 0$ and calls f *orthogonal to g*. Evidently, orthogonality is a symmetric relation. If $V \subset PC^1_{2\pi}(\mathbb{R}, \mathbb{C})$ is any subset, define the *orthogonal space $V^\perp$ of V* by

---

[1] Often the factor $\frac{1}{2\pi}$ is omitted in this definition, and in the related definition of $\|f\|_2$. We insert it for reasons of normalization.

$$V^\perp = \{x \mid x \in PC^1_{2\pi}(\mathbb{R}, \mathbb{C}), x \perp v, \text{ all } v \in V\}.$$

Exercise 182 Show that $V^\perp$ is a vector subspace of $PC^1_{2\pi}(\mathbb{R}, \mathbb{C})$.

Recall that the Kronecker delta symbol is defined by $\delta_{ij} = 0$ if $i \neq j$, and $\delta_{ii} = 1$, where $i, j \in \mathbb{Z}$.

For the following orthogonality relations, we need this exercise about primitive functions.

Exercise 183 Let $m, n \in \mathbb{Z}$. Use the trigonometric equations

$$\sin(mx)\sin(nx) = \frac{1}{2}(\cos((m-n)x) - \cos((m+n)x))$$

$$\sin(mx)\cos(nx) = \frac{1}{2}(\sin((m-n)x) + \sin((m+n)x))$$

$$\cos(mx)\cos(nx) = \frac{1}{2}(\cos((m-n)x) + \cos((m+n)x))$$

to prove the following primitive function expressions:

$$\int \sin(mx)\sin(nx)\,dx = \begin{cases} \frac{1}{2}\left(\frac{\sin((m-n)x)}{m-n} - \frac{\sin((m+n)x)}{m+n}\right) & \text{for } m \neq \pm n, \\ \pm\frac{1}{2}\left(x - \frac{\sin(2mx)}{2m}\right) & \text{for } m = \pm n. \end{cases}$$

$$\int \sin(mx)\cos(nx)\,dx = \begin{cases} -\frac{1}{2}\left(\frac{\cos((m-n)x)}{m-n} + \frac{\cos((m+n)x)}{m+n}\right) & \text{for } m \neq \pm n, \\ -\frac{1}{2}\frac{\cos(2mx)}{2m} & \text{for } m = \pm n. \end{cases}$$

$$\int \cos(mx)\cos(nx)\,dx = \begin{cases} \frac{1}{2}\left(\frac{\sin((m-n)x)}{m-n} + \frac{\sin((m+n)x)}{m+n}\right) & \text{for } m \neq \pm n, \\ \frac{1}{2}\left(x + \frac{\sin(2mx)}{2m}\right) & \text{for } m = \pm n. \end{cases}$$

Proposition 327 (Trigonometric Orthogonality Relations) *Let* $k, l \in \mathbb{Z}$. *Then:*

(i) $\int_0^{2\pi} \cos(kx)\cos(lx)\,dx = \int_0^{2\pi} \sin(kx)\sin(lx)\,dx = \pi\delta_{kl}$ *for k or l* $\neq 0$.

(ii) $\int_0^{2\pi} \cos(kx)\sin(lx)\,dx = 0$.

(iii) $\int_0^{2\pi} e_k = 2\pi\delta_{0k}$.

(iv) $\|e_k\|_2 = 1$, *and* $e_k \perp e_l$ *for* $k \neq l$. *The basis* $(e_n)_{n=-N,-N+1,...N-1,N}$ *of* U_N *is an orthonormal basis, i.e., its vectors have norm 1 and are mutually orthogonal with respect to the Hermitian form* $\langle ?, ? \rangle$.

Proof These equations are direct applications of the orthogonality formulas in exercise 183. $\qquad\square$

38.4 Fourier's Theorem

Definition 228 *Let* $n \in \mathbb{Z}$. *Then the map*

$$\hat{?}_n : PC^1_{2\pi}(\mathbb{R}, \mathbb{C}) \to \mathbb{C} : f \mapsto \hat{f}_n = \langle f, e_n \rangle = \frac{1}{2\pi} \int_0^{2\pi} f(x) \cdot e^{-inx} \, dx$$

is $\mathbb{C}$-*linear, and the value* $\hat{f}_n$ *is called the* n-*th* Fourier coefficient *of* f. *For* $N \in \mathbb{N}$, *the* $\mathbb{C}$-*linear map*

$$\mathcal{F}_N : PC^1_{2\pi}(\mathbb{R}, \mathbb{C}) \to U_N : f \mapsto \mathcal{F}_N(f) = \sum_{-N}^{N} \hat{f}_n \cdot e_n = \sum_{-N}^{N} \langle f, e_n \rangle \cdot e_n$$

is idempotent: $(\mathcal{F}_N)^2 = \mathcal{F}_N$, *and onto* U_N. *The image* $\mathcal{F}_N(f)$ *of a function* $f \in PC^1_{2\pi}(\mathbb{R}, \mathbb{C})$ *is called the* N-*th partial sum of the Fourier series of* f, *while the universally guaranteed limit*

$$\mathcal{F}_\infty(f) = \sum_{\lim} \hat{f}_n \cdot e_n$$

of the partial sums $\mathcal{F}_N(f)$ *of* f *in* U_∞ *is called the* Fourier series *of* f. *The Fourier series map*

$$\mathcal{F}_\infty : PC^1_{2\pi}(\mathbb{R}, \mathbb{C}) \to U_\infty$$

is also a $\mathbb{C}$-*linear function for the component-wise vector-space structure on* U_∞.

The final question is whether we can reconstruct functions from their Fourier series. We need the following auxiliary construction: If $f \in PC^1_{2\pi}(\mathbb{R}, \mathbb{C})$, then there may be some points of discontinuity. What values at those points can we derive from the information given by Fourier series information at those points? By virtue of integrating over the whole period, we can evidently not expect to obtain the delicate information about the behavior of f at such critical points. But we have this information: If x is a point of the partition $P \in Part(0, 2\pi)$ for f, then there is a smooth function f_1 defined on the closed interval ending with x, and there is a smooth function f_2 defined on the closed interval beginning at x. Evidently, $f_1(x) = \lim_{y<x, y \to x} f_1(y) = \lim_{y<x, y \to x} f(y)$, while $f_2(x) = \lim_{y>x, y \to x} f_2(y) = \lim_{y>x, y \to x} f(y)$, so the values $f_1(x)$ and $f_2(x)$ depend only on f and not on the smooth auxiliary functions f_1 and f_2. Define the value $^\alpha f(x) = (f_1(x) + f_2(x))/2$. For a point which is not a splitting point of the partition I, define $^\alpha f(x) = f(x)$. This function $^\alpha f \in PC^1_{2\pi}(\mathbb{R}, \mathbb{C})$ is what one can recover from the Fourier series.

Exercise 184 Show that the map

$$^\alpha? : PC^1_{2\pi}(\mathbb{R}, \mathbb{C}) \to PC^1_{2\pi}(\mathbb{R}, \mathbb{C}) : f \mapsto {}^\alpha f$$

is a linear idempotent map. The sub-vector space $CC^1_{2\pi}(\mathbb{R}, \mathbb{C})$ of continuous piece-wise smooth 2π-periodic functions is left pointwise invariant by this map.

Proposition 328 (Fourier's Theorem) *If $f \in PC^1_{2\pi}(\mathbb{R}, \mathbb{C})$, and if $x \in \mathbb{R}$, then $\lim_{N \to \infty} \mathcal{F}_N(f)(x)$ exists, and equals ${}^\alpha f(x)$. This limit is denoted by $\sum_{-\infty}^{\infty} \hat{f}_n \cdot e_n(x)$. So we have*

$$^\alpha f(x) = \sum_{-\infty}^{\infty} \hat{f}_n \cdot e_n(x).$$

In other words, we have a function $\sum_{-\infty}^{\infty} \hat{f}_n \cdot e_n$, which evaluates to $\sum_{-\infty}^{\infty} \hat{f}_n \cdot e_n(x)$ at each $x \in \mathbb{R}$. In particular, if f is also continuous, then

$$f = \sum_{-\infty}^{\infty} \hat{f}_n \cdot e_n.$$

This means that $\mathcal{F}_\infty : CC^1_{2\pi}(\mathbb{R}, \mathbb{C}) \to U_\infty$ is injective and that f is calculated from $\mathcal{F}_\infty(f) = \sum_{\lim} \hat{f}_n \cdot e_n$ by the limit function $\sum_{-\infty}^{\infty} \hat{f}_n \cdot e_n$. Sometimes, these two objects are confused, but we stress that the difference is real, not just notational. However, for continuous piece-wise smooth functions, the difference does not matter.

Proof We shall not give a complete proof, but describe essential steps thereof, which can be stated in a compact way. We first need an informative expression of the remainder $\mathcal{F}_N(f)(x) - {}^\alpha f(x)$ at a given point x. Then this remainder is shown to converge to zero by use of what is called Riemann's lemma. We restrict our proof sketch to a C^1 function f in order to avoid special case manipulations for non-differentiable points. To describe the remainder, we need this function

$$g(t) = \frac{f(x+t) - f(x)}{2\sin(t/2)} \text{ for } t \in [0, 2\pi],$$

which is continuous on $[0, 2\pi]$ and converges to $\lim_{t \to 0} \frac{f(x+t)-f(x)}{2t/2} \cdot \frac{2t/2}{2\sin(t/2)} = f'(0) \cdot 1$ according to what we know from the proof of proposition 266. Then it is shown that the remainder has this form:

$$\mathcal{F}_N(f)(x) - f(x) = \frac{1}{\pi} \int_0^{2\pi} g(t) \sin\left(t(N + \tfrac{1}{2})\right) dt.$$

Riemann's lemma states that $\lim_{\lambda \to 0} \int_a^b g(t) \sin(\lambda t) dt = 0$ if $g : [a, b] \to \mathbb{C}$ is piecewise continuous. So the remainder converges to 0 as $N \to \infty$. Here is the proof of Riemann's lemma. We may clearly suppose that g is continuous and then treat the general case by adding up the continuous parts of g. Substituting t by $t = \tau + \frac{\pi}{\lambda}$ we have $\sin(\lambda t) = \sin(\lambda \tau + \pi) = -\sin(\lambda \tau)$. Setting $L_\lambda = \int_a^b g(t)\sin(\lambda t)dt$, we obtain $L_\lambda = -\int_{a-\frac{\pi}{\lambda}}^{b-\frac{\pi}{\lambda}} g(\tau + \frac{\pi}{\lambda})\sin(\lambda \tau)dt$. Choosing λ large enough to obtain $a < b - \frac{\pi}{\lambda} < b$, we have

$$2L_\lambda = \int_{b-\frac{\pi}{\lambda}}^{b} g(t)\sin(\lambda t)dt - \int_{a-\frac{\pi}{\lambda}}^{a} g(t+\tfrac{\pi}{\lambda})\sin(\lambda t)dt +$$

$$\int_a^{b-\frac{\pi}{\lambda}} \left(g(t) - g(t+\tfrac{\pi}{\lambda})\right)\sin(\lambda t)dt.$$

Setting $s(\lambda) = \sup\{|g(t)-g(t')| \mid t,t' \in [a,b], |t-t'| \le \frac{\pi}{\lambda}\}$ and $y = \sup\{|g(t)| \mid t \in [a,b]\}$, we obtain

$$2|L_\lambda| \le 2y\frac{\pi}{\lambda} + s(\lambda)(b-a).$$

But $s(\lambda) \to 0$ if $\lambda \to \infty$ since $|g(t) - g(t')|$ vanishes on the diagonal $t = t'$ of $[a,b] \times [a,b]$, and becomes arbitrary small in a small neighborhood thereof. Therefore $L_\lambda \to 0$ if $\lambda \to \infty$. $\qquad\square$

There is a beautiful a posteriori justification for the Fourier series $\mathcal{F}_\infty(f)$ of a function f in geometrical terms. Given $N \in \mathbb{N}$, consider the orthogonal subspace $U_N^\perp$. Let $f \in PC_{2\pi}^1(\mathbb{R}, \mathbb{C})$ and set $G_N(f) = f - \mathcal{F}_N(f)$ (an element of $Ker(\mathcal{F}_N)$, by the idempotency of $\mathcal{F}_N$). Clearly, $G_N(f) \perp e_n$ for $|n| \le N$, i.e., $G_N(f) \in U_N^\perp$. Therefore $f = G_N(f) + \mathcal{F}_N(f)$ is an *orthogonal decomposition* of f, i.e., $\mathcal{F}_N(f) \in U_N$ and $G_N(f) \in U_N^\perp$. But evidently, $U_N \cap U_N^\perp = 0$. This means that we have a direct *orthogonal* decomposition

$$PC_{2\pi}^1(\mathbb{R}, \mathbb{C}) = U_N \oplus U_N^\perp.$$

The special role of $\mathcal{F}_N(f)$ is made evident by this result:

Proposition 329 *With the above notations, if $f \in PC_{2\pi}^1(\mathbb{R}, \mathbb{C})$, and if $F \in U_N$, then $\|f - F\| \ge \|f - \mathcal{F}_N(f)\|$ and equality holds iff $F = \mathcal{F}_N(f)$. Moreover, $\lim_{N \to \infty} \|f - \mathcal{F}_N(f)\| = 0$, or, equivalently, we have* Parseval's equation, *a kind of infinite "Pythagorean theorem":*

$$\sum_{n=-\infty}^{\infty} (\hat{f}_n)^2 = \|f\|^2.$$

Proof We have $\|f - F\|^2 = \|f - \mathcal{F}_N(f)\|^2 + \|F - \mathcal{F}_N(f)\|^2 + 2Re(\langle f - \mathcal{F}_N(f), F - \mathcal{F}_N(f)\rangle)$. But $(f - \mathcal{F}_N(f)) \perp (F - \mathcal{F}_N(f))$, therefore $\|f - F\|^2 = \|f - \mathcal{F}_N(f)\|^2 + \|F - \mathcal{F}_N(f)\|^2$, i.e., $\|f - F\| \ge \|f - \mathcal{F}_N(f)\|$ and equality holds iff $F = \mathcal{F}_N(f)$. For the proof of Parseval's equation, we refer to [14], section 4.7. $\qquad\square$

Intuitively, this proposition states that the N-th partial sum of the Fourier series of f is the nearest point to f in U_N.

38.5 Restatement in Terms of the Sine and Cosine Functions

This last part of the general theory deals with the restatement of Fourier's theorem in terms of trigonometric functions, and in particular deals with Fourier series of real-valued periodic functions.

The alternate representation of Fourier series stems from a base change on U_N induced by base changes on the subspaces $D_n = \mathbb{C}e_n \oplus \mathbb{C}e_{-n}$ for $n > 0$ and $D_0 = \mathbb{C}e_0$. The new basis on D_n is defined by the well-known Euler formulas, see also proposition 257:

$$\cos(nx) = \frac{e^{inx} + e^{-inx}}{2} \quad \text{and} \quad \sin(nx) = \frac{e^{inx} - e^{-inx}}{2i}.$$

In other words, given a Fourier series

$$\sum_{n=-N}^{N} c_n e^{inx},$$

the question is whether it can be rewritten as

$$\sum_{n=0}^{N} a_n \cos(nx) + b_n \sin(nx).$$

Here are the corresponding formulas in terms of the base change

$$(e_n(x), e_{-n}(x)) \mapsto (\cos(nx), \sin(nx))$$

for $n \neq 0$, and $(e_0 = 1) \mapsto (2\cos(0x) = 2)$. The latter is in conformance with the general formula, which, for all $n \geq 0$, reads

$$a_n = c_n + c_{-n} \quad \text{and} \quad b_n = i(c_n - c_{-n}),$$

in particular, $a_0 = 2c_0$ and $b_0 = 0$. Conversely, one has, for all $n \geq 0$,

$$c_n = \frac{a_n - ib_n}{2} \quad \text{and} \quad c_{-n} = \frac{a_n + ib_n}{2}.$$

This yields, for every D_n, $n > 0$,

$$c_n e^{inx} + c_{-n} e^{-inx} = a_n \cos(nx) + b_n \sin(nx)$$

With these base change translation formulas, the functions f are approximated by

$$\sum_{n=-N}^{N} c_n e^{inx} = \frac{a_0}{2} + \sum_{n=1}^{N} a_n \cos(nx) + b_n \sin(nx),$$

while the limit representation is

$$\mathcal{F}_\infty(f) = \sum_{\text{lim}} c_n e_n = \frac{a_0}{2} + \sum_{\text{lim}} a_n \cos(n \cdot ?) + b_n \sin(n \cdot ?),$$

or for the functions

$${}^{\alpha}f(x) = \sum_{n=-\infty}^{\infty} c_n e^{inx} = \frac{a_0}{2} + \sum_{n=1}^{\infty} a_n \cos(nx) + b_n \sin(nx).$$

By the uniqueness of the coefficients c_n and their integral formula $c_n = \hat{f}_n$ (see definition 228), f is real iff $c_{-n} = \overline{c_n}$ for all $n \geq 0$. This condition is equivalent to a_n and b_n being real, as can be seen from the transformation formulas given above. The formula $c_{-n} = \overline{c_n}$ is the classical Fourier statement for real-valued 2π-periodic piecewise smooth functions. For real-valued functions, there is a frequently used equivalent statement using phase quantities, which follows from the trigonometric relation $\sin(a + b) = \cos(a)\sin(b) + \sin(a)\cos(b)$:

$$\frac{a_0}{2} + \sum_{n=1}^{\infty} a_n \cos(nx) + b_n \sin(nx) = \frac{a_0}{2} + \sum_{n=1}^{\infty} A_n \sin(nx + P_n)$$

with

$$A_n = \sqrt{a_n^2 + b_n^2},$$

$$P_n = \begin{cases} \arctan(b_n/a_n) & \text{if } a_n \neq 0, \\ 0 & \text{if } b_n = 0 \text{ and } a_n = 0, \\ \frac{\pi}{2} & \text{if } b_n > 0 \text{ and } a_n = 0, \\ -\frac{\pi}{2} & \text{if } b_n < 0 \text{ and } a_n = 0. \end{cases}$$

The sequence $(A_n)_n$ is called the *amplitude spectrum*, and the sequence $(P_n)_n$ is called the *phase spectrum* of f. The function $A_n \sin(nx + P_n)$ is called the *n-th partial* or *harmonic* of the Fourier series, the first partial is also called the *fundamental* of the series.

Example 180 From given amplitude and phase spectra, a periodic function can be constructed. Let $a_0 = 0$, $(A_n)_n = (1/n)_n$ and $(P_n)_n = (0)_n$. The amplitude spectrum $(A_n)_n$ is shown in figure 38.3.

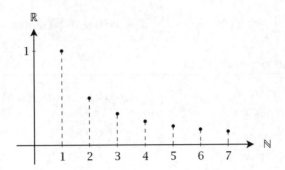

Fig. 38.3. An amplitude spectrum $(A_n)_n$ where $A_n = 1/n$.

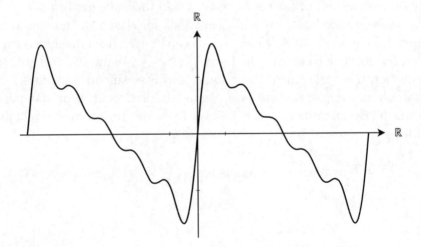

Fig. 38.4. The function described by $f(x) = \sum_{n=1}^{5} \frac{1}{n} \sin(nx)$.

The function $f(x) = \sum_{n=1}^{5} A_n \sin(nx + P_n)$, i.e., the Fourier series with five terms, is shown in figure 38.4. It is apparent that f is an approximation of a sawtooth curve.

Keeping the amplitude spectrum $(A_n)_n$, but replacing $(P_n)_n$ by $(P'_n)_n = (n^2)_n$, we get the function reproduced in figure 38.5.

This technique of proceeding from amplitude and phase spectra to the corresponding function is called *Fourier synthesis*.

If we have a more general period p instead of 2π, then the formulas are these: Call $\nu = 1/p$ the frequency of $f \in PC_p^1(\mathbb{R}, \mathbb{C})$. Then we have the representation

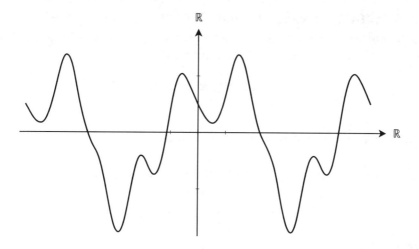

Fig. 38.5. The function described by $f(x) = \sum_{n=1}^{5} \frac{1}{n} \sin(nx + n^2)$. This function looks quite different from figure 38.4, but its amplitude spectrum is the same, while its phase spectrum differs.

$$^{\alpha}f(x) = \sum_{n=-\infty}^{\infty} c_n e^{i2\pi nvx}$$

$$= \frac{a_0}{2} + \sum_{n=1}^{\infty} a_n \cos(2\pi nvx) + b_n \sin(2\pi nvx).$$

And for real-valued functions:

$$^{\alpha}f(x) = \frac{a_0}{2} + \sum_{n=1}^{\infty} A_n \sin(2\pi nvx + P_n)$$

where the coefficients and the members of the amplitude and phase spectra are calculated by

$$a_n = 2v \int_0^p f(x) \cos(2\pi nvx)\, dx \quad n \geq 0,$$

$$b_n = 2v \int_0^p f(x) \sin(2\pi nvx)\, dx \quad n > 0,$$

$$A_n = \sqrt{a_n^2 + b_n^2},$$

$$P_n = \begin{cases} \arctan(b_n/a_n) & \text{if } a_n \neq 0, \\ 0 & \text{if } b_n = 0 \text{ and } a_n = 0, \\ \frac{\pi}{2} & \text{if } b_n > 0 \text{ and } a_n = 0, \\ -\frac{\pi}{2} & \text{if } b_n < 0 \text{ and } a_n = 0. \end{cases}$$

Example 181 We take up the example of figure 38.2 (b), the rectangle curve, and compute the Fourier coefficients a_n and b_n. Consider a particular 1-periodic rectangle function (figure 38.7 (d)):

$$f(x) = \begin{cases} 1 & \text{for } n \le x < \frac{1}{2} + n, n \in \mathbb{Z}, \\ -1 & \text{for } \frac{1}{2} + n \le x < 1 + n, n \in \mathbb{Z}. \end{cases}$$

Then $v = 1/p = 1$, and we have the following equations:

$$a_n = 2 \int_0^1 f(x) \cos(2\pi nx)\, dx \quad n \ge 0,$$

$$b_n = 2 \int_0^1 f(x) \sin(2\pi nx)\, dx \quad n > 0.$$

For $n = 0$ in particular:

$$a_0 = 2 \int_0^{\frac{1}{2}} \cos(2\pi 0x)\, dx + 2 \int_{\frac{1}{2}}^1 -\cos(2\pi 0x)\, dx$$

$$= 2 \int_0^{\frac{1}{2}} -1\, dx + 2 \int_{\frac{1}{2}}^1 1\, dx = 0$$

and, in general, for $n > 0$:

$$a_n = 2 \int_0^{\frac{1}{2}} -\cos(2\pi nx)\, dx + 2 \int_{\frac{1}{2}}^1 \cos(2\pi nx)\, dx$$

$$= \frac{1}{\pi n}(\sin(2\pi n) - 2\sin(\pi n)),$$

$$b_n = 2 \int_0^{\frac{1}{2}} -\sin(2\pi nx)\, dx + 2 \int_{\frac{1}{2}}^1 \sin(2\pi nx)\, dx$$

$$= \frac{s}{\pi n}(\cos(\pi n) - 1).$$

We calculate the coefficients a_n and b_n for $n = 1, 2, 3$:

$$a_1 = 0, \quad b_1 = -\frac{4}{\pi},$$

$$a_2 = 0, \quad b_2 = 0,$$

$$a_3 = 0, \quad b_3 = -\frac{4}{3\pi}.$$

Generally, $a_n = 0$ for all n and

$$b_n = \begin{cases} -\frac{4}{\pi n} & \text{if } n \text{ odd,} \\ 0 & \text{if } n \text{ even.} \end{cases}$$

We have thus the approximation

$$\mathcal{F}_3(f)(x) = \frac{a_0}{2} + \sum_{n=1}^{3} a_n \cos(2\pi n x) + b_n \sin(2\pi n x)$$

$$= -\frac{4}{\pi} \sin(2\pi x) - \frac{4}{3\pi} \sin(6\pi x).$$

The corresponding amplitude and phase spectra are given by:

$$A_n = \begin{cases} \frac{4}{\pi n} & \text{if } n \text{ odd,} \\ 0 & \text{if } n \text{ even,} \end{cases}$$

$$P_n = \begin{cases} \pi & \text{if } n \text{ odd,} \\ 0 & \text{if } n \text{ even.} \end{cases}$$

The amplitude spectrum $(A_n)_n$ is shown in figure 38.6.

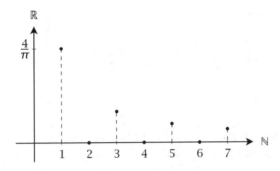

Fig. 38.6. Amplitude spectrum $(A_n)_n$.

The Fourier series for f is given by

$$^{\alpha}f(x) = \sum_{n=0}^{\infty} \frac{4}{\pi(2n+1)} \sin(2\pi(2n+1)x + \pi).$$

See figure 38.7 for $\mathcal{F}_3(F)$, $\mathcal{F}_5(f)$ and $\mathcal{F}_7(f)$. The wobbling at the ridges is an indicator of the discontinuities of the approximated function f. As terms of higher order are added, the wobbling gets more intensive.

This technique of calculating the coefficients a_n and b_n (or, equivalently, the frequency spectrum $(A_n)_n$ and phase spectrum $(P_n)_n$) is called *Fourier analysis*. In digital signal processing (DSP), where the focus is on time-dependent functions, Fourier analysis is also known as passing from the time domain to the frequency domain.

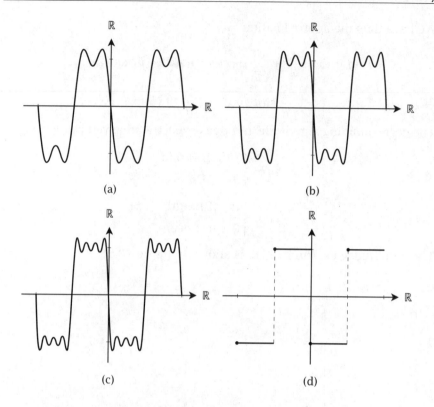

Fig. 38.7. Approximations to the rectangle function f: (a) $\mathcal{F}_3(f)$, (b) $\mathcal{F}_5(f)$, (c) $\mathcal{F}_7(f)$, and (d) f itself.

38.6 Finite Fourier Series and Fast Fourier Transform

For practical purposes, the general Fourier theory may appear to be over-loaded for two reasons: First, concrete functions which are to be be represented by Fourier series are virtually never periodic, and second, the functions to be represented are given only as a finite set of values. How are such obstructions overcome?

Non-periodic functions can be interpreted as having a period that tends to infinity. This means that we have to extend the theory from finite to infinite periods. This branch leads to the so-called *Fourier transform* $\hat{f}$ of a function f, an instance of which was already alluded to in the Fourier coefficients $\hat{f}_n$. We give a short overview of this theory in section 38.8.

The finiteness constraint leads to finite Fourier series, a theory which is also applicable to the computerized Fourier theory as it is implemented in the Fast Fourier Transform (FFT) algorithm. We shall briefly describe this portion of the practical application of Fourier's theory, because it plays an omnipresent role in the theory of digital signal processing, and more specifically in image and sound analysis for multimedia science. For example, the popular MP3 audio compression algorithms are based on Fourier theory.

The concrete situation is this: One determines the values $f(x)$ on a finite set $A \subset \mathbb{R}$ of arguments of a function $f : \mathbb{R} \to \mathbb{C}$ and would like to find a finite Fourier series S (an element of U_N) such that the series coincides with the function on all measured arguments, i.e., $f(x) = S(x)$, for all $x \in A$. Of course, the measured function is by no means periodic, i.e., the finite measurements cannot imply any (finite) period. The idea therefore is to interpret the measured values as if they they were derived from a periodic function, and then apply the Fourier method to these data.

As we shall apply this setup to the FFT algorithm, we need to have a very special distribution of the arguments from A as follows. To fix the ideas, we assume that $A = Z_N = \{\frac{k}{N} \mid k = 0, 1, \ldots N - 1\}$—but this is simply a normalized setup which could be adopted to a realistic size without any difficulty. It means that one measures N values $f_k = f(k/N) \in \mathbb{C}$ with a *sampling period* of $\Delta = 1/N$, corresponding to a *sampling frequency* of $1/\Delta = N$. Moreover, the FFT algorithm requires that $N = 2^M$, but at present, we can get by with a general N and specialize later on. One then looks for a representation

$$f_k = \sum_{n=0}^{N-1} y_n e^{i2\pi n \frac{k}{N}}, \quad k = 0, 1, 2, \ldots N - 1,$$

with frequency $v = 1$. This means that the fundamental frequency is defined by the period 1, the total length of the N sampling periods $N \cdot \Delta = 1$. In other words, the function f is cooked down to a periodic function of period $p = 1$, or, in other words, the trace of the function f on those N measured points is extrapolated in a 1-periodic way to the left and to the right of the closed unit interval $[0, 1]$. This representation needs a word of caution concerning the intervening powers $n = 0, 1, \ldots N - 1$ of the fundamental function $e^{i2\pi \frac{k}{N}}$. We know from the discussion above that a trigonometric series is real-valued iff the coefficients c_n are conjugate under change of index sign. The above formula looks as if it contained positive powers only. However, this impression is erroneous, since we have $e^{i2\pi\frac{n}{N}} \cdot e^{i2\pi\frac{N-n}{N}} = 1$, i.e., the second factor is the inverse $e^{i2\pi\frac{-n}{N}}$ of the first one, and is related to $-n$ instead of n, as required in the symmetric Fourier series formula. In most practical cases, the number N is even, $N = 2M$. The highest frequency in the sinusoidal representation is then $N/2 = M$, half the sampling frequency. This frequency is also called the *Nyquist* frequency.

The remainder of this discussion is devoted to the solution of the above formula, i.e., the proof that there is exactly one sequence $(y_n)_{n=0,1,...N-1}$ yielding the required function values f_k. To this end, denote by ϵ_n the functions $\epsilon_n(k) = e^{i2\pi n\frac{k}{N}}$ defined on $\mathbb{Z}_N$ (in fact: if $k = k'$ mod N, then $\epsilon_n(k) = \epsilon_n(k')$). We also work on $\mathbb{Z}_N$ concerning the given function f, i.e., we are given a set function $f : \mathbb{Z}_N \to \mathbb{C} : k \mapsto f_k$, an element in the N-dimensional $\mathbb{C}$-vector space $\mathbb{C}^{\mathbb{Z}_N}$. On this space, we have a Hermitian form (a discrete version of the form $\langle ?, ? \rangle$ known from the Fourier theory)

$$\langle g, h \rangle = \frac{1}{N} \sum_{k=0}^{N-1} g(k)\overline{h(k)}$$

with associated norm $\|h\| = \frac{1}{N} \sum_{k=0}^{N-1} |g(k)|^2$.

Proposition 330 *For any two $m, l = 0, 1, \ldots N - 1$, we have*

$$\langle \epsilon_m, \epsilon_l \rangle = \delta_{ml} \quad (Kronecker).$$

Exercise 185 Give a proof of proposition 330. For the case $m \neq l$, use the fact that $1 + q + q^2 + \ldots q^{N-1} = \frac{1-q^N}{1-q}$ and apply this to $q = e^{i2\pi\frac{m-l}{N}}$.

The following corollary yields the representation of f at N given arguments in terms of the given basis as required above.

Corollary 331 *On $\mathbb{C}^{\mathbb{Z}_N}$, the sequence $(\epsilon_n)_{n=0,1,...N-1}$ is an orthonormal basis, and each ϵ_n is a group homomorphism $\epsilon_n : \mathbb{Z}_N \to U$, where U is the multiplicative group of complex numbers z with norm $|z| = 1$. A function $f \in \mathbb{C}^{\mathbb{Z}_N}$ has the representation*

$$f = \sum_{n=0}^{N-1} y_n \epsilon_n,$$

with uniquely determined Fourier coefficients

$$y_n = \langle f, \epsilon_n \rangle = \frac{1}{N} \sum_{k=0}^{N-1} f_k \cdot e^{-i2\pi n \frac{k}{N}},$$

and we have Parseval's equation

$$\|f\|^2 = \sum_{n=0}^{N-1} |y_n|^2 = \sum_{n=0}^{N-1} |\langle f, \epsilon_n \rangle|^2.$$

Example 182 Given is a function f defined at points $0, 1, \ldots 7$ by $f_k = 1$ for $k \in \{0, 1, 2, 3\}$ and $f_k = 0$ for $k \in \{4, 5, 6, 7\}$. The finite Fourier transform assumes that f is effectively a function of period 8. Figure 38.8 shows f and its periodic continuation.

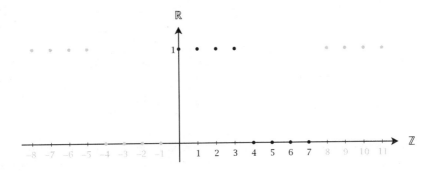

Fig. 38.8. The function f from example 182 (black), and its periodic continuation (gray).

The Fourier transform of f involves calculating the coefficient y_k, for $k \in \{0, 1, \ldots 7\}$:

$$y_n = \frac{1}{8} \sum_{k=0}^{7} f_k \cdot e^{-i2\pi n \frac{k}{8}} = \frac{1}{8}\left(1 + e^{-i\pi n \frac{1}{4}} + e^{-i\pi n \frac{1}{2}} + e^{-i\pi n \frac{3}{4}}\right).$$

This translates into the following equations, after applying some trigonometric transformations:

$$y_0 = \frac{1}{2}, \quad y_1 = \frac{1}{8} + i\left(-\frac{1}{8} - \frac{1}{4\sqrt{2}}\right),$$

$$y_2 = 0, \quad y_3 = \frac{1}{8} + i\left(\frac{1}{8} - \frac{1}{4\sqrt{2}}\right),$$

$$y_4 = 0, \quad y_5 = \frac{1}{8} + i\left(-\frac{1}{8} + \frac{1}{4\sqrt{2}}\right),$$

$$y_6 = 0, \quad y_7 = \frac{1}{8} + i\left(\frac{1}{8} + \frac{1}{4\sqrt{2}}\right).$$

Recall that a complex number z can be written as $z = |z|e^{i \cdot arg(z)}$, where $-\pi < arg(z) \leq \pi$. In figure 38.9, both spectra $|y_k|$ and $arg(y_k)$ are shown. The previous discussion has emphasized, that y_5, y_6 and y_7 are the complex conjugates of y_3, y_2, and y_1, respectively. This can be seen in the amplitude spectrum, where the symmetry about the point 4 is obvious, and in the phase spectrum, where the values for points greater than 4 are the negatives of the points less than 4.

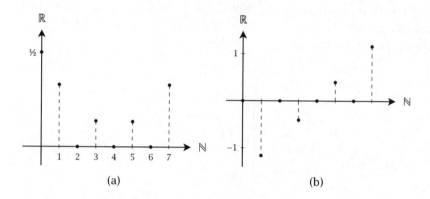

Fig. 38.9. The amplitude spectrum $|y_k|$ (a), and phase spectrum $abs(y_k)$ (b), of f.

38.7 Fast Fourier Transform (FFT)

The FFT algorithm is a way to speed up the calculation of Fourier co-efficients for finite Fourier analysis. In order to describe the notion of calculation speed in a precise way, one introduces the Landau symbols:

Definition 229 (Landau's O Symbol) *For two sequences $c = (c_n)_n$ and $d = (d_n)_n \in \mathbb{R}^{\mathbb{N}}$, one says that d is of order O of c, in signs $d \leq_O c$, iff there is a positive real number λ and a natural number L such that $d_n \leq \lambda c_n$ for $n > L$. This relation is reflexive and transitive, but not antisymmetric. We define $O(c) = \{d \mid d \leq_O c\}$, i.e., d is of order O of c, iff $d \in O(c)$. The equivalence relation $\leq_O \cap \leq_O^{-1}$ is denoted by $\sim_O$. The equivalence class of a sequence c under $\sim_O$ is denoted by $\Theta(c)$. The $\sim_O$-equivalence classes are called* growth classes. *The relation $\leq_O$ induces a partial ordering on growth classes, which is denoted by $\leq$. In particular,*

$\Theta(c) = \Theta(d)$ *iff there are positive real numbers* λ, μ *and a natural number* L *such that* $\lambda c_n \le d_n \le \mu c_n$ *for* $n > L$.

Exercise 186 Show that $\Theta(256 \cdot n^2 + 50 \cdot n) < \Theta(n^3)$. Show that for every polynomial $p \in \mathbb{R}[X]$, $\Theta(p(n)) < \Theta(e^n)$. Show that for every non-constant polynomial $p = a_k X^k + a_{k-1} X^{k-1} + \ldots a_0 \in \mathbb{R}[X]$ with positive leading coefficient a_k, we have $\log(n) <_o p(n)$. Show that for any two polynomials $p, q \in \mathbb{R}[X]$ with positive leading coefficients, we have $\Theta(p) \le \Theta(q)$ iff $deg(p) \le deg(q)$. Show that $O(n \log(n)) < O(p(n))$ for a quadratic polynomial with positive leading coefficient.

Let us now estimate the growth class of the number of arithmetic operations needed to calculate the Fourier coefficients

$$y_n = \frac{1}{N} \sum_{k=0}^{N-1} f_k \cdot e^{-i2\pi n \frac{k}{N}}.$$

We suppose that the quantities $f_0, f_1 \ldots f_{N-1}$ as well as the basic value $\epsilon(N) = e^{-i2\pi \frac{1}{N}}$ are given. To calculate all y_n, one needs at most $N - 2$ multiplications to calculate the powers $\epsilon(N)^2, \ldots \epsilon(N)^{N-1}$, and for each n, one needs N multiplications to get the $f_k \cdot \epsilon(N)^{kn}$, further $N-1$ additions, and one multiplication by $\frac{1}{N}$, so we need at most $q(N) = 2N^2 + N$ arithmetic operations to get all coefficients y_n. The FFT algorithm manages to reduce this growth class $O(q(N)) = O(N^2)$ to the class $O(\log(N)N) \subsetneq O(N^2)$ for the special values $N = 2^n$ of dyadic powers.

Proposition 332 *Let* $N = 2^n$. *Suppose that the quantity* $\epsilon(N) = e^{-i2\pi \frac{1}{N}}$ *as well as the values* $f_0, f_1 \ldots f_{N-1}$ *of* $f \in \mathbb{C}^{\mathbb{Z}_N}$ *are given. Then the calculation of the Fourier coefficients* $y_n = \langle f, \epsilon_n \rangle$ *takes at most* $4 \cdot n \cdot 2^n = 4N \log_2(N)$ *operations, i.e., if Fourier(N) denotes the growth function of the minimal number of arithmetic operations (addition, multiplication) needed to calculate all the* y_n, *then we have*

$$\Theta(Fourier(N)) \le \Theta(N \log(N)).$$

To begin with, the case $N = 2$ is straightforward. In fact,

$$y_0 = \frac{1}{2}(f_0 + f_1), \qquad y_1 = \frac{1}{2}(f_0 + (-1)f_1),$$

which needs less than $4 \cdot 1 \cdot 2^1 = 8$ operations. The proof relies on a lemma which manages the recursive step from *Fourier(N)* to *Fourier$(2N)$*.

Lemma 333 *Given all the quantities $\epsilon(N)$, we have*

$$Fourier(2N) \le 2Fourier(N) + 8N.$$

With this lemma, and the induction hypothesis that for $N = 2^{n-1}$, one needs at most $4 \cdot (n-1) \cdot 2^{(n-1)}$ operations, it follows that for $N = 2^n$, we need at most $2 \cdot 4 \cdot (n-1) \cdot 2^{(n-1)} + 8 \cdot 2^{n-1} = 4 \cdot n \cdot 2^n$ operations, and the proposition follows.

But the lemma is more than an auxiliary result, its very content is implemented in the calculation routines, so we should make it explicit in its own right, instead of hiding it in a proof section. The main idea is a recursion by splitting the calculation of the $2N$ Fourier coefficients y_n of $f \in \mathbb{C}^{\mathbb{Z}_{2N}}$ into two subcalculations of auxiliary functions $f^+, f^- \in \mathbb{C}^{\mathbb{Z}_N}$, defined by

$$f_k^+ = f_{2k} \quad k = 0, 1, 2, \ldots N,$$
$$f_k^- = f_{2k+1} \quad k = 0, 1, 2, \ldots N.$$

By induction hypothesis, it is assumed that we have already calculated the coefficients

$$y_0^+, y_1^+, \ldots y_{N-1}^+,$$
$$y_0^-, y_1^-, \ldots y_{N-1}^-.$$

of f^+ and f^-, respectively, each requiring not more than $Fourier(N)$ arithmetic operations. Then we have

$$y_k = \frac{1}{2}(y_k^+ + \epsilon(2N)^k \cdot y_k^-), \quad k = 0, 1, \ldots 2N - 1 \qquad (*)$$

where one sets $y_{k+N}^{\pm} = y_k^{\pm}$.

Exercise 187 Give a proof of the preceding formula.

Now, knowing the $2N$ coefficients $y_0^{\pm}, y_1^{\pm}, \ldots y_{N-1}^{\pm}$, the above formula needs three operations (two multiplications and one addition) for each $0 \le k < 2N$, then at most $2N$ multiplications for the powers $\epsilon(2N)^k$, and not more than $2Fourier(N)$ calculations for the two auxiliary functions. This means that

$$Fourier \le 2N \cdot 3 + 2N + 2 \cdot Fourier(N) = 2Fourier(N) + 8N,$$

and we are done.

Example 183 Consider the case of a function given by 8 values $f_k, 0 \leq k < 8$. According to the recursion described above, the Fourier coefficients for two smaller functions, f^+ and f^- must be found, where

$$
\begin{aligned}
f_i^+ &= f_{2i}, & i = 0, 1, 2, 3 \\
f_i^- &= f_{2i+1}, & i = 0, 1, 2, 3
\end{aligned}
$$

Continuing the process, these functions are again subdivided, yielding

$$
\begin{aligned}
f_j^{++} &= f_{2j}^+ &= f_{4j}, & j = 0, 1 \\
f_j^{+-} &= f_{2j+1}^+ &= f_{4j+2}, & j = 0, 1 \\
f_j^{-+} &= f_{2j}^- &= f_{4j+1}, & j = 0, 1 \\
f_j^{--} &= f_{2j+1}^- &= f_{4j+3}, & j = 0, 1
\end{aligned}
$$

Now the function is broken down into parts with only two values each— the calculation of the coefficients can now begin by using

$$
\begin{aligned}
y_0^{++} &= \tfrac{1}{2}(f_0^{++} + f_1^{++}), & y_1^{++} &= \tfrac{1}{2}(f_0^{++} - f_1^{++}) \\
y_0^{+-} &= \tfrac{1}{2}(f_0^{+-} + f_1^{+-}), & y_1^{+-} &= \tfrac{1}{2}(f_0^{+-} - f_1^{+-}) \\
y_0^{-+} &= \tfrac{1}{2}(f_0^{-+} + f_1^{-+}), & y_1^{-+} &= \tfrac{1}{2}(f_0^{-+} - f_1^{-+}) \\
y_0^{--} &= \tfrac{1}{2}(f_0^{--} + f_1^{--}), & y_1^{--} &= \tfrac{1}{2}(f_0^{--} - f_1^{--})
\end{aligned}
$$

After having completed the calculations for $N = 2$, one can proceed to the case $N = 4$ in which the new coefficients are calculated from the old coefficients, following equation $(*)$. This yields

$$
\begin{aligned}
y_0^+ &= \tfrac{1}{2}(y_0^{++} + \epsilon(4)^0 \cdot y_0^{+-}), & y_1^+ &= \tfrac{1}{2}(y_1^{++} + \epsilon(4)^1 \cdot y_1^{+-}) \\
y_2^+ &= \tfrac{1}{2}(y_0^{++} + \epsilon(4)^2 \cdot y_0^{+-}), & y_3^+ &= \tfrac{1}{2}(y_1^{++} + \epsilon(4)^3 \cdot y_1^{+-}) \\
y_0^- &= \tfrac{1}{2}(y_0^{-+} + \epsilon(4)^0 \cdot y_0^{--}), & y_1^- &= \tfrac{1}{2}(y_1^{-+} + \epsilon(4)^1 \cdot y_1^{--}) \\
y_2^- &= \tfrac{1}{2}(y_0^{-+} + \epsilon(4)^2 \cdot y_0^{--}), & y_3^- &= \tfrac{1}{2}(y_1^{-+} + \epsilon(4)^3 \cdot y_1^{--})
\end{aligned}
$$

Note that the indexes of the coefficients have been reduced by the rule $y_{k+N}^\pm = y_k^\pm$. Now the final step can be performed:

$$
\begin{aligned}
y_0 &= \tfrac{1}{2}(y_0^+ + \epsilon(8)^0 \cdot y_0^-), & y_1 &= \tfrac{1}{2}(y_1^+ + \epsilon(8)^1 \cdot y_1^-) \\
y_2 &= \tfrac{1}{2}(y_2^+ + \epsilon(8)^2 \cdot y_2^-), & y_3 &= \tfrac{1}{2}(y_3^+ + \epsilon(8)^3 \cdot y_3^-) \\
y_4 &= \tfrac{1}{2}(y_0^+ + \epsilon(8)^4 \cdot y_0^-), & y_5 &= \tfrac{1}{2}(y_1^+ + \epsilon(8)^5 \cdot y_1^-) \\
y_6 &= \tfrac{1}{2}(y_2^+ + \epsilon(8)^6 \cdot y_2^-), & y_7 &= \tfrac{1}{2}(y_3^+ + \epsilon(8)^7 \cdot y_3^-)
\end{aligned}
$$

This can be visualized in a so-called butterfly diagram, as in figure 38.10.

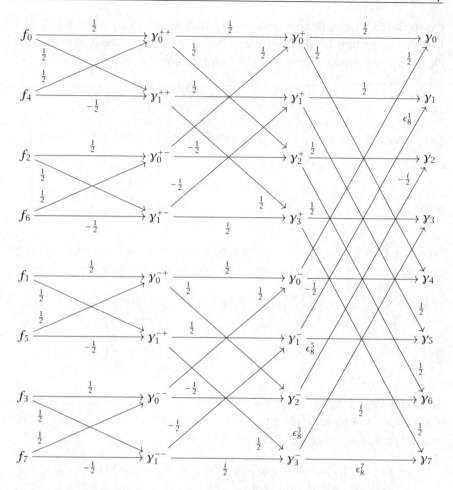

Fig. 38.10. Schema of the Fast Fourier Transform for $N = 8$. For the sake of clarity, the expressions of the form $\epsilon(4)^k$ as well as $\epsilon(8)^{2j}$ have been evaluated, whereas for odd k the abbreviations $\epsilon_8^k = \frac{1}{2}\epsilon(8)^k$ have been used.

Each y can be reached from two other nodes a and b. To calculate a factor y, one has to multiply a with the value next to the arrow joining a and y and add that to the result of multiplying b with the value next to the arrow joining a and y.

By replacing all y's with their values, making use of the fact that $\epsilon(4) = e^{-i2\pi\frac{1}{4}} = -i$, and expressing all intermediate functions by the original f:

$$y_0 = \frac{1}{2}(\frac{1}{2}(\frac{1}{2}(f_0 + f_4) + \frac{1}{2}(f_2 + f_6)) + \epsilon(8)^0\frac{1}{2}(\frac{1}{2}(f_1 + f_5) + \frac{1}{2}(f_3 + f_7)))$$

$$y_1 = \frac{1}{2}(\frac{1}{2}(\frac{1}{2}(f_0 - f_4) - \frac{i}{2}(f_2 - f_6)) + \epsilon(8)^1\frac{1}{2}(\frac{1}{2}(f_1 - f_5) - \frac{i}{2}(f_3 - f_7)))$$

$$y_2 = \frac{1}{2}(\frac{1}{2}(\frac{1}{2}(f_0 + f_4) - \frac{1}{2}(f_2 + f_6)) + \epsilon(8)^2\frac{1}{2}(\frac{1}{2}(f_1 + f_5) - \frac{1}{2}(f_3 + f_7)))$$

$$y_3 = \frac{1}{2}(\frac{1}{2}(\frac{1}{2}(f_0 - f_4) + \frac{i}{2}(f_2 - f_6)) + \epsilon(8)^3\frac{1}{2}(\frac{1}{2}(f_1 - f_5) + \frac{i}{2}(f_3 - f_7)))$$

$$y_4 = \frac{1}{2}(\frac{1}{2}(\frac{1}{2}(f_0 + f_4) + \frac{1}{2}(f_2 + f_6)) + \epsilon(8)^4\frac{1}{2}(\frac{1}{2}(f_1 + f_5) + \frac{1}{2}(f_3 + f_7)))$$

$$y_5 = \frac{1}{2}(\frac{1}{2}(\frac{1}{2}(f_0 - f_4) - \frac{i}{2}(f_2 - f_6)) + \epsilon(8)^5\frac{1}{2}(\frac{1}{2}(f_1 - f_5) - \frac{i}{2}(f_3 - f_7)))$$

$$y_6 = \frac{1}{2}(\frac{1}{2}(\frac{1}{2}(f_0 + f_4) - \frac{1}{2}(f_2 + f_6)) + \epsilon(8)^6\frac{1}{2}(\frac{1}{2}(f_1 + f_5) - \frac{1}{2}(f_3 + f_7)))$$

$$y_7 = \frac{1}{2}(\frac{1}{2}(\frac{1}{2}(f_0 - f_4) + \frac{i}{2}(f_2 - f_6)) + \epsilon(8)^7\frac{1}{2}(\frac{1}{2}(f_1 - f_5) + \frac{i}{2}(f_3 - f_7)))$$

Applying this result to example 182, where $f_0 = f_1 = f_2 = f_3 = 1$, and $f_4 = f_5 = f_6 = f_7 = 0$, and observing that $\epsilon(8) = e^{-i2\pi\frac{1}{8}} = \frac{1}{\sqrt{2}} - \frac{i}{\sqrt{2}}$ we get:

$$y_0 = \frac{1}{2}(\frac{1}{2}(\frac{1}{2} + \frac{1}{2}) + \epsilon(8)^0\frac{1}{2}(\frac{1}{2} + \frac{1}{2})) = \frac{1}{2}$$

$$y_1 = \frac{1}{2}(\frac{1}{2}(\frac{1}{2} - \frac{i}{2}) + \epsilon(8)^1\frac{1}{2}(\frac{1}{2} - \frac{i}{2})) = \frac{1}{8} + i(-\frac{1}{8} - \frac{1}{4\sqrt{2}})$$

$$y_2 = \frac{1}{2}(\frac{1}{2}(\frac{1}{2} - \frac{1}{2}) + \epsilon(8)^2\frac{1}{2}(\frac{1}{2} - \frac{1}{2})) = 0$$

$$y_3 = \frac{1}{2}(\frac{1}{2}(\frac{1}{2} + \frac{i}{2}) + \epsilon(8)^3\frac{1}{2}(\frac{1}{2} + \frac{i}{2})) = \frac{1}{8} + i(\frac{1}{8} - \frac{1}{4\sqrt{2}})$$

$$y_4 = \frac{1}{2}(\frac{1}{2}(\frac{1}{2} + \frac{1}{2}) + \epsilon(8)^4\frac{1}{2}(\frac{1}{2} + \frac{1}{2})) = 0$$

$$y_5 = \frac{1}{2}(\frac{1}{2}(\frac{1}{2} - \frac{i}{2}) + \epsilon(8)^5\frac{1}{2}(\frac{1}{2} - \frac{i}{2})) = \frac{1}{8} + i(-\frac{1}{8} + \frac{1}{4\sqrt{2}})$$

$$y_6 = \frac{1}{2}(\frac{1}{2}(\frac{1}{2} - \frac{1}{2}) + \epsilon(8)^6\frac{1}{2}(\frac{1}{2} - \frac{1}{2})) = 0$$

$$y_7 = \frac{1}{2}(\frac{1}{2}(\frac{1}{2} + \frac{i}{2}) + \epsilon(8)^7\frac{1}{2}(\frac{1}{2} + \frac{i}{2})) = \frac{1}{8} + i(\frac{1}{8} + \frac{1}{4\sqrt{2}})$$

These results are identical to those obtained in example 182.

38.8 The Fourier Transform

For a thorough treatment of the Fourier transform, we refer to advanced texts for this subjects, e.g. [40]. However, we need to sketch this more advanced topic briefly, not only because of its theoretical prominence, but also because it is basic to the following chapter on wavelets. We shall therefore not include the proofs of the results of this section.

Recall from section 38.5 that the exponential version of the Fourier series of a 2π-periodic function is given by

$$^\alpha f(x) = \sum_{n=-\infty}^{\infty} c_n e^{i2\pi n\nu x}.$$

More generally, if the function f is piecewise C^1, continuous, and has period p with corresponding frequency $\nu = 1/p$, then we have the representation

$$f(x) = \sum_{n=-\infty}^{\infty} c_n e^{i2\pi n\nu x},$$

where

$$c_n = \frac{1}{p} \int_{-p/2}^{p/2} f(x) e^{-i2\pi n\frac{1}{p}x}\, dx.$$

In order to deduce a formula corresponding to the limit period $p \to \infty$, one writes

$$f(x) = \frac{1}{\sqrt{2\pi}} \sum_{n=-\infty}^{\infty} \mathcal{F}_p(f)(n\tfrac{2\pi}{p}) e^{i(n\frac{2\pi}{p})x} \cdot \frac{2\pi}{p},$$

with

$$\mathcal{F}_p(f)(\xi) = \frac{1}{\sqrt{2\pi}} \int_{-p/2}^{p/2} f(x) e^{-i\xi x}\, dx.$$

This infinite sum represents an approximation with integration arguments $n \cdot \frac{2\pi}{p}$ of step width $\frac{2\pi}{p}$ to an integral

$$f(x) = \frac{1}{\sqrt{2\pi}} \int_{-\infty}^{\infty} \mathcal{F}(f)(\xi) e^{i\xi x}\, d\xi,$$

with

$$\mathcal{F}(f)(\xi) = \frac{1}{\sqrt{2\pi}} \int_{-\infty}^{\infty} f(x) e^{-i\xi x}\, dx,$$

and where the integration interval is in fact the whole real line and not a proper compact interval as in the integral defined before. In view of the fact that this infinite integration interval stems from the infinite sum $\sum_{n=-\infty}^{\infty} = \lim_{N\to\infty} \sum_{-N}^{N}$ this motivates the following definition:

Definition 230 *If $g : \mathbb{R} \to \mathbb{C}$ is integrable on every closed interval of $\mathbb{R}$, we write $\int_{-\infty}^{\infty} g(x)\, dx$ for the limit $\lim_{b\to\infty} \int_{-b}^{b} g(x)\, dx$, if this limit exists. The number $\int_{-\infty}^{\infty} g(x)\, dx \in \mathbb{R}$ is called the* improper integral *of g.*

Here is an important class of functions which admit the improper integral:

Definition 231 *For a positive natural number r, denote by $\mathcal{L}^r(\mathbb{R})$ the $\mathbb{C}$-vector subspace of functions $f : \mathbb{R} \to \mathbb{C}$ such that $|f|^r$ is integrable on every closed interval $[a,b] \subset \mathbb{R}$ and such that the improper integral $\int_{-\infty}^{\infty} |f(x)|^r\, dx$ exists. Such functions are also said to be absolutely r-integrable on $\mathbb{R}$, and for $r = 1$, absolutely integrable on $\mathbb{R}$. In $\mathcal{L}^r(\mathbb{R})$, we have the $\mathbb{C}$-vector subspace $\mathcal{N}$ of functions f such that $\{x \mid f(x) \neq 0\}$ has measure 0 (see definition 197). Denote $L^r(\mathbb{R}) = \mathcal{L}^r(\mathbb{R})/\mathcal{N}$. When talking about absolutely r-integrable functions, one usually identifies functions f and g when $f - g \in \mathcal{N}$. The following statements should be understood in this sense of considering the classes of functions in $L^r(\mathbb{R})$.*

The interest in these functions resides in the following proposition:

Proposition 334 *If $g \in L^1(\mathbb{R})$, then the improper integral $\int_{-\infty}^{\infty} g(x)\, dx$ exists.*

Observe that with $g \in L^r(\mathbb{R})$, and $\xi \in \mathbb{R}$, we have also $g \cdot e^{i \cdot \xi \cdot ?} \in L^r(\mathbb{R})$ since $|g \cdot e^{i \cdot \xi \cdot ?}|^r = |g|^r$. This makes possible the following definition:

Definition 232 *Let $f \in L^1(\mathbb{R})$. Then the* Fourier transform *of f is the function $\mathcal{F}(f) : \mathbb{R} \to \mathbb{C}$, also denoted by $\hat{f}$, which evaluates to*

$$\mathcal{F}(f)(\xi) = \frac{1}{\sqrt{2\pi}} \int_{-\infty}^{\infty} f(x) e^{-i\xi x}\, dx.$$

One also has the adjoint transform *$\mathcal{F}^*(f) : \mathbb{R} \to \mathbb{C}$, defined by*

$$\mathcal{F}^*(f)(x) = \mathcal{F}(f)(-x) = \frac{1}{\sqrt{2\pi}} \int_{-\infty}^{\infty} f(\xi) e^{ix\xi}\, d\xi.$$

The two transforms are related by the Fourier integral theorem:

Proposition 335 (Fourier Integral Theorem) *If $f \in L^2(\mathbb{R})$, then so are the Fourier transform $\mathcal{F}(f)$ and the adjoint Fourier transform $\mathcal{F}^*(f)$. Moreover, in $L^2(\mathbb{R})$,*

$$f(x) = \mathcal{F}^*(\mathcal{F}(f))(x)$$
$$= \frac{1}{\sqrt{2\pi}} \int_{-\infty}^{\infty} \mathcal{F}(f)(\xi) e^{ix\xi}\, d\xi$$
$$= \frac{1}{2\pi} \int_{-\infty}^{\infty} \int_{-\infty}^{\infty} f(u) e^{i\xi(x-u)}\, du\, d\xi.$$

This means that the adjoint is the inverse of $\mathcal{F}$, $\mathcal{F}^* = \mathcal{F}^{-1}$, and $\mathcal{F}$: $L^2(\mathbb{R}) \xrightarrow{\sim} L^2(\mathbb{R})$ is an automorphism. Moreover, if we denote by $\langle f, g \rangle = \int_{-\infty}^{\infty} f(x) \cdot \overline{g(x)}\, dx$ the Hermitian form[2] for elements $f, g \in L^2(\mathbb{R})$, then the Fourier transform and its inverse are isometries, i.e.,

$$\langle \mathcal{F}(f), \mathcal{F}(g) \rangle = \langle \mathcal{F}^*(f), \mathcal{F}^*(g) \rangle = \langle f, g \rangle,$$

and therefore

$$\langle \mathcal{F}(f), g \rangle = \langle f, \mathcal{F}^*(g) \rangle.$$

Example 184 The Fourier transform of the step function

$$f(x) = \begin{cases} 1 & \text{if } -b \le x \le b, \\ 0 & \text{else}, \end{cases}$$

is sought (figure 38.11).

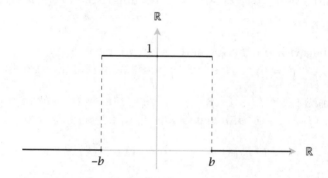

Fig. 38.11. A step function of width $2b$.

We have

$$\mathcal{F}(f)(\xi) = \frac{1}{\sqrt{2\pi}} \int_{-\infty}^{\infty} f(x) e^{-i\xi x}\, dx.$$

Since f is 0 outside the interval $[-b, b]$, the integral becomes

---

[2] The Hölder inequality $\left| \int_{-\infty}^{\infty} f(x)g(x)\, dx \right| \le \|f\|_2 \cdot \|g\|_2$, with the norm $\|f\|_2 = \sqrt{\langle f, f \rangle}$, guarantees that the form is a finite number. See [3, p. 116].

$$\mathcal{F}(f)(\xi) = \frac{1}{\sqrt{2\pi}} \int_{-b}^{b} f(x) e^{-i\xi x}\, dx$$

$$= \frac{1}{\sqrt{2\pi}} \int_{-b}^{b} e^{-i\xi x}\, dx \qquad \text{(since } f(x) = 1 \text{ in } [-b, b])$$

$$= \frac{1}{\sqrt{2\pi}} \left(\frac{i}{\xi} e^{-i\xi x} \right) \Big|_{-b}^{b}$$

$$= \frac{1}{\sqrt{2\pi}} \left(\frac{i}{\xi} e^{-i\xi b} - \frac{i}{\xi} e^{i\xi b} \right)$$

$$= \frac{1}{\sqrt{2\pi}} \frac{2}{\xi} \left(\frac{e^{i\xi b} - e^{-i\xi b}}{2i} \right) \qquad \text{(multiplying by } \tfrac{2i}{2i})$$

$$= \sqrt{\frac{2}{\pi}} \frac{\sin(b\xi)}{\xi} \qquad \text{(Euler formula for sin)}$$

The function $\frac{\sin(x)}{x}$ occurs frequently in DSP. It is not defined at $x = 0$, but, since $\lim_{x \to 0, x < 0} = \lim_{x \to 0, x > 0} = 1$, we can extend its domain to $x = 0$ by defining it to be equal to 1 there. Thus extended, it is called $\text{sinc}(x)$ and looks like figure 38.12.

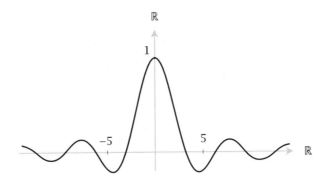

Fig. 38.12. The sinc function.

CHAPTER 39

Wavelets

39.1 Introduction

The Fourier representation of functions, be it by Fourier series or by the Fourier transform, is based on periodic functions which, by definition, are not very realistic in the sense that real functions never take values different from zero outside a finite interval. For example, acoustic sounds have a finite duration, so the defining air pressure differs only for a finite time from the normal pressure. In other words, such periodic functions are smeared over an infinite time, whereas the function to be described is localized in a small finite window. The success of wavelet theory was established in 1984 with a paper by Pierre Goupillaud, Jean Morlet and Alex Grossman (Cycle-Octave and Related Transforms in Seismic Signal Analysis. Geoexploration, 23:85–102, 1984), following the idea of constructing general functions from wavelets, i.e., small, localized "excitation packages" $\psi : \mathbb{R} \to \mathbb{C}$, i.e., either their support $supp(\psi)$—by definition the closure of $\{x \mid \psi(x) \neq 0\}$—is compact, or at least their values tend to zero very fast (in a specific technical sense). Figure 39.1 shows the graphs of some the following wavelets:

1. *Haar's* wavelet:

$$H(x) = \begin{cases} 1 & \text{if } 0 \leq x < \frac{1}{2}, \\ -1 & \text{if } \frac{1}{2} \leq x < 1, \\ 0 & \text{else.} \end{cases}$$

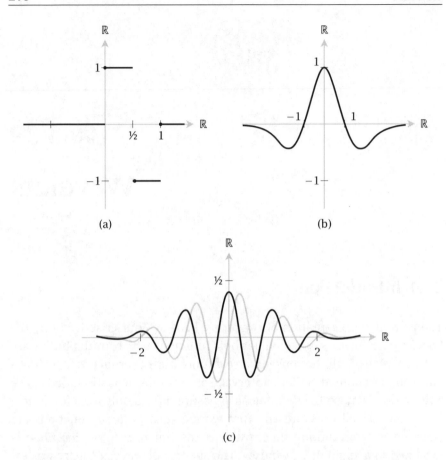

Fig. 39.1. Types of wavelets: (a) Haar's wavelet $H(x)$, (b) Murenzi's Mexican Hat wavelet $\psi_{Mex}(x)$, (c) Morlet's wavelet with $\nu = 1$, $Re(\psi_{Mor,\nu}(x))$ (black), $Im(\psi_{Mor,\nu}(x))$ (gray).

2. *Murenzi's Mexican Hat* wavelet:

$$\psi_{Mex}(x) = (1 - x^2)e^{-x^2/2}.$$

3. *Morlet's* wavelet ($\nu > 0.8$):

$$\psi_{Mor,\nu}(x) = \frac{1}{\sqrt{2\pi}}e^{-x^2/2}e^{i2\pi\nu x}.$$

4. *Meyer's* wavelet is given indirectly by its Fourier transform

$$\hat{\psi}_{Mey,\nu}(\xi) = \frac{1}{\sqrt{2\pi}}e^{i\xi/2}(w(\xi) + w(-\xi)),$$

where

$$w(\xi) = \begin{cases} \sin\left(\frac{\pi}{2}v\left(\frac{3\xi}{2\pi} - 1\right)\right) & \text{if } \frac{2\pi}{3} \leq \xi \leq \frac{4\pi}{3}, \\ \cos\left(\frac{\pi}{2}v\left(\frac{3\xi}{4\pi} - 1\right)\right) & \text{if } \frac{4\pi}{3} \leq \xi \leq \frac{8\pi}{3}, \\ 0 & \text{else.} \end{cases}$$

The idea is that one should be able to describe a sufficiently tame function f by an integral. This is similar to the Fourier transform, where the "coefficients" involve the exponential function, but here a fixed wavelet ψ plays this role. This idea is however only of theoretical use and must be adapted to a calculus which can be handled by computers. This latter program is met with the so-called frame theory, a theory which rests on the fact that one may find a denumerable basis, much like the powers of $e^{i2\pi vx}$ in Fourier theory, which permits a representation of the original function in the form of a series. We shall first give a concise presentation of the general theory and then specialize to the frame theory, in particular the theory involving the Haar wavelet.

39.2 The Hilbert Space $L^2(\mathbb{R})$

We shall not develop full Hilbert space theory here, but introduce the characteristic features of Hilbert spaces for the specific case of $L^2(\mathbb{R})$. To begin with, the $\mathbb{C}$-vector space $L^2(\mathbb{R})$ is provided with the Hermitian form

$$\langle f, g \rangle = \int_{-\infty}^{\infty} f(x)\overline{g(x)}\, dx$$

and the associated norm $\|f\| = \sqrt{\langle f, f \rangle}$ introduced in proposition 335 of chapter 38, and which has the properties (i)-(vi), and (viii) (triangle inequality) of sorite 326, whereas property (vii) of that sorite is replaced by a stronger property

(vii') $\|f\| \geq 0$, and $\|f\| = 0$ iff $f = 0$.

(Attention, this is in $L^2(\mathbb{R})$ and means that $f \in \mathcal{N}$.)

The space $L^2(\mathbb{R})$ becomes a metric space by the usual distance $d(f, g) = \|f - g\|$. The name "Hilbert space" is associated with the fact that (1) $(L^2(\mathbb{R}), d)$ is a complete metric space (see definition 204 in chapter 33) and (2) the metric d is deduced from a norm $\|x\| = \sqrt{\langle x, x \rangle}$, which stems from a Hermitian form $\langle ?, ? \rangle$. For such a space, one extends the concept of a basis from linear algebra to a *Schauder basis*.

Definition 233 *A* Schauder basis, *or simply a basis, of* $L^2(\mathbb{R})$ *is a sequence* $(b_i)_{i \in \mathbb{N}}$ *of functions* $b_i \in L^2(\mathbb{R})$ *such that every function* $f \in L^2(\mathbb{R})$ *can be written as a convergent series* $\sum_i \beta_i \cdot b_i = \lim_{N \to \infty} \sum_{i=0,1,\dots N} \beta_i \cdot b_i$, *where* $(\beta_i)_{i \in \mathbb{N}}$ *is a unique sequence of complex numbers. In particular, any finite subsequence of such a basis consists of linearly independent functions.*

A Schauder basis $(\beta_i)_i$ *is called* orthonormal Schauder basis *iff in the given Hermitian form we have* $\langle b_i, b_j \rangle = \delta_{ij}$ *for all* $i, j \in \mathbb{N}$.

It can be shown[1] that, for an orthonormal Schauder basis, the representation as a convergent sum is the same if one changes the indexes by a permutation $\pi : \mathbb{N} \overset{\sim}{\to} \mathbb{N}$, i.e., $\sum_i \beta_i \cdot b_i = \sum_i \beta_{\pi(i)} \cdot b_{\pi(i)}$.

We have already encountered this type of basis in Fourier theory, see the orthogonality relations in proposition 327 and Fourier's theorem 328.

Here is a slightly different description of $L^2(\mathbb{R})$:

Proposition 336 *Suppose that we are given an orthonormal Schauder basis* $(e_i)_i$ *of* $L^2(\mathbb{R})$. *Then the map*

$$f \mapsto \langle f, e. \rangle = (\langle f, e_i \rangle)_i \in \mathbb{C}^{\mathbb{N}}$$

is a linear isomorphism $\langle ?, e. \rangle : L^2(\mathbb{R}) \overset{\sim}{\to} l^2$ *onto the complex vector space* l^2 *of sequences* $(c_i)_i \in \mathbb{C}^{\mathbb{N}}$ *such that* $\sum_i |c_i|^2 < \infty$. *If we define a Hermitian form*

$$\langle (c_i)_i, (d_i)_i \rangle = \sum_i c_i \cdot \overline{d_i}$$

on l^2, *then* $\langle ?, e. \rangle$ *is an isometry of Hilbert spaces.*

Proof Clearly, the map $\langle ?, e. \rangle$ is a linear injection by the definition of a Schauder basis. Now, by the Cauchy criterion a series $x = \sum_i c_i \cdot e_i$ is convergent iff the partial sums $\sum_{N \le i \le M} c_i \cdot e_i$ tend to zero as $N, M \to \infty$, and this means $\sum_{N \le i \le M} |c_i|^2 \to 0$, because the basis vectors e_i are orthonormal. This is equivalent to the statement $(\langle f, e_i \rangle)_i \in l^2$. On the other hand, the condition $\sum_i |c_i|^2 < \infty$ implies that on l^2, the Hermitian form $\langle (c_i)_i, (d_i)_i \rangle$ is defined. In fact, for all finite sums $\sum_{i=0,\dots N} c_i \cdot \overline{d_i}$, we have the Schwarz inequality $| \sum_{i=0,\dots N} c_i \cdot \overline{d_i} |^2 \le \sum_{i=0,\dots N} |c_i|^2 + \sum_{i=0,\dots N} |d_i|^2 \le \sum_i |c_i|^2 + \sum_i |d_i|^2 < \infty$. Therefore the series $\sum_i c_i \cdot \overline{d_i}$ is absolutely convergent and the Hermitian form is defined. $\square$

A last fact, concerning Hilbert subspaces, must be mentioned. By definition, a *Hilbert subspace* $V \subset L^2(\mathbb{R})$ is more than a vector subspace, one

---

[1] Consult any book on functional analysis, for example [34].

also asks for completeness, i.e., V, with the induced Hermitian form, must be complete, which is equivalent to closedness, see exercise 166. Here is the crucial fact about orthogonal Hilbert subspaces:

Proposition 337 *If H is a Hilbert space with Hermitian form $\langle ?, ? \rangle$, then for every Hilbert subspace $S \subset H$, we have a direct decomposition*

$$H = S \oplus S^{\perp},$$

where $S^{\perp}$ is the Hilbert subspace of vectors orthogonal to all of S.

Proof Let $x \in H$. Let $\xi = \inf\{d(x, s) | s \in S\}$. Then there is a sequence $(s_i)_i$ of elements of S, such that $\xi = \lim_{i \to \infty} \|s_i - x\|$. But the sequence $(s_i)_i$ is Cauchy. In fact, we have the general equation $\|u + v\|^2 + \|u - v\|^2 = 2(\|u\|^2 + \|v\|^2)$, whence (with $x - s_m = u, s_n - x = v$)

$$\|s_m - s_n\|^2 = 2(\|x - s_m\|^2 + \|x - s_n\|^2) - 4\|x - \frac{1}{2}(s_m + s_n)\|^2.$$

But since $\frac{1}{2}(s_m + s_n) \in S$, we have $\|x - \frac{1}{2}(s_m + s_n)\|^2 \leq \xi^2$, and there is an index N such that for $\varepsilon > 0$, $\|x - s_m\|^2 < \frac{1}{4}\varepsilon + \xi^2$ for $m > N$. Taking $m, n > N$, we get

$$2(\|x - s_m\|^2 + \|x - s_n\|^2) - 4\|x - \frac{1}{2}(s_m + s_n)\|^2 < \varepsilon,$$

whence the claim. By the completeness of S, $(s_i)_i$ converges to $p \in S$. Let us show that $S \perp (x - p)$. Let $t \neq 0$ in S (the case $S = 0$ is trivial). Suppose $\langle t, x - p \rangle \neq 0$. Replacing t by $i \cdot t$, we may suppose that $\langle t, x - p \rangle \neq 0$ is not imaginary. Then we have $P(\lambda) = \|\lambda \cdot t + (x - p)\|^2 = \lambda^2 \cdot \|t\|^2 + 2\lambda \cdot Re(\langle t, x - p \rangle) + \|x - p\|^2$. Its minimal value is achieved where the derivative $P'(\lambda)$ vanishes. So we have the equation $\lambda = -\frac{Re(\langle t, x - p \rangle)}{\|t\|^2}$ for $\lambda \neq 0$. This means that p is not the nearest point, a contradiction, whence $S \perp (x - p)$. The same argument also yields unicity of the solution p, since for another solution p', the line $\lambda \cdot p + (1 - \lambda) \cdot p'$ would contain a point nearer the the two p, p'. So $H = S + S^{\perp}$, and the evident intersection $S \cap S^{\perp} = 0$ yields the direct sum representation $H = S \oplus S^{\perp}$. $\quad\square$

The theory of wavelets is based on the idea of calculating the scalar product of a function $f \in L^2(\mathbb{R})$ with a double-parametric family $(T^b a \bullet \psi)_{b \in \mathbb{R}, a \in \mathbb{R}^*}$ of "deformations" of a fixed wavelet ψ by means of the scalar product family $\langle f, T^b a \bullet \psi \rangle$ of "wavelet coefficients" of f. Its efficiency is due to the fact that, by use of the so-called frame construction method, one may find an orthonormal basis $(T^{b_i} a_i \bullet \psi)_{i \in \mathbb{N}}$ within the family of deformations $(T^b a \bullet \psi)_{b,a}$.

Before we proceed, recall the notations $GL(M)$ and $GA(M)$ for the general linear group and the general affine group over a module M, from sorite 186 and definition 168 in volume 1.

We construct a *group action of* GA *on* $L^2(\mathbb{R})$, i.e., by definition[2], a group homomorphism

$$\bullet : GA(\mathbb{R}) \to GL(L^2(\mathbb{R}))$$

defined by $(T^b a \bullet f)(x) = \frac{1}{\sqrt{|a|}} f(\frac{x-b}{a})$, where the transformation $T^b a = T^b \circ a$ is the affine map sending $r \in \mathbb{R}$ to $T^b a(r) = b + ar$.

Exercise 188 Verify that $\bullet$ is a group action.

Proposition 338 *If* $T^b a \in GA(\mathbb{R})$, *then*

$$T^b a \bullet ? : L^2(\mathbb{R}) \xrightarrow{\sim} L^2(\mathbb{R})$$

is an isometry, i.e., for any $f, g \in L^2(\mathbb{R})$, *we have*

$$\langle T^b a \bullet f, T^b a \bullet g \rangle = \langle f, g \rangle.$$

In particular, $(T^b a \bullet f) \perp (T^b a \bullet g)$ *iff* $f \perp g$, *and* $\|T^b a \bullet f\| = \|f\|$.

Proof Since we have a group action, the linear map $T^b a \bullet ?$ is a bijection. Moreover, we have

$$\langle T^b a \bullet f, T^b a \bullet g \rangle = \int_{-\infty}^{\infty} \frac{1}{\sqrt{|a|}} f\left(\frac{x-b}{a}\right) \overline{\frac{1}{\sqrt{|a|}} g\left(\frac{x-b}{a}\right)} dx$$

$$= \int_{-\infty}^{\infty} f(\eta) \overline{g(\eta)} d\eta \qquad \text{with } \eta = \frac{x-b}{a}$$

$$= \langle f, g \rangle.$$

$\square$

Definition 234 *A function* $\psi \in L^2(\mathbb{R})$ *is called a* wavelet *if*

$$0 < c_\psi = 2\pi \int_{-\infty}^{\infty} \frac{|\hat{\psi}(\xi)|^2}{|\xi|} d\xi < \infty,$$

where $\hat{\psi}$ *is the Fourier transform of* ψ, *see definition 232.*

Given two real numbers a *and* b, *with* $a \neq 0$, *the* ψ-*wavelet coefficient at index* a, b *of* $f \in L^2(\mathbb{R})$ *is the scalar product*

---

[2] A (left) group action of a group G on a set X is a homomorphism $\mu : G \to S(X)$ into the symmetry group of X, or, equivalently, a map $v : G \times X \to X$ such that (1) $v(e, x) = x$ for all $x \in X$ and the neutral element $e \in G$, and (2) $v(h, v(g, x)) = v(hg, x)$ for all $x \in X, h, g, \in G$. The equivalence is induced by $v(g, x) = \mu(g)(x)$.

$$L_\psi f(a,b) = \frac{1}{\sqrt{c_\psi}} \langle f, T^b a \bullet \psi \rangle = \frac{1}{\sqrt{c_\psi |a|}} \int_{-\infty}^{\infty} f(x) \cdot \overline{\psi}\left(\frac{x-b}{a}\right) dx,$$

which defines a function $L_\psi f \in L^2(\mathbb{R}^2)$.

There are many wavelets, more precisely, the set $W \subset L^2(\mathbb{R})$ of wavelets is *dense*, i.e., by definition, $\overline{W} = L^2(\mathbb{R})$. In other words, there are wavelets in every neighborhood of each function $f \in L^2(\mathbb{R})$, see [24] for a proof.

It is theoretically important (and comforting) to know that the family of ψ-wavelet coefficients $L_\psi f(a,b)$ is rich enough to rebuild f. The formula is as follows. If $f \in L^2(\mathbb{R})$ and ψ is a wavelet, then

$$f(t) = \frac{1}{\sqrt{c_\psi}} \int_{-\infty}^{\infty} \int_{-\infty}^{\infty} L_\psi f(a,b) \cdot T^b a \bullet \psi \frac{da\, db}{a^2}.$$

See [24] for a proof. This theoretical result, together with the density of wavelets and the density of the denumerable set $\mathbb{Q}$ of rational numbers in $\mathbb{R}$ suggests that it might be possible to reconstruct f using only a countable set of wavelets $T^{b_i} a_i \bullet \psi$. This is what frame theory is about, and what gives us back the advantage of Fourier's approach against the doubly real-valued set of indexes a, b of wavelet coefficients.

39.3 Frames and Orthonormal Wavelet Bases

Frames are families $(T^{b_i} a_j \bullet \psi)_{i,j \in \mathbb{Z}}$ of deformed wavelets $T^{b_i} a_j \bullet \psi$ for special values of the deformation indexes a_j and b_i.

Definition 235 *Given two real numbers $a_0 > 1$ and $b_0 > 0$, a frame index family is a family*

$$FI(a_0, b_0) = ((a_0^m, nb_0 a_0^m))_{m,n \in \mathbb{Z}},$$

of elements in $\mathbb{R}^* \times \mathbb{R}$. A frame function system for a wavelet ψ and $FI(a_0, b_0)$ is the family of wavelets

$$(\psi, a_0, b_0) = \left(\psi_{m,n}^{a_0, b_0}\right)_{m,n \in \mathbb{Z}}, \text{ where } \psi_{m,n}^{a_0, b_0} = T^{nb_0 a_0^m} a_0^m \bullet \psi$$

indexed by two integers m and n, which define the action of the affine transformation $T^{nb_0 a_0^m} a_0^m$ for the deformation index pair $(a_0^m, nb_0 a_0^m) \in FI(a_0, b_0)$.

A wavelet ψ is called orthonormal *if the family* $(\psi, 2, 1)$ *defines an orthonormal Schauder basis of* $L^2(\mathbb{R})$. *In this case, we also write*

$$\psi_{m,n} = \psi_{m,n}^{2,1} = T^{n2^m} 2^m \bullet \psi, \quad i.e.,$$

$$\psi_{m,n}(x) = 2^{-m/2} \psi(2^{-m}x - n).$$

Remark 33 Observe that in Hilbert space theory, the invariance of infinite series of orthogonal vectors under permutations of the index set $\mathbb{N}$, guarantees that we may take any bijection of $\mathbb{N} \xrightarrow{\sim} \mathbb{Z} \times \mathbb{Z}$ to define the double sum of the representation of f. Such a basis solves our problem of selecting a denumerable deformation index set $(a_i, b_i)_i$ such that f is represented by the wavelet coefficients $L_\psi f(a_i, b_i)$.

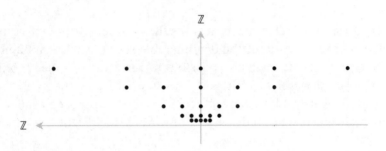

Fig. 39.2. Part of the frame index family $FI(2, 1)$ for $m, n \in \{-2, \ldots 2\}$. The diagram has been rotated counter-clockwise by $90°$.

The following corollary is immediate from the isometric property of deformations:

Corollary 339 *If* $(\psi^k)_{k \in \mathbb{Z}}$ *is a family of orthonormal functions in* $L^2(\mathbb{R})$, *and if* $m, n \in \mathbb{Z}$, *then so is* $(\psi_{m,n}^k)_k$.

For example, let $\chi \in L^2(\mathbb{R})$ be the characteristic function of the half-open interval $[0, 1[$, i.e.,

$$\chi(x) = \begin{cases} 1 & \text{if } x \in [0, 1[, \\ 0 & \text{else.} \end{cases}$$

Then, with the notations of definition 235, we have

$$\chi_{n,k}(x) = 2^{-n/2}\chi(2^{-n}x - k)$$

$$= \begin{cases} 2^{-n/2} & \text{if } x \in [2^n k, 2^n(k+1)[, \\ 0 & \text{else.} \end{cases}$$

The system $(\psi^k = \chi_{0,k})_k$ is orthonormal, and so is the system $(\psi^k_n)_k$ with $\psi^k_n = (\chi_{0,k})_{n,0} = \chi_{n,k}$, by corollary 339. The deformations of χ are related by the *two-scale equation*:

$$\chi_{m,n} = \frac{1}{\sqrt{2}}(\chi_{m-1,2n} + \chi_{m-1,2n+1}). \tag{$*$}$$

We need the characteristic function since it is strongly related to the Haar wavelet H. In fact, we have this *two-scale relation*:

$$H_{m,n} = \frac{1}{\sqrt{2}}(\chi_{m-1,2n} - \chi_{m-1,2n+1}). \tag{$**$}$$

Next, we start the construction of the Haar wavelet basis using a chain of auxiliary Hilbert subspaces $V_n \subset L^2(\mathbb{R})$.

Proposition 340 *For $n \in \mathbb{Z}$, the map*

$$v_n : l^2 \to L^2(\mathbb{R}) : (a_k)_k \mapsto \sum_k a_k \chi_{n,k}$$

is a linear isometry of Hilbert spaces onto the Hilbert subspace $V_n = \mathrm{Im}(v_n)$, in particular $(\chi_{n,k})_k$ is an orthonormal Schauder basis of V_n. We have a commutative diagram of isomorphisms of Hilbert spaces

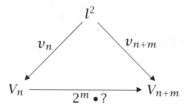

The sequence of spaces $(V_n)_{n \in \mathbb{Z}}$ has these properties:

(i) *For all $n \in \mathbb{Z}$, $V_n \subset V_{n-1}$.*

(ii) *$\overline{\bigcup_n V_n} = L^2(\mathbb{R})$, i.e., for every $f \in L^2(\mathbb{R})$, there is a sequence $(f_n)_{n \in \mathbb{N}}$ with $f_n \in V_{-n}$, which converges to f, i.e., $\lim_{n \to \infty} f_n = f$.*

(iii) *$\bigcap_n V_n = 0$.*

(iv) *If $W_n = V_n^\perp \subset V_{n-1}$ is the orthogonal space of V_n in V_{n-1}, then the isometry $2^m \bullet ?$ induces an isomorphism $W_n \overset{\sim}{\to} W_{n+m}$ of Hilbert spaces.*

Proof (i) The linear map v_n is an isometry because the canonical basis $(e_k)_k$, with $e_k = (0, \ldots 0, 1, 0 \ldots 0)$ having a 1 at index k and 0 else, is mapped bijectively onto the orthonormal sequence $(\chi_{n,k})_k$. Moreover, $2^m \bullet \chi_{n,k} = (2^m \circ T^{k2^n} 2^n) \bullet \chi = T^{k2^{m+n}} 2^{m+n} \chi = \chi_{m+n,k}$. The inclusion $V_n \subset V_{n-1}$ follows from the two-scale equation ($*$).

Claim (ii) is a standard result in functional analysis, we do not prove it here.

As to claim (iii), if $f \in \bigcap_n V_n$, then f must be constant over any interval $[k2^n, (k+1)2^n[$ of length 2^n for all $n \to \infty$, i.e., x must be a constant in $L^2(\mathbb{R})$, i.e., it must vanish.

For claim (iv), observe that the deformation isomorphism $2^m \bullet ?$ maps V_{n-1} onto V_{m+n-1} and also the subspace $V_n \subset V_{n-1}$ onto $V_{m+n} \subset V_{m+n-1}$. Since $2^m \bullet ?$ is an isometry, the orthogonal spaces W_n, W_{m+n} also do correspond.   □

This implies that $W_n \perp W_m$ for all $m \neq n$. In order to construct a Haar wavelet basis of $L^2(\mathbb{R})$, we first construct one for W_0. Then we observe that the Hilbert space isomorphism $2^m \bullet ? : W_0 \xrightarrow{\sim} W_m$ induced by the deformation $2^m \bullet ?$ on wavelets generates a basis for W_m. We then take any $f \in L^2(\mathbb{R})$ and use the limit representation $\lim_{n \to \infty} f_n = f$ with $f_n \in V_{-n}$. We then rewrite $f_{n+1} = f_n + g_n$, with $g_n \in W_{-n}$. Therefore $f = f_0 + \sum_{n \in \mathbb{N}} g_n$. Similarly, one can show that $f_0 = \sum_{n \in \mathbb{N}} g_{-n+1}$, and therefore $f = \sum_{n \in \mathbb{Z}} g_n$, i.e., $L^2(\mathbb{R}) = \bigoplus_{n \in \mathbb{Z}} W_n$ in the sense of converging infinite sums.

Proposition 341 *The system of deformed Meyer wavelets is orthonormal.*

The proof technique of this proposition is far beyond our modest context. See [24] for a proof. However, the essential technique can also be traced for the Haar wavelet. This is what we shall discuss now.

Lemma 342 *The system of deformed Haar wavelets $(H_{0,k})_{k \in \mathbb{Z}}$ is an orthonormal basis of W_0.*

We need the proof idea for the fast wavelet transform, so let us make it explicit here: The system is evidently an orthonormal system of shifted wavelets in V_{-1}, by the two-scale relation ($**$) with $m = 0$. Why is it also a basis of W_0? Take $g \in W_0 = V_0^{\perp} \subset V_{-1}$. Then by proposition 340,

$$g = \sum_{k \in \mathbb{Z}} a_k \chi_{-1,k}, \quad (a_k)_k \in l^2$$

$$= \sum_{k \in \mathbb{Z}} (a_{2k} \chi_{-1,2k} + a_{2k+1} \chi_{-1,2k+1}).$$

But, since g is in the space orthogonal to V_0, and thus orthogonal to the basis functions $\chi_{0,k}$ of V_0,

$$0 = \langle g, \chi_{0,k} \rangle$$

$$= \int_{-\infty}^{\infty} \sum_{l \in \mathbb{Z}} (a_{2l} \chi_{-1,2l} + a_{2l+1} \chi_{-1,2l+1}) \overline{\chi_{0,k}}$$

$$= \int_{-\infty}^{\infty} (a_{2k} \chi_{-1,2k} + a_{2k+1} \chi_{-1,2k+1}) \overline{\chi_{0,k}}$$

$$= \int_{k}^{k+1} (a_{2k} \chi_{-1,2k} + a_{2k+1} \chi_{-1,2k+1}) \overline{\chi_{0,k}}$$

$$= \int_{k}^{k+1} (a_{2k} 2^{-1/2} + a_{2k+1} 2^{-1/2})$$

$$= 2^{-1/2} (a_{2k} + a_{2k+1}).$$

This implies

$$a_{2k} + a_{2k+1} = 0.$$

Therefore

$$a_{2k} \chi_{-1,2k} + a_{2k+1} \chi_{-1,2k+1} = a_{2k} \chi_{-1,2k} - a_{2k} \chi_{-1,2k+1}$$

$$= a_{2k} (\chi_{-1,2k} - \chi_{-1,2k+1})$$

$$= a_{2k} \sqrt{2} H_{0,k},$$

thus, finally,

$$g = \sum_{k \in \mathbb{Z}} \sqrt{2} a_{2k} H_{0,k}.$$

Corollary 343 *For $n \in \mathbb{Z}$, the system of deformed Haar wavelets $(H_{n,k})_{k \in \mathbb{Z}}$ is an orthonormal basis of W_n.*

This follows from lemma 342 and proposition 340. And here is the promised Haar basis, a fact which follows from corollary 343 and the result $L^2(\mathbb{R}) = \bigoplus_{n \in \mathbb{Z}} W_n$ mentioned above.

Proposition 344 *The system of deformed Haar wavelets $(H_{n,k})_{n,k \in \mathbb{Z}}$ is an orthonormal basis of $L^2(\mathbb{R})$.*

39.4 The Fast Haar Wavelet Transform

We shall now discuss an algorithm for computing the coefficients of a function in the Haar wavelet basis $(H_{n,k})_{n,k \in \mathbb{Z}}$.

We start from a function f whose approximation f_N on V_N is given by

$$f_N = \sum_{k \in \mathbb{Z}} a_{Nk} \chi_{N,k}.$$

In practical cases, one has evidently only finitely many different coefficients, or still more concretely, finitely many coefficients different from 0. The fast Haar wavelet transform is an algorithm which tends to reduce the components from the basis $(\chi_{N,k})_k$ of V_N and to replace them by Haar wavelet components. We know from proposition 340 that the decreasing chain $\ldots \supset V_n \supset V_{n+1} \supset \ldots$ tends towards 0, so we want to restrict the $\chi_{n,k}$ components to spaces V_n with successively increasing index n. The two-scale formulas (∗) and (∗∗) are used in the algorithm as follows.

These formulas define a new basis of the space V_{m-1} in terms of the space V_m and of the space W_m, in fact:

$$\chi_{m-1,2n} = \frac{1}{\sqrt{2}} (\chi_{m,n} + H_{m,n}),$$

$$\chi_{m-1,2n+1} = \frac{1}{\sqrt{2}} (\chi_{m,n} - H_{m,n}).$$

So one starts rewriting f_N as

$$f_N = \sum_k a_{N+1,k} \chi_{N+1,k} + \sum_k c_{N+1,k} H_{N+1,k}$$

where the coefficients $a_{N+1,k}$ and $c_{N+1,k}$ are calculated from the two-scale formulas. The general step is this: suppose that the coefficients

$$a_{m,k} \text{ and } c_{m,k} \text{ for } m = n, n-1, \ldots N+1$$

are calculated, then we obtain the coefficients $a_{m+1,k}$ and $c_{m+1,k}$ by the two-scale formulas via the linear expressions

$$a_{m+1,k} = \frac{1}{\sqrt{2}} (a_{m,2k} + a_{m,2k+1}),$$

$$c_{m+1,k} = \frac{1}{\sqrt{2}} (a_{m,2k} - a_{m,2k+1}),$$

which are calculated in finite time if the original data are zero for all but finitely many coefficients.

The preceding procedure is coined multiscale analysis (MSA) and was introduced by Stéphane Mallat in 1989 and Yves Meyer in 1990.

Let us recapitulate the MSA described here with respect to the chain $\ldots V_{n+1} \subset V_n \subset V_{n-1} \ldots$ of subspaces, which fill out $L^2(\mathbb{R})$. We are given an approximation f_N of $f \in L^2(\mathbb{R})$ by deformations of character-istic functions of a certain granularity, i.e., $f_N \in V_N$. We therefore write $f_N = \sum_k a_{N,k} \chi_{N,k}$. We now use the successive orthogonal decomposition $V_N = V_{N+1} \oplus W_{N+1}$, $V_{N+1} = V_{N+2} \oplus W_{N+2}$, etc., until $V_{m-1} = V_m \oplus W_m$ for a certain $m > N$. This yields a corresponding successive decomposition of f_N by $f_N = f_{N+1} + g_{N+1}$, $f_{N+1} = f_{N+2} + g_{N+2}$, etc., until $f_{m-1} = f_m + g_m$ with $f_m \in V_m, g_m \in W_m$. We may visualize this decomposition by

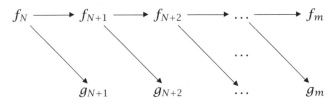

This yields the decomposition $f_N = f_m + \sum_{i=N+1}^{m} g_i$, where for increasing m, the summand f_m tends to be constant over successively increasing intervals, while the summand captures more and more the non-constant contributions to f_N. The two-scale formulas yield the successive split-off of g-components (c-coefficients) from the f-components (a-coefficients).

Example 185 Consider a function f_0 on the interval $[0, 16[$ which is de-fined as a sum of characteristic functions $f_0 = \sum_{k=0}^{15} a_{0,k} \chi_{0,k}$ (see fig-ure 39.3).

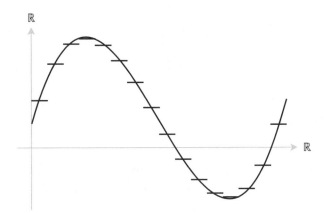

Fig. 39.3. The function f_0 which is to be transformed (one may think of f_0 as being a sequence of samples read from a CD, for instance).

In this example we show how f_0 is approximated by a sum of Haar-wavelets. Using the formulas given above one can calculate the coefficients $a_{1,k}$, $c_{1,k}$ and split the function f_0 into $f_1 = \sum_{k=0}^{15} a_{1,k} \chi_{1,k}$ and $g_1 = \sum_{k=0}^{15} c_{1,k} H_{1,k}$ (see figure 39.4).

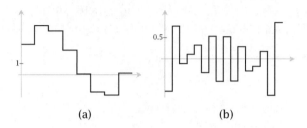

(a) (b)

Fig. 39.4. The functions f_1 (a) and g_1 (b), which actually is the first approximation of f_0.

Continuing this process, one gets the coefficients $a_{2,k}$, $c_{2,k}$, which yield $f_2 = \sum_{k=0}^{15} a_{2,k} \chi_{2,k}$ and $g_2 = \sum_{k=0}^{15} c_{2,k} H_{2,k}$ with the approximation to f_0 now being $\overline{g_2} = g_1 + g_2$ (see figure 39.5).

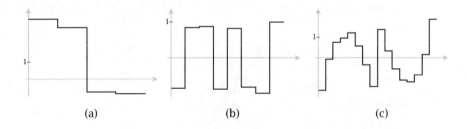

(a) (b) (c)

Fig. 39.5. The functions f_2 (a) and g_2 (b), as well as the approximation $\overline{g_2} = g_1 + g_2$ (c).

Figures 39.6 and 39.7 show the next approximation steps. For $n > 4$, all coefficients $a_{n,k}$, $c_{n,k}$ vanish, i.e, $f_n = 0$ and $g_n = 0$.

Exercise 189 Calculate the first four Haar approximation levels of

$$f_3 = 0.5\chi_{3,0} - 2\chi_{3,1} + \chi_{3,4}.$$

Haar wavelets exemplify in a simple way the principles behind wavelet approximation. However, Haar approximations of continuous functions are

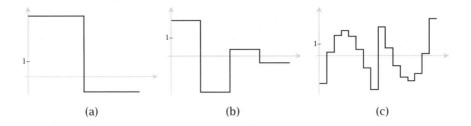

Fig. 39.6. The functions f_3 (a), g_3 (b), and the approximation $\overline{g_3} = g_1 + g_2 + g_3$ (c).

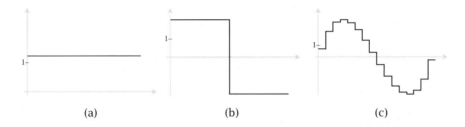

Fig. 39.7. The functions f_4 (a), g_4 (b), and the approximation $\overline{g_4} = g_1 + g_2 + g_3 + g_4$ (c).

step functions and, thus, not continuous. Other wavelets, like Mirenzi's or Morlet's, are continuous functions, hence better suited for approximating smooth functions. However, the calculation of transforms based on such wavelets is more complicated and does not fit into the modest scope of an introductory exposition.

CHAPTER **40**

Fractals

40.1 Introduction

This chapter gives a first insight into the fascinating world of fractals. These geometric objects have been unleashed to science and popularized by the mathematician Benoit Mandelbrot around 1982, mainly through his book *The Fractal Geometry of Nature* [27], following a preliminary study of the Julia set. The novelty of this mathematical approach was hinted at in the book's title, claiming that nature is based on a special type of geometry. This time, however, nature was not to be restricted to planetary trajectories cast into quadratic equations by classical geometry. It rather included complex inorganic shapes, such as snow flakes or clouds, but also living nature, such as plants, in particular shapes of leaves. This is perhaps the reason why this kind of geometry has attracted the interest of scientists and artists alike. Mathematically, we encounter a much more prosaic scenery. We want to present a thoroughly mathematical setup of the basic ideas of a particular kind of fractals. It is based on contractions on metric spaces and their fixpoints. However, in this theory, the points being mapped by contractions are not original geometric points, but big objects, i.e., compact sets in given natural metric spaces. Therefore, fixpoints for fractals are compact sets having bizarre shapes, for example those snow flakes, clouds or fern leaves mentioned above. Our presentation also includes a first glance at the notion of the dimension of a complex geometric object, which is more refined than that known from classical geometry.

Before we come to the specific theory, let us complete some elementary aspects of metric spaces. A review of section 33.2 for the elementary facts about metric spaces will be helpful. We have learned that each metric space (X, d) induces a canonical topology on X whose open sets are the subsets $O \subset X$ such that for every element $x \in O$, there is an open ball $B_\varepsilon(x) = \{y \mid d(x, y) < \varepsilon\}$ contained in O. A closed set in X is defined as the complement in X of an open set. If (X, c) and (Y, d) are two metric spaces, then a map $f : X \to Y$ is *continuous* iff it fulfills the following three equivalent conditions (completely analogous to the definition of continuity for subsets of $\mathbb{R}^n$, compare to lemma 238; in fact, those sets bear a topology which is also induced by a metric, i.e., the metric defined on the englobing $\mathbb{R}^n$):

(i) *The inverse image $f^{-1}(O)$ of every open set $O \subset Y$ is open in X.*

(ii) *The inverse image $f^{-1}(C)$ of every closed set $C \subset Y$ is closed in X.*

(iii) *For every $x \in X$ and $\varepsilon > 0$, there is a $\delta > 0$ such that $f(B_\delta(x)) \subset B_\varepsilon(f(x))$.*

It is easy to check that the metric spaces, together with continuous maps and their set-theoretic compositions define a category **Metr**. A typical continuous map $f : X \to Y$ is an isometry, by definition a map such that $d(f(x), f(y)) = c(x, y)$ for all $x, y \in X$. In fact, we may then choose $\delta = \varepsilon$ in the criterion (iii) for continuity.

In this chapter, (X, d) will always denote a complete metric space.

40.2 Hausdorff-Metric Spaces

The theory of fractals deals with compact sets $K \subset X$ of a complete metric space (X, d). Recall the three-fold characterization of a compact set in $\mathbb{R}^n$ given in proposition 236. The equivalence of (i) and (ii) in this proposition is true without change for any complete metric space and its associated topology as defined in section 33.2 under the second axiom of countability which requires that there is countable basis of the topology. This means that there is countable set of open sets, such that every every open set is a union of sets from this basis. Reread the proof, and everything will work mutatis mutandis.

Lemma 345 *If $f : X \to Y$ is a continuous map between metric spaces, then the image $f(K)$ of a compact subset $K \subset X$ is compact.*

Exercise 190 Give a proof of lemma 345 using the second characterization (ii) of compactness given in proposition 236.

We are now going to define a new complete metric space, which is derived from X. Its underlying set is

$$\mathcal{H}(X) = \{K \mid K \text{ a non-empty compact subset of } X\}.$$

The metric on $\mathcal{H}$ is given as follows. Let $A, B \in \mathcal{H}$. Then for any $x \in X$, the map

$$d_x : B \to \mathbb{R} : b \mapsto d(x, b)$$

is continuous, and therefore, by lemma 345 and exercise 138, has a minimum

$$d(x, B) = \min\{d(x, b) \mid b \in B\}.$$

Analogously, the function $d(?, A) : B \to \mathbb{R} : b \mapsto d(b, A)$ is continuous, and we may define

$$d(A, B) = \max\{d(a, B) \mid a \in A\}.$$

Evidently, $d(A, B)$ is not symmetric in general, so one sets

$$h(A, B) = \max(d(A, B), d(B, A)).$$

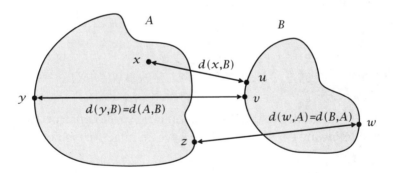

Fig. 40.1. In this case, clearly $d(A, B) > d(B, A)$, so $h(A, B) = d(A, B)$.

Exercise 191 Verify the claims concerning the continuity of the distance functions d_x and $d(?, A)$.

Proposition 346 *The function*

$$h = h(d) : \mathcal{H}(X) \times \mathcal{H}(X) \to \mathbb{R}$$

is a metric, called the Hausdorff metric, *and* $\mathcal{H}(X)$ *is complete. If* $(C_i)_i$ *is a Cauchy sequence in* $\mathcal{H}(X)$, *then it converges to the compact set* C *whose points are the limits* $\lim_{i \to \infty} c_i$ *of Cauchy sequences in* X *with* $c_i \in C_i$, *for all* i. *The metric space* $(\mathcal{H}(X), h)$ *is called the* Hausdorff-metric space of (X, d).

Proof Clearly, $h(A, A) = 0$, and, by definition, $h(A, B) = h(B, A)$. If $A \neq B$, then (up to an exchange of roles) we may suppose that there is $a \in A - B$. Then $d(a, B) > 0$, by the compactness of B. So $d(A, B) = \max\{d(x, B) \mid x \in A\} \geq d(a, B) > 0$. Therefore also $h(A, B) > 0$. To prove the triangle inequality, take three compact sets A, B and C. We first show that $d(A, C) \leq d(A, B) + d(B, C)$ and, symmetrically, $d(C, A) \leq d(C, B) + d(B, A)$. For $a \in A$, we have

$$
\begin{aligned}
d(a, C) &= \min\{d(a, c) \mid c \in C\} \\
&\leq \min\{d(a, b) + d(b, c) \mid b \in B, c \in C\}, \qquad \text{for all } b \in B \\
&\leq \min\{d(a, b) \mid b \in B\} + \max\{\min\{d(b, c) \mid c \in C\} \mid b \in B\} \\
&= d(a, B) + \max\{d(b, C) \mid b \in B\} \\
&= d(a, B) + d(B, C).
\end{aligned}
$$

Whence $d(A, C) \leq d(A, B) + d(B, C)$. Then we have

$$
\begin{aligned}
h(A, C) &\leq \max(d(A, B) + d(B, C), d(C, B) + d(B, A)) \\
&\leq \max(d(A, B), d(B, A)) + \max(d(B, C), d(C, B)) \\
&= h(A, B) + h(B, C).
\end{aligned}
$$

The completeness of $\mathcal{H}(X)$ requires a long technical proof, which we must omit here. Refer to [5], section 2.7, for details.   □

The perfect naturality of the Hausdorff-metric space construction results from the following consideration: We have an injective isometry

$$i_X : X \to \mathcal{H}(X) : x \mapsto \{x\},$$

i.e., $h(i(x), i(y)) = d(x, y)$, and i_X is therefore continuous and gives us back the space (X, d) by restriction of the Hausdorff-metric space to the image $Im(i_X)$. Moreover, we know from lemma 345 that for any continuous map $f : X \to Y$ between metric spaces, the induced set map $\mathcal{H}(f) : \mathcal{H}(X) \to \mathcal{H}(Y) : K \mapsto f(K)$ is defined. The next result is entitled "naturality of the Hausdorff metric". This refers to the commutation of the $\mathcal{H}$-construction with composition of maps, with the identity, and also to the commutative of proposition 347. The systematic background of this type of commutativity has been dealt with in the chapter 36 on category theory under the title of natural transformations.

Proposition 347 (Naturality of the Hausdorff Metric) *With the above notations, if $f : X \to Y$ is continuous, then so is*

$$\mathcal{H}(f) : \mathcal{H}(X) \to \mathcal{H}(Y)$$

Moreover, if $g : Y \to Z$ is a second continuous map between metric spaces, then we have

$$\mathcal{H}(g \circ f) : \mathcal{H}(g) \circ \mathcal{H}(f),$$

and $\mathcal{H}(Id_X) = Id_{\mathcal{H}(X)}$. Finally, we have this commutative diagram

$$
\begin{array}{ccc}
X & \xrightarrow{i_X} & \mathcal{H}(X) \\
{\scriptstyle f}\downarrow & & \downarrow{\scriptstyle \mathcal{H}(f)} \\
Y & \xrightarrow{i_Y} & \mathcal{H}(Y)
\end{array}
$$

of continuous maps. In particular, $\mathcal{H} : Top(X, Y) \to Top(\mathcal{H}(X), \mathcal{H}(Y))$ is an injection.

Proof Let $f : X \to Y$ be a continuous map between the complete metric spaces (X, d) and (Y, e). We show that $\mathcal{H}(f) : \mathcal{H}(X) \to \mathcal{H}(Y)$ is also continuous by use of the criterion (iii) for continuity described in the introduction of this chapter. So let $\varepsilon > 0$, and take compact sets $K, L \subset X$. We have to find $\delta > 0$ such that $h(K, L) < \delta$ implies $h(f(K), f(L)) < \varepsilon$. It is sufficient to find δ such that $d(K, L) < \delta$ implies $e(f(K), f(L)) < \varepsilon$. The function $g(x) = e(f(x), f(L))$ is continuous and for each $x \in L$, we have $g(x) = 0$. For every $x \in L$ take a $\delta(x) > 0$ such that $d(z, x) < \delta(x)$ implies $g(z) \leq \varepsilon$. Then take a finite cover of L by $U_{\delta(x_i)/2}(x_i)$ and then take the minimum δ of all $\delta(x_i)/2$. Then, if for a z, we have $d(z, y) < \delta$ for a $y \in L$, then, if $y \in U_{\delta(x_{i_0})/2}(x_{i_0})$, we have $d(z, x_{i_0}) < \delta(x_i)$, whence $g(z) < \varepsilon$. So, if $d(K, L) < \delta$, then for all $k \in K$, $d(k, L) < \delta$, whence $g(k) < \varepsilon$, i.e., $e(f(k), f(L)) < \varepsilon$, whence $e(f(K), f(L)) < \varepsilon$.

If $g : Y \to Z$ is a second continuous map, then for a compact $K \subset X$, $\mathcal{H}(g \circ f)(K) = (g(f(K)) = \mathcal{H}(g)(\mathcal{H}(f)(K)) = \mathcal{H}(g) \circ \mathcal{H}(f)(K)$. Clearly $\mathcal{H}(Id_X) = Id_{\mathcal{H}(X)}$, so $\mathcal{H}$ is a functor on the category of metric spaces with continuous maps. Moreover, for a singleton $x \in X$, we have $\mathcal{H}(f)(i_X(x)) = \mathcal{H}(f)(\{x\}) = \{f(x)\} = i_X(f(x))$, and the diagram commutes. □

40.3 Contractions on Hausdorff-Metric Spaces

The Hausdorff metric construction is not only natural on metric spaces, but also conserves contractions. More precisely:

Proposition 348 *Let (X, d) and (Y, e) be metric spaces and denote by Contra(X, Y) the set of contractions $f : X \to Y$ as defined in proposition 295. Then we have Contra$(X, Y) \subset Top(X, Y)$ and the Hausdorff map $\mathcal{H}$ transforms contractions into contractions, i.e., we have an injection*

$$\mathcal{H} : Contra(X, Y) \to Contra(\mathcal{H}(X), \mathcal{H}(Y)).$$

Proof We already know that isometries are continuous. Now, a contraction f is similar to an isometry, only that there is an even stronger condition $e(f(x), f(y)) \le k \cdot d(x, y)$ compared to $e(f(x), f(y)) = d(x, y)$ for an isometry. Therefore $Contra(X, Y) \subset Top(X, Y)$. On the other hand, let $f : X \to Y$ be a contraction with $e(f(x), f(y)) \le k \cdot d(x, y)$ for all $x, y \in X$. Then for a compact set $L \subset X$ and a point $x \in X$, we have $e(f(x), f(L)) = \min\{e(f(x), f(l)) \mid l \in L\} \le \min\{k \cdot d(x, l) \mid l \in L\} = k \cdot d(x, L)$. So $e(f(K), f(L)) = \max\{e(f(x), f(L)) \mid x \in K\} \le \max\{k \cdot d(x, L) \mid x \in K\} = k \cdot d(K, L)$. Therefore $h(f(K), f(L)) \le k \cdot h(K, L)$, and we are done. $\square$

In particular, if $X = Y$, we set $Contra(X) = Contra(X, X)$, and if $c \in Contra(X)$, then, by the commutativity of the diagram from proposition 347, we have the commutative diagram

$$
\begin{array}{ccc}
X & \xrightarrow{\ i_X\ } & \mathcal{H}(X) \\
{\scriptstyle c}\downarrow & & \downarrow{\scriptstyle \mathcal{H}(c)} \\
X & \xrightarrow{\ i_X\ } & \mathcal{H}(X)
\end{array}
$$

which connects two contractions by the injection i_X. We then write $i_X : c \to \mathcal{H}(c)$. More generally, if $f : X \to Y$ is a continuous map, and if $c_X \in Contra(X)$ and $c_Y \in Contra(Y)$ are such that the diagram

$$
\begin{array}{ccc}
X & \xrightarrow{\ f\ } & Y \\
{\scriptstyle c_X}\downarrow & & \downarrow{\scriptstyle c_Y} \\
X & \xrightarrow{\ f\ } & Y
\end{array}
$$

commutes, then we write $f : c_X \to c_Y$ and call this a *morphism of contractions*. Clearly, morphisms of contractions can be composed, composition is associative, and the identity Id_X is a morphism $c \to c$. We may therefore speak about the "category of contractions", for the precise meaning of this wording, please refer to chapter 36.

Proposition 349 *If $f : c_X \to c_Y$ is a morphism of contractions, and if $Fix(c_X)$ and $Fix(c_Y)$ are the unique fixpoints of these contractions as guaranteed by proposition 295, then we have*

$$f(Fix(c_X)) = Fix(c_Y).$$

In particular, the fixpoint $Fix(c)$ maps to the fixpoint $Fix(\mathcal{H}(c)) = \{Fix(c)\}$ under the canonical embedding $i_X : X \to \mathcal{H}(X)$. This result tells us that the Hausdorff transition does not create interesting new fixpoints from a given contraction. However, there is a dramatic generalization when we consider the set of all contractions $Contra(\mathcal{H}(X))$ and its subset $\mathcal{H}(Contra(X))$.

Lemma 350 *The set $Contra(X)$ is closed under composition, i.e., if $c_1, c_2 \in Contra(X)$, then also $c_1 \circ c_2 \in Contra(X)$ (it is a semigroup, i.e., a monoid without neutral element), and so is the subset $\mathcal{H}(Contra(X))$ of $Contra(\mathcal{H}(X))$, because of the naturality of the $\mathcal{H}$-operator.*

Proof If $c_1, c_2 \in Contra(X)$ with constants $0 < u_1, u_2 < 1$, respectively, such that

$$d(c_1(x), c_1(y)) \le u_1 \cdot d(x, y) \quad \text{and} \quad d(c_2(x), c_2(y)) \le u_2 \cdot d(x, y),$$

then $d((c_1 \circ c_2)(x), (c_1 \circ c_2)(y)) \le u_1 \cdot d(c_2(x), c_2(y)) \le u_1 u_2 \cdot d(x, y)$ with $0 < u_1 u_2 < 1$. □

On the Hausdorff-metric space $\mathcal{H}(X)$, the contraction set is significantly richer by the following lemma:

Lemma 351 *If $c, d \in Contra(\mathcal{H}(X))$, then the sum map $c \boxplus d : \mathcal{H}(X) \to \mathcal{H}(X)$, defined by $(c \boxplus d)(K) = c(K) \cup d(K)$, is also a contraction. We have the following structure[1] on $Contra(\mathcal{H}(X))$: If $c, d, e \in Contra(\mathcal{H}(X))$, then*

(i) $c \boxplus d = d \boxplus c$.

(ii) $(c \boxplus d) \boxplus e = c \boxplus (d \boxplus e) = c \boxplus d \boxplus e$.

---

[1] This is called a semiring, since it is both a commutative additive and a multiplicative semigroup which are related by distributivity.

(iii) $(c \circ d) \circ e = c \circ (d \circ e) = c \circ d \circ e.$

(iv) $(c \boxplus d) \circ e = (c \circ e) \boxplus (d \circ e),$ and $c \circ (d \boxplus e) = (c \circ d) \boxplus (c \circ e).$

Proof We first show that for any compact $B, D, E \subset X$, we have $d(B, D \cup E) \leq d(B, D)$. In fact,

$$d(B, D \cup E) = \max\{d(b, D \cup E) \mid b \in B\}$$
$$= \max\{\min\{d(b, D), d(b, E)\} \mid b \in B\}$$
$$\leq \max\{d(b, D) \mid b \in B\} = d(B, D).$$

Then we have

$$d(B \cup C, D \cup E) \leq \max\{d(B, D \cup E), d(C, D \cup E)\}$$
$$\leq \max\{d(B, D), d(C, E)\}$$
$$\leq \max\{h(B, D), h(C, E)\},$$

The last expression is symmetric under exchange of B, D and C, E, so we have $h(B \cup C, D \cup E) \leq \max\{h(B, D), h(C, E)\}$. Now, taking the two above yields

$$h((c_1 \boxplus c_2)(K), (c_1 \boxplus c_2)(L)) = h(c_1(K) \cup c_2(K), c_1(L) \cup c_2(L))$$
$$\leq \max\{h(c_1(K), c_1(L)), h(c_2(K), c_2(L))\}$$
$$\leq \max\{u_1, u_2\} \cdot h(K, L),$$

if u_1 and u_2 are the constants of contraction of c_1 and c_2, respectively, whence $c_1 \boxplus c_2$ is a contraction. Showing properties (i) through (iv) reduces to straightforward set-theoretic calculations. $\qquad\square$

This finally gives us the necessary structure for the definition of so-called deterministic fractals on X.

Definition 236 *Given a complete metric space (X, d), denote by*

$$Frac(X) = \langle \mathcal{H}(Contra(X)) \rangle$$

the semiring in $Contra(\mathcal{H}(X))$ which is generated[2] by the semigroup $\mathcal{H}(Contra(X))$. By lemma 351, this is the set of all finite sums $\boxplus_i \mathcal{H}(c_i)$, where $c_i \in Contra(X)$. The elements $f \in Frac(X)$ are called (deterministic) *fractals on X, while the fixpoint $Fix(f)$ of such a fractal f is called its* attractor. *If we identify the attractor of f with the constant contraction, we have the evident constant morphism $f \to Fix(f)$ of fractals.*

Observe that we do not identify a fractal with its attractor, but rather stress the generating contraction as its structural essence.

---

[2] This is the smallest semiring in $Contra(X)$ which contains $\mathcal{H}(Contra(X))$.

Lemma 352 *If $h = T^t \circ g \in Aff_{\mathbb{R}}(\mathbb{R}^n)$ is an affine endomorphism of the Euclidean n-space, then h is a contraction iff its linear part g is so, and the latter is true if the Euclidean norm $\|g\| < 1$. In this case, the fixpoint of h is*

$$Fix(T^t \circ g) = \left(\sum_{i=0}^{\infty} g^i\right)(t).$$

Proof We know that $d(T^t \circ g(x), T^t \circ g(y)) = \|T^t \circ g(x) - T^t \circ g(y)\| = \|g(x - y)\| \leq \|g\| \cdot \|x - y\| = \|g\| \cdot d(x, y)$, so this proves the statements about the affine contractions. If such a contraction is given, then $(T^t \circ g)((\sum_{i=0}^{\infty} g^i)(t)) = t + (\sum_{i=1}^{\infty} g^i)(t) = (\sum_{i=0}^{\infty} g^i)(t)$. $\qquad\qquad\square$

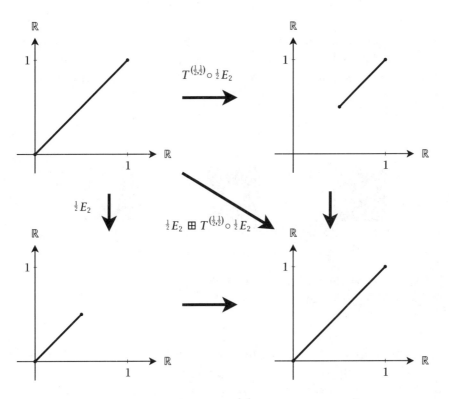

Fig. 40.2. The application of $\frac{1}{2}E_2 \boxplus T^{(\frac{1}{2}, \frac{1}{2})} \circ \frac{1}{2}E_2$ to Δ. The result at the lower right is $\Delta = \frac{1}{2}E_2(\Delta) \cup T^{(\frac{1}{2}, \frac{1}{2})} \circ \frac{1}{2}E_2(\Delta)$.

Example 186 A first simple, but very instructive example of a fractal involves the diagonal line $\Delta = \{\lambda \cdot (1, 1) \mid \lambda \in [0, 1]\}$ of the unit square in $\mathbb{R}^2$, see figure 40.2. It is the attractor of the fractal $\frac{1}{2}E_2 \boxplus T^{(\frac{1}{2},\frac{1}{2})} \circ \frac{1}{2}E_2$, where E_2 is the 2×2 unit matrix. The fact that the involved affine homomorphisms are contractions follows from lemma 352. Observe that the fixpoints of the induced contractions involved are $Fix(\frac{1}{2}E_2) = (0,0)$ and $Fix(T^{(\frac{1}{2},\frac{1}{2})} \circ \frac{1}{2}E_2) = (\frac{1}{2},\frac{1}{2})$.

Example 187 The Sierpinski carpet is very similar to the diagonal from the previous example. We start with the unit square $\blacksquare = \{(x,y) \mid 0 \le x \le 1, 0 \le y \le 1\}$. The fractal is given by

$$Sierpinski = \left(I \boxplus T^{(\frac{1}{3},0)} \boxplus T^{(0,\frac{1}{3})} \boxplus T^{(\frac{2}{3},0)} \boxplus T^{(0,\frac{2}{3})} \boxplus T^{(\frac{2}{3},\frac{1}{3})} \boxplus T^{(\frac{1}{3},\frac{2}{3})} \boxplus T^{(\frac{2}{3},\frac{2}{3})}\right) \circ \frac{1}{3}E_2.$$

In other words, the content of the unit square is scaled by $\frac{1}{3}$ and then placed at the north, south, west, east, northeast, northwest, southeast, southwest of the unit square. Figure 40.3 illustrates the first three applications of *Sierpinski*.

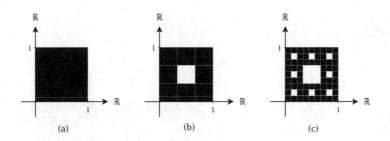

(a)   (b)   (c)

Fig. 40.3. (a) The unit square $\blacksquare$, (b) the first step *Sierpinski*($\blacksquare$), (c) the second step *Sierpinski*^{2}($\blacksquare$).

An approximation of the attractor *Fix(Sierpinski)* is shown in figure 40.4, but, of course, in contrast to the diagonal from example 186, the attractor itself cannot effectively be drawn, since it is infinitely "riddled with holes".

Example 188 The famous *Koch curve* is constructed using the same procedure as the previous two examples. To make things more clear, the transformation is split into four parts:

$Koch_1 = \frac{1}{3}E_2$: A scaling by $\frac{1}{3}$.

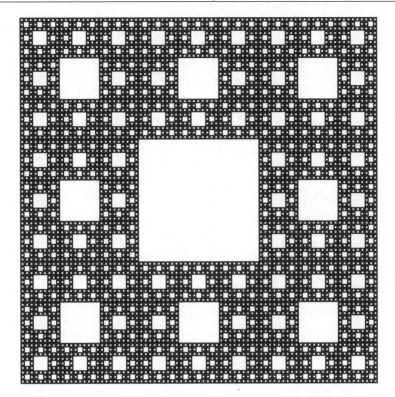

Fig. 40.4. The (approximated) attractor of the *Sierpinski* fractal. In fact it is the fifth iterate *Sierpinski*^{5}(■).

$Koch_2 = T^{(\frac{1}{3},0)}R_{60}\frac{1}{3}E_2$: A scaling by $\frac{1}{3}$, followed by a rotation R_{60} by $60°$ and then a translation by the vector $(\frac{1}{3},0)$.

$Koch_3 = T^{(\frac{1}{2},\frac{\sqrt{3}}{6})}R_{-60}\frac{1}{3}E_2$: A scaling by $\frac{1}{3}$, followed by a rotation by $-60°$ and then a translation by the vector $(\frac{1}{2},\frac{\sqrt{3}}{6})$.

$Koch_4 = T^{(\frac{2}{3},0)}\frac{1}{3}E_2$: A scaling by $\frac{1}{3}$, followed by a translation by the vector $(\frac{2}{3},0)$.

Then the fractal whose attractor is the Koch curve is

$$Koch = Koch_1 \boxplus Koch_2 \boxplus Koch_3 \boxplus Koch_4.$$

The first three iterations, starting with the unit interval $u = [0,1]$, are shown in figure 40.5. To make the four parts of the transformation more easily recognizable, their respective results are set in black for $Koch_1$, dark gray for $Koch_2$, middle gray for $Koch_3$ and light gray for $Koch_4$.

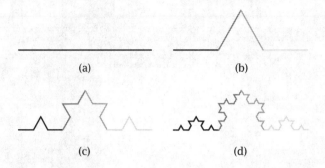

Fig. 40.5. (a) The unit interval $u = [0,1] = Koch^0(u)$, (b) $Koch^1(u)$, (c) $Koch^2(u)$, and (d) $Koch^3(u)$.

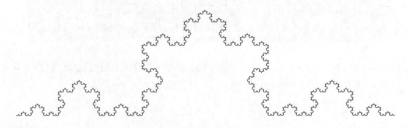

Fig. 40.6. The fifth iterate $Koch^5(u)$.

40.4 Fractal Dimension

The classification of fractals is far from being settled. By definition, a classification involves describing all "essentially" different fractals and grouping "similar" fractals into classes. More precisely, one may ask for the description of all equivalence classes of fractals under the relation $a \sim b$ (here $a \in Frac(X)$ and $b \in Frac(Y)$, where (X,d) and (Y,c) are complete metric spaces) iff there is an isomorphism $f : a \to b$ such that d and the transported metric $c_f(x,y) = c(f(x),f(y))$ on X are equivalent. The latter means that there are two positive constants u and v such that for all $x, y \in X$,

$$u \cdot d(x, y) \le c_f(x, y) \le v \cdot d(x, y).$$

Exercise 192 Prove that equivalence among metrics is in fact an equivalence relation.

A classical method for determining such equivalence classes is the exhibition of numerical invariants, i.e., numbers associated with fractals which do not change within an equivalence class. It is then hoped that one may find enough such invariants such that their totality is fine enough to distinguish classes by their associated invariants. For example, in the elementary case of finite sets X, the invariant $card(X) \in \mathbb{N}$ completely describes the equivalence class of such a set under set bijections (equipollence): X is equipollent to Y iff $card(X) = card(Y)$.

For a fractal a, a first such numerical invariant is its dimension $dim(a)$. This is a non-negative real number, which need not be defined for every fractal, but if it is defined for a on (X, d), and if the fractal b on (Y, c) is equivalent to a in the above sense, then $dim(b) = dim(a)$, which means that the dimension is a numeric invariant for the said equivalence relation.

The concept of a geometric dimension is motivated by two observations: To begin with, when we have a curved line A in $\mathbb{R}^n$, the dimension should be 1. Similarly, we expect that a subset $A \subset \mathbb{R}^n$, which is a patchwork of smoothly deformed pieces of a plane, has dimension equal to 2.

The second point is that the covering argument for such a patchwork can also be considered as "casting" of a "wild set" into a patchwork of "standard" covering charts, like line intervals, open/closed balls or cubes, etc. The ingenious idea in the concept of dimension, which we shall now describe, is that both observations can be merged into one and the same criterion. This criterion is, described intuitively, that the covering by "standard" charts exhibits a specific behavior when one successively decreases their admitted size. For example, if one covers a square S by adjacent small squares S_i having side length equal to, say, $\frac{1}{10}$ of the side length of S, then one needs 10^2 such squares, whereas if one covers an interval I by adjacent intervals of length $\frac{1}{10}$ of the length of I, one needs 10^1 of them. So the growth of the number of covering standard charts is an indicator of the dimension: 2 for a square, 1 for an interval. This combinatorial growth factor is the idea behind the following definition, which—as we shall see below—covers a number of fractals which are far from being standard geometric forms.

The number $N(n)$ of standard charts grows like n^{dim} if you cover your object by charts of $\varepsilon = 1/n$ the object size, or, equivalently, $N(\varepsilon)$ behaves like ε^{-dim}. If the dimension of an object is defined in this way, the dimension of a fractal is a real number and need not be a natural number. This, then, is the origin of the name "fractal": It describes geometric forms having fractional dimension.

Definition 237 *Let $K \in X$ be a non-empty compact set in a metric space (X, d). For a positive real number ε, call $\mathcal{N}(K, \varepsilon)$ the minimal natural number N such that there is a covering of K by N closed balls $\overline{B}_\varepsilon(x_i), i = 1, 2, \ldots N$. Since K is compact, this number always exists. For a positive real number ε, the ε-dimension of K is defined by*

$$dim(K, \varepsilon) = \frac{\log(\mathcal{N}(K, \varepsilon))}{\log(1/\varepsilon)},$$

i.e.,

$$\mathcal{N}(K, \varepsilon) = \varepsilon^{-dim(K, \varepsilon)}.$$

We say that K has fractal dimension $dim(K)$ iff $\lim_{\varepsilon \to 0} dim(K, \varepsilon)$ exists and is equal to $dim(K)$, i.e.,

$$dim(K) = \lim_{\varepsilon \to 0} dim(K, \varepsilon).$$

There is an important subtlety in this definition. It is obvious that the ε-dimension $dim(K, \varepsilon)$ depends on the basic choice of closed balls for the covering number $\mathcal{N}(K, \varepsilon)$. If, instead, we had taken closed cubes, the number would change. However, it would not change dramatically, but this difference must be dealt with, when one calculates the limit. Now, suppose that some change of the covering standard charts causes a change from $\mathcal{N}(K, \varepsilon)$ to $c \cdot \mathcal{N}(K, \varepsilon)$ for a positive constant c. Then the ε-dimension changes to

$$dim^*(K, \varepsilon) = \frac{\log(c \cdot \mathcal{N}(K, \varepsilon))}{\log(1/\varepsilon)} = \frac{\log(c)}{\log(1/\varepsilon)} + dim(K, \varepsilon),$$

which evidently converges to the limit of $dim(K, \varepsilon)$ since the first summand converges to zero. So we are happy that the quite arbitrary choice of the type of "standard" chart is in fact not relevant for the definition of the dimension.

There is another point which matters in this definition: The limit process uses all ε, while, in reality, such a condition is far from computationally effective. Happily, one has this result, which also uses closed cubes as easier "standard" charts instead of closed balls in the case of $X = \mathbb{R}^n$ with the standard Euclidean distance.

Proposition 353 (Box Counting Theorem) *Let $K \subset \mathbb{R}^m$ be a non-empty compact set. Define $\mathcal{N}_n(K)$ as the number of closed cubes ("boxes") of side length $\frac{1}{2^n}$, $B = \left[k_1 \cdot \frac{1}{2^n}, (k_1 + 1) \cdot \frac{1}{2^n}\right] \times \left[k_2 \cdot \frac{1}{2^n}, (k_2 + 1) \cdot \frac{1}{2^n}\right] \times \ldots \left[k_m \cdot \frac{1}{2^n}, (k_m + 1) \cdot \frac{1}{2^n}\right]$, $k_i \in \mathbb{Z}$, which have a non-empty intersection with K. If $dim(K)$ exists, then we have*

$$dim(K) = \frac{1}{\log(2)} \cdot \lim_{n \to \infty} \frac{\log(\mathcal{N}_n(K))}{n}.$$

Proof This theorem is a statement regarding the transition from counting balls in a finite covering of K by balls to counting small cubes in a finite covering of K by cubes. But it is also a theorem about the replacement of the limit for $\varepsilon \to 0$ by the limit for a sequence of size parameters $\varepsilon_n \to 0$. We shall not prove the following technical (but not really difficult) fact concerning the transition to a sequence of size parameters: It states that if we are given a compact set $K \subset X$ and a sequence $\varepsilon_n = c \cdot r^n$ with $c > 0$ and $0 < r < 1$, then the dimension D of K is defined iff the limit $D = \lim_{n \to \infty} dim(K, c \cdot r^n)$ exists. The necessity of this fact is clear, its sufficiency essentially follows from the fact that $\mathcal{N}(K, \varepsilon)$ is an increasing function of ε. For the details, refer to [5], section 5.1. Admitting this result, observe that a ball of radius $\frac{1}{2^n}$ hits at most 2^m of our boxes of length $\frac{1}{2^{n-1}}$ (to see this intuitively, you may draw a picture for $m = 1, 2, 3$). Therefore, $2^{-m}\mathcal{N}_{n-1}(K) \leq \mathcal{N}(K, \frac{1}{2^n})$. If a ball has radius $\frac{1}{2^n}$, then by the theorem of Pythagoras, it can include a cube of largest length $\frac{1}{2^{l(n)}}$, where $l(n)$ is the smallest integer with $\frac{1}{2^{l(n)}} \leq \frac{1}{2^{n-1}\sqrt{m}}$, i.e., $l(n) \geq n - 1 + \frac{1}{2}\log(m)$. Clearly, $\lim_{n \to \infty} \frac{l(n)}{n} = 1$. Therefore we also have $\mathcal{N}(K, \frac{1}{2^n}) \leq \mathcal{N}_{l(n)}(K)$ and

$$2^{-m}\mathcal{N}_{n-1}(K) \leq \mathcal{N}(K, \tfrac{1}{2^n}) \leq \mathcal{N}_{l(n)}(K) \leq \mathcal{N}_{2n}(K)$$

for large n. Hence

$$\frac{1}{\log(2)} \lim_{n \to \infty} \frac{\log(\mathcal{N}_n(K))}{n} = \frac{1}{\log(2)} \lim_{n \to \infty} \frac{\log(2^{-m}\mathcal{N}_n(K))}{n}$$

$$= \lim_{n \to \infty} dim(K, 1/2^n)$$

$$= \frac{1}{\log(2)} \lim_{n \to \infty} \frac{2\log(\mathcal{N}_{2n}(K))}{2n}$$

$$= \frac{1}{\log(2)} \lim_{n \to \infty} \frac{\log(\mathcal{N}_n(K))}{n}$$

defines the fractal dimension. □

Example 189 Let $K = [0,1]^3 \subset \mathbb{R}^3$ be the unit cube. Then we have $\mathcal{N}_1(K) = 8, \mathcal{N}_2(K) = 64, \ldots \mathcal{N}_n(K) = 8^n$. Therefore

$$\frac{1}{\log(2)} \cdot \frac{\log(\mathcal{N}_n(K))}{n} = \frac{1}{\log(2)} \cdot \frac{n\log(8)}{n} = 3,$$

whence $dim(K) = 3$, as expected.

Example 190 We calculate the dimension of the Koch curve using the original definition of dimension by closed balls (in this case, closed disks). By the preceding discussion, we can use as ε the diameter of the closed disks, instead of their radius, this will make the calculation a little simpler.

With $K = Fix(Koch)$, and referring to figure 40.7 (a), with diameter $\varepsilon = \frac{1}{3}$, $\mathcal{N}(K, \frac{1}{3}) = 4$, thus

$$dim(K, \tfrac{1}{3}) = \frac{\log(4)}{\log(3)}.$$

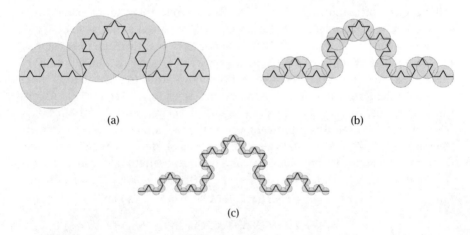

(a)   (b)

(c)

Fig. 40.7. Minimal coverings of the Koch curve, by disks of diameter (a) $1/3$, (b) $1/9$, and (c) $1/27$.

Parts (b) and (c) of the figure show the situation for $\varepsilon = \frac{1}{9}$ and $\varepsilon = \frac{1}{27}$, respectively. The corresponding values for $dim(K, \varepsilon)$ are

$$dim(K, \tfrac{1}{9}) = \frac{\log(16)}{\log(9)} \text{ and } dim(K, \tfrac{1}{27}) = \frac{\log(64)}{\log(27)}.$$

In general, it can be seen that

$$dim(K, 3^{-n}) = \frac{\log(4^n)}{\log(3^n)} = \frac{n\log(4)}{n\log(3)} = \frac{\log(4)}{\log(3)}.$$

Thus, the dimension of the Koch curve is

$$dim(K) = \frac{\log(4)}{\log(3)} \approx 1.26186.$$

A further interesting feature is the length of the Koch curve. The length of the unit interval is, of course, 1. The length of the first iterate is $4/3$, of the second $16/9$. Thus, the length of the attractor is $\lim_{n\to\infty} \frac{4^n}{3^n}$, which diverges to infinity. On the other hand, the area under the curve is finite, because the whole curve fits into the unit square. We have here an example of the "infinitely long coastline" of an island, often alluded to in popular literature on fractals.

Example 191 In order to calculate the dimension of the Sierpinski carpet, it is convenient to adapt the box counting theorem to use boxes of side length $1/3^n$ instead of $1/2^n$. The definition of dimension is accordingly adapted to:

$$dim(K) = \frac{1}{\log(3)} \cdot \lim_{n\to\infty} \frac{\log(\mathcal{N}_n(K))}{n}.$$

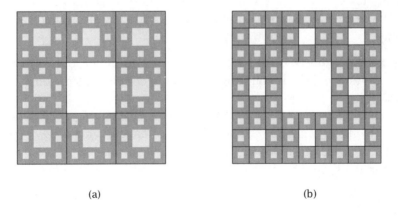

(a)   (b)

Fig. 40.8. Coverings of the Sierpinski carpet by squares of side length $1/3$ (a), and $1/9$ (b).

Setting $K = Fix(Sierpinski)$, we see from figure 40.8, that $\mathcal{N}_1(K) = 8$, $\mathcal{N}_2(K) = 64$, and, in general, $\mathcal{N}_n(K) = 8^n$. With the above formula

$$dim(K) = \frac{1}{\log(3)} \cdot \lim_{n \to \infty} \frac{\log(8^n)}{n}$$

$$= \frac{1}{\log(3)} \cdot \lim_{n \to \infty} \frac{n \log(8)}{n}$$

$$= \frac{1}{\log(3)} \cdot \log(8)$$

$$= 3 \cdot \frac{\log(2)}{\log(3)}$$

$$\approx 1.8928.$$

Exercise 193 The Sierpinski gasket (or triangle) is defined by the following contraction:

$$Gasket = \left(I \boxplus T^{(\frac{1}{4}, \frac{\sqrt{3}}{4})} \boxplus T^{(\frac{1}{2}, 0)} \right) \circ \frac{1}{2} E_2.$$

Draw a few iterations to get an approximation of $Fix(Gasket)$, starting with a triangle of side length 1.

Use the box counting theorem to determine $dim(Fix(Gasket))$.

Exercise 194 All examples so far have been about compact subsets of $\mathbb{R}^2$. The principles of this chapter, however, readily apply to $\mathbb{R}^3$ (and $\mathbb{R}^n$ for any integer $n > 0$, for that matter). Figure 40.9 shows the attractor of *Sponge*, the Sierpinski carpet extended to $\mathbb{R}^3$, also called "Menger sponge".

Proceed analogously to the Sierpinski carpet to define the contraction *Sponge* and use the box counting theorem to compute its dimension.

Proposition 354 *Let (X, d) and (Y, c) be complete metric spaces, and a $\in$ Frac(X) and b $\in$ Frac(Y). Suppose that we have a fractal isomorphism $f : a \to b$ which also induces an equivalence of metrics c and d_f, (i.e., a $\sim$ b in the sense of the introduction of this section). Then dim(Fix(a)) = dim(Fix(b)).*

Proof The proof of this theorem is very technical, we have to omit it here and again refer to [5], section 5.1.   □

Example 192 Let S be a shearing by 2 and R a rotation by 30° defined by the matrixes

$$S = \begin{pmatrix} 1 & 2 \\ 0 & 1 \end{pmatrix} \quad \text{and} \quad R = \begin{pmatrix} \frac{\sqrt{3}}{2} & \frac{1}{2} \\ -\frac{1}{2} & \frac{\sqrt{3}}{2} \end{pmatrix},$$

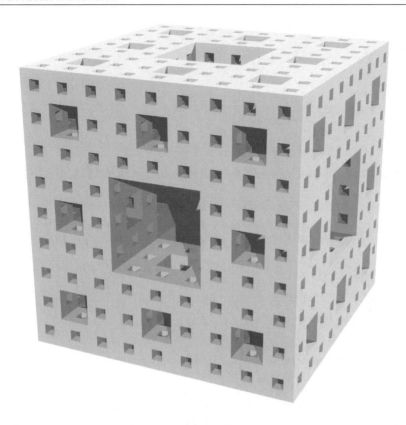

Fig. 40.9. This is not the Borg spaceship, but an approximation of *Fix(Sponge)*.

and let T be the isomorphism $T = RS$.

We have the metric spaces $(\mathbb{R}^2, d)$ and $(\mathbb{R}^2, c)$, where d is the usual Euclidean metric on $\mathbb{R}^2$, and $c(x, y) = d(T^{-1}x, T^{-1}y)$. Then c is a metric (the reader should verify this), and, since $c_T(x, y) = c(Tx, Ty) = d(T^{-1}Tx, T^{-1}Ty) = d(x, y)$, both metrics, d and c, are equivalent.

According to proposition 354, the dimensions of K and $T(K)$, where $K = Fix(Koch)$, are equal (figure 40.10).

A last example will close this chapter and illustrate how fractals can be used to model forms that occur in nature. The fractal function for *Barnsley's fern* is defined through these four functions from $\mathbb{R}^2$ to $\mathbb{R}^2$:

(a)   (b)

Fig. 40.10. The Koch curve K (a), and its transformation $T(K)$ (b).

$$fern_1((x,y)) = \begin{pmatrix} 0.85 & 0.04 \\ -0.04 & 0.85 \end{pmatrix} \begin{pmatrix} x \\ y \end{pmatrix} + \begin{pmatrix} 0 \\ 1.60 \end{pmatrix},$$

$$fern_2((x,y)) = \begin{pmatrix} -0.15 & 0.28 \\ 0.26 & 0.24 \end{pmatrix} \begin{pmatrix} x \\ y \end{pmatrix} + \begin{pmatrix} 0 \\ 0.44 \end{pmatrix},$$

$$fern_3((x,y)) = \begin{pmatrix} 0.20 & -0.26 \\ -0.23 & 0.22 \end{pmatrix} \begin{pmatrix} x \\ y \end{pmatrix} + \begin{pmatrix} 0 \\ 1.60 \end{pmatrix},$$

$$fern_4((x,y)) = \begin{pmatrix} 0 & 0 \\ 0 & 0.16 \end{pmatrix} \begin{pmatrix} x \\ y \end{pmatrix}.$$

Then $Fern_i = \mathcal{H}(fern_i)$, and $Fern = Fern_1 \boxplus Fern_2 \boxplus Fern_3 \boxplus Fern_4$. The compact start set can be chosen quite arbitrarily, however, the single point $(0,0)$ is enough to generate the beautiful picture in figure 40.11.

Fig. 40.11. The (approximated) attractor *Fix*(*Fern*).

CHAPTER 41

Neural Networks

41.1 Introduction

The term "Neural Networks" denotes an information processing paradigm which differs fundamentally from ordinary programming: instead of using a single powerful central processing unit one deals with a multitude of interconnected units. This model is based on the structure of biological brains which are huge networks of neurons ("brain cells"), see figure 41.1.

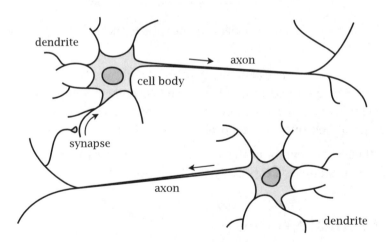

Fig. 41.1. Schematic rendering of two biological neurons, each of which transmits electrochemically encoded information through its axon and receives information on its dendritic ramifications.

As mentioned above, artificial or *formal neurons* are modeled after biological neurons: every neuron receives signals from other neurons, does some processing, and sends signals to other neurons to which it is connected.

The processing performed by formal neurons consists of several stages: first, a weighted sum is calculated from the *inputs I_i* and the synaptic *weights w_i*. This sum is passed to a *activation function a*, which determines how strong the "reaction" to the weighted sum should be. This result is changed (i.e., scaled or transformed etc.) by an *output function o*. The result of the output function, finally, is the value which is sent off to other formal neurons (see figure 41.2).

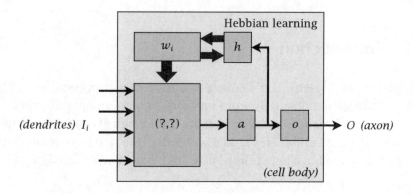

Fig. 41.2. The inner workings of a formal neuron.

A special feature of neural networks is their ability to "learn": Neural networks can be modified to better respond to a given set of inputs. Such training typically involves changing of the neurons' synaptic weights depending on their output. This process is called "Hebbian learning".

Since their introduction in the 1940s neural networks have been used in research and industry, especially for AI and robotics.

41.2 Formal Neurons

The theory of artificial neural networks (ANN) is an information processing theory, which is based on the exchange of time-dependent data streams. The mathematical model we pursue in this chapter deals with discrete time in $\mathbb{Z}$ and is based on data values in $\mathbb{R}$. Here is the basic framework of data streams.

Definition 238 *For $n \in \mathbb{N}$, the n-stream domain is the $\mathbb{R}$-vector space $D^n = (\mathbb{R}^{\mathbb{Z}})^n$. For $n = 1$, we also call it the* stream domain D. *An element $x = (x_t)_{t \in \mathbb{Z}} \in D^n$ is called an n-stream. Its evaluation x_t at "time" $t \in \mathbb{Z}$ is denoted by $x(t)$. We use the notation $p_i : D^n \to D$ for the i-th projection.*

By the canonical identification $(\mathbb{R}^{\mathbb{Z}})^n \xrightarrow{\sim} \mathbb{R}^{\mathbb{Z} \times n} \xrightarrow{\sim} (\mathbb{R}^n)^{\mathbb{Z}}$, an n-stream x is regarded either as an n-dimensional vector $x = (x_1, \ldots x_n)$ of streams $x_i \in D$ or else as a stream of n-dimensional vectors $x(t) \in \mathbb{R}^n$. Depending on the given context we shall make use of the appropriate view without special emphasis.

For $n \geq 1$, there is a canonical injection

$$\mathbb{R}^n \to D^n : \xi \mapsto x(t) = \xi, \text{ for all } t \in \mathbb{Z},$$

identifying a vector $\xi \in \mathbb{R}^n$ with what is called a constant n- stream, *which we then also denote by ξ. An n-stream $x \in D^n$ is called* initially constant, *iff there is $t_0 \in \mathbb{Z}$ such that $x(t) = x(t_0)$ for all $t \leq t_0$; it is called* eventually constant *iff there is $t_0 \in \mathbb{Z}$ such that $x(t) = x(t_0)$ for all $t \geq t_0$.*

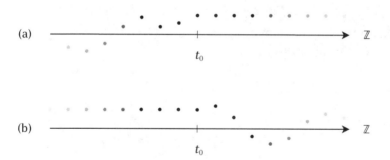

Fig. 41.3. (a) Eventually constant 1-stream, (b) initially constant 1-stream.

For $\lambda \in \mathbb{Z}$, one has a linear map, called the shift operator,

$$^{\lambda}? : D^n \xrightarrow{\sim} D^n$$

sending x to $^{\lambda}x$ such that $^{\lambda}x(t) = x(t + \lambda)$, the n-stream shifted by λ. A n-stream x is called periodic *iff there is an integer period $\tau > 0$ such that $^{\tau}x = x$.*

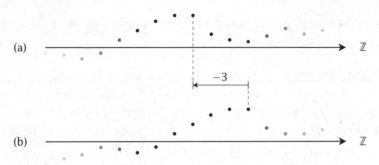

Fig. 41.4. (a) A 1-stream x, (b) the shifted 1-stream ^{-3}x.

Sorite 355 *With the notations of definition 238, we have these facts:*

(i) *The shift operator $^\lambda?$ defines a group action*

$$\mathbb{Z} \to GL(D^n),$$

i.e., $^0x = x$ and $^\mu(^\lambda x) = {}^{\mu+\lambda}x$, for all $x \in D^n$ and $\lambda, \mu \in \mathbb{Z}$.

(ii) *The subspace $\mathbb{R}^n \subset D^n$ of constant streams is the subspace of those streams x left invariant by every shifting, i.e., $^\lambda x = x$, all $\lambda \in \mathbb{Z}$. In other words: $\mathbb{R}^n = \bigcap_{\lambda \in \mathbb{Z}} {}^\lambda D^n$, where $^\lambda D^n$ is the vector subspace of D^n consisting of all λ-periodic streams; observe that $^0D^n = D^n$.*

(iii) *The stream domain projections p_i and the shift operators commute, i.e., the diagrams*

$$
\begin{array}{ccc}
{}^\mu D^n & \xrightarrow{\ \lambda?\ } & {}^\mu D^n \\
{\scriptstyle p_i}\downarrow & & \downarrow{\scriptstyle p_i} \\
{}^\mu D & \xrightarrow{\ \lambda?\ } & {}^\mu D
\end{array}
$$

shiftops
commute for all $i = 1, \ldots n$ and $\lambda, \mu \in \mathbb{Z}$.

To define a formal neuron, we shall deal with functions between n-stream domain spaces. The common construction is as follows. One is given a function $f : \mathbb{R}^n \to \mathbb{R}^m$ and applies this function at every time $t \in \mathbb{Z}$ to n-streams, i.e., one defines an induced function $f : D^n \to D^m$ by $f(x)(t) = f(x(t))$, the symbol f being unchanged, if no confusion is likely. Clearly such a function commutes with every shift operator (the reader should verify this as an exercise). For example, if $(?,?) : \mathbb{R}^n \times \mathbb{R}^n \to \mathbb{R}$ is the standard scalar product, we obtain a bilinear function $b : D^n \times D^n \to D$. But observe that the shift operators $^\lambda?$, with $\lambda \neq 0$, are not of this type of induced functions. We will also deal with other more general functions $h : D^n \to D^m$ between stream domains. A common one is defined by a family $h = (h(t))_{t \in \mathbb{Z}}$ of functions $h(t) : \mathbb{R}^n \to \mathbb{R}^m$, i.e. $h(x)(t) = h(t)(x(t))$.

Before proceeding to the definition of a formal neuron, it is of advantage to describe some simple processes in the language of diagrams and n-streams. Consider the way a neuron processes its input. All the signals that ever reach the neuron as inputs or leave the neuron can be interpreted as n-streams. If the weights can change over time as is the case in Hebbian learning, they must be described as n-streams, too. The weighted sum calculated from the inputs is now the stream that is the result of the scalar product of the input and weight n-streams. The entire process performed by a neuron can be represented by the diagram

$$D^n \times D^n \xrightarrow{(?,?)} D \xrightarrow{a} D \xrightarrow{o} D$$

From the n-streams describing weights and inputs a stream of output values is generated: A pair (w, x) of weight n-stream w and input n-stream x is mapped to the output stream $o(a(w, x))$.

Hebbian learning, which changes the weight vector at "time" t depending on the weight vector and an output value at "time" $t - 1$ can be represented like this:

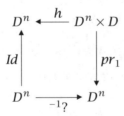

Here the identity Id and the projection pr_1 simply ensure that the the n-stream of weight represented in the upper line is identical to the one in the lower line. The time shift by -1 describes the fact that any weight which is changed by h is the successor in time of the weight which h uses as an argument.

Keep in mind that the arrows in this diagram *do not* depict the flow of data, but instead define certain relations between the sets.

Definition 239 *A formal neuron is a triple of functions*

$$N = (a : \mathbb{R} \to \mathbb{R}, o : \mathbb{R} \to \mathbb{R}, h : D^n \times D \to D^n),$$

where a is called the activation function, *o is called the* output function, $h = (h(t))_t$ *is called the* Hebbian learning function, *and n is called the* fan-in dimension *of N. If the symbol N is given, then one writes* $n = dim_N$, $a = act_N$, *and* $o = out_N$.

The process diagram $\mathcal{P}_N$ *of a formal neuron N is the diagram of set functions*

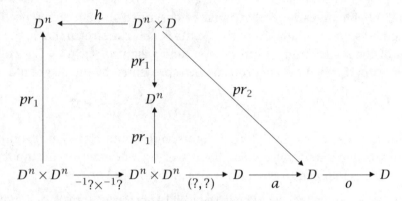

with the standard scalar product $(?, ?)$ *as explained above. The process diagram gives rise to its limit*[1]

$$S_N = \lim \mathcal{P}_N.$$

This is called the state space *of the formal neuron N.*

Let $\alpha : S_N \to D$ *be the projection into the codomain D of the output function o in* $\mathcal{P}_N$; α *is called the* axon *or* output *of N. For each* $i = 1, \ldots n$, *we have the projection* $\delta^i : S_N \to D$ *induced by the projection* $D^n \times D^n \xrightarrow{pr_2} D^n \xrightarrow{pr_i} D$ *from the 2n-stream space in the left lower corner of* $\mathcal{P}_N$; δ_i *is called the* i-th dendrite *or* input *of N. The associated diagram of set functions*

---

[1] Remember that for the category of sets the limit can be constructed by selecting the subset of the Cartesian product of all n-streams which contains exactly those n-streams which form a solution of the "equations" described by the diagram (see proposition 309).

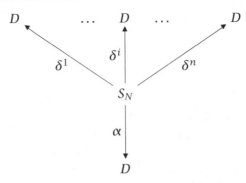

is denoted by $\blacktriangleright_N$ *and is called the* input-output shape, *or* IO-shape, *of* N. *An element* $s \in S_N$ *is called a* state *of* N, *and for a state* s, *its vector* $\delta(s) = (\delta^1(s), \ldots \delta^n(s)) \in D^n$ *is called its* input vector, *the value* $\alpha(s) \in D$ *its* output, *and the first projection* $\omega(s)$ *of* s *to* D^n *in the lower left corner* $D^n \times D^n$ *of the process diagram is called the* weight *of* s. *If in a neuron, the weight is fixed to* w, *one denotes by* N^w *the corresponding restricted structure, including the restricted limit* $S_N^w \subset S_N$, *and the corresponding IO-shape* $\blacktriangleright_N^w$.

Example 193 If $dim_N = 0$, then the process diagram reduces to

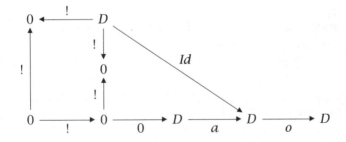

and the limit consists of exactly one element: From the lowest horizontal level, we have the image $0 \in D$ of the 0 map into D. Its image $a(0)$, and the image $o(a(0))$, together with the image $a(0)$ under the identity on D concludes the description of the unique limit element. Therefore, the only output element in the axon of such a neuron is the element $o(a(0))$. This means that a zero-dimensional neuron N is essentially defined by a single stream $o = o(a(0)) \in D$. Therefore a zero-dimensional neuron is also called an *input neuron*, since its state space is a singleton, yielding a single axonal output stream o. Such a neuron's IO-shape $\blacktriangleright_N$ is denoted by $\triangleright o$. An input neuron typically represents the sensorial input for the

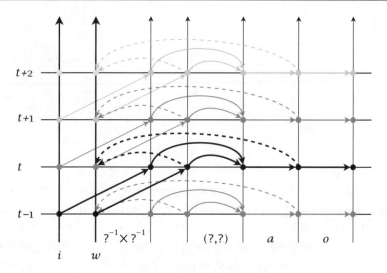

Fig. 41.5. This figure shows how the elements of the various n-streams are handled in a neuron. The diagonal arrows represent the time-shift operator: they transfer input and weights from $t-1$ to t. The calculation of the weighted sum is symbolized by the pair of curved arrows going to the right. The dashed arrows represent the Hebb learning function, which uses the weights from the previous time step and the current result of the activation function to change the current weights.

neural processing, such as it is provided by the retina (visual sense) or cochlea (auditory sense).

Example 194 Neurons may also just pipe information without further changes. Such a neuron N is defined by $dim_N = 1$ and $a_N = o_N = Id_D$, whereas h is the constant map with value 1. What does the state space look like? Let $x \in D$ be any stream. Then it consists of all these elements at the places of our diagram:

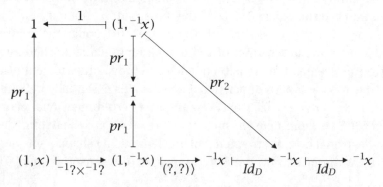

We denote this neuron by Thru, and its IO-shape $\blacktriangleright$Thru by $D \xrightarrow{\text{Thru}} D$, so that its elements are the pairs $x \xrightarrow{\text{Thru}} {}^{-1}x$. This means that the Thru neuron just shifts input and output time by -1, i.e., the output stream at time t is the input stream at time $t - 1$.

Example 195 Suppose that the activation function of a neuron N with dimension $n = dim_N$ is the negative translation by $\theta \in \mathbb{R}$, $a_N = T^{-\theta}$. A *sigmoid function* is a monotone (non-decreasing) function $\sigma : \mathbb{R} \to [0, 1] \subset \mathbb{R}$ such that $\lim_{x \to -\infty} \sigma(x) = 0$ and $\lim_{x \to \infty} \sigma(x) = 1$, see figure 41.9. If the output function of N is the sigmoid function σ, then the neuron is denoted by $\text{Perc}_{n,\theta,\sigma}$, and it is called a *$\sigma$-perceptron with threshold θ*. In particular, if $\sigma = \chi_{[0,\infty[} = \eta$, the *Heaviside function*, then $\text{Perc}_{n,\theta,\eta}$ is also called a *perceptron* or a *McCulloch-Pitts neuron* $\text{MCPerc}_{n,\theta}$. In this case, for any pair $(w, x) \in D^n \times D^n$ of a weight w and an input x of a state s, the output of that state at time $t + 1$ is 0 if $(w(t), x(t)) < \theta$, and 1 else.

Example 196 A special case is the invariant weight learning function $h_N = pr_1 : \mathbb{R}^n \times \mathbb{R} \to \mathbb{R}^n$, which, as the name suggests, does not change the weights. This means that in the state space, the weights must be constant. Therefore the state space S_N of such a neuron contains the non-empty subspace CS_N^w of constant inputs and outputs and weight w. While any input $x \in D^n$ is allowed, the corresponding output is $o_N(a_N((w, x)))$, and, in the case of a McCulloch-Pitts neuron $\text{MCPerc}_{n,\theta}$, the constant state space $CS_{\text{MCPerc}_{n,\theta}}^w$ for w is described by the IO-pairs $(x, \eta((w, x) - \theta))$, or, equivalently, by the input vectors $x \in \mathbb{R}^n$ alone. Identifying $CS_{\text{MCPerc}_{n,\theta}}^w$ with $\mathbb{R}^n$, the McCulloch-Pitts neuron $\text{MCPerc}_{n,\theta}$ defines a partition of the input space $\mathbb{R}^n$ by the two fibers $CS_{\text{MCPerc}_{n,\theta}}^{w,+} = \{x \mid \eta((w, x) - \theta) = 1\}$ and $CS_{\text{MCPerc}_{n,\theta}}^{w,-} = \{x \mid \eta((w, x) - \theta) = 0\}$ of the output function. For any subset $X \subset \mathbb{R}^n$ of constant inputs, we then write $X^+ = X \cap CS_{\text{MCPerc}_{n,\theta}}^{w,+}$ and $X^- = X \cap CS_{\text{MCPerc}_{n,\theta}}^{w,-}$, supposing that the perceptron parameters n, θ are clear. Furthermore, as a shorthand notation, we write the weights of McCulloch-Pitts neuron as a superscript: $\text{MCPerc}_{2,0.5}^{-1,0.5}$ is a neuron having two inputs with assigned weights -1 and 0.5, respectively.

In logical applications, the interesting subset is $X = Q^n = 2^n$ of bit sequences of length n. The subsets $Q^{n,+}$ and $Q^{n,-}$ represent those n-bit sequences with output 1 and 0, respectively. So this situation deals with the problem of representing logical functions $f : Q^n \to Q$ by means of perceptron state spaces for given weight w and threshold θ.

Example 197 Consider the neuron $\text{MCPerc}_{2,-0.5}^{(-1,1)}$. We want to show that it represents the implication IMPLIES : $Q^2 \to Q$. We simply calculate the result for all four input combinations:

$$\eta(((0,0),(-1,1)) - (-0.5)) = \eta(0.5) \qquad\qquad = 1$$
$$\eta(((0,1),(-1,1)) - (-0.5)) = \eta(1.5) \qquad\qquad = 1$$
$$\eta(((1,0),(-1,1)) - (-0.5)) = \eta(-0.5) \qquad\qquad = 0$$
$$\eta(((1,1),(-1,1)) - (-0.5)) = \eta(0.5) \qquad\qquad = 1$$

The results are in complete accordance with the truth table for the logical operation 'IMPLIES' (cf. section 1.1 in volume 1).

Exercise 195 Show that the negation NOT : $Q \to Q$ is represented by the neuron $\text{MCPerc}_{1,-0.5}^{-1}$, the conjunction AND : $Q^2 \to Q$ by $\text{MCPerc}_{2,1.5}^{(1,1)}$, and the disjunction OR : $Q^2 \to Q$ by $\text{MCPerc}_{2,0.5}^{(1,1)}$.

41.2.1 Geometric Interpretation of Perceptron Processing

By an easy calculation it can be shown that there is no perceptron that represents the XOR function, which returns 1 exactly for $(0,1)$ and $(1,0)$. In general, it can be shown that of all possible functions $f : Q^n \to Q$, only a small number can be represented by a perceptron. Let us briefly give a geometric interpretation of the situation. We suppose $w \neq 0$, otherwise everything becomes trivial.

For a given weight w, the value of activation function $x \mapsto (w,x) - \theta$ is checked for being non-negative or negative by the Heaviside output function. The separating value is $x \mapsto (w,x) - \theta = 0$. The solutions of this equation are described as follows. Since $(w,w) \neq 0$, there is a real number r such that $(w, r \cdot w) = \theta$, i.e., we find a vector d such that $(w,d) = \theta$. Therefore the equation becomes

$$(w, x - d) = 0, \text{ i.e., } w \perp (x - d).$$

This means that the solutions are those x, which, when shifted by $-d$, are contained in the subspace $w^\perp \subset \mathbb{R}^n$. The latter is the kernel of the non-zero linear form $(w,?) : \mathbb{R}^n \to \mathbb{R}$. This means that $w^\perp$ is an $(n-1)$-dimensional subspace of $\mathbb{R}^n$, i.e., a *hyperplane of* $\mathbb{R}^n$, see definition 216 of section 24.1 in volume 1. Therefore the solution set of the activation function $(w, x - d) = 0$ is the shifted hyperplane $H = T^{+d}w^\perp$. We can now regard the space $\mathbb{R}^n$ as split into two subsets H^+ and H^-, where our hyperplane H is the boundary between the closed set H^+ and the open complement H^-, see figure 41.6. This is clear from the fact that $(w, ? - d)$ is continuous.

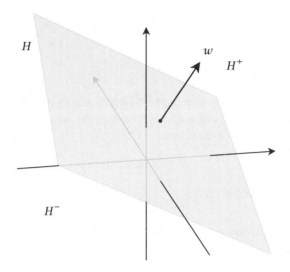

Fig. 41.6. A hyperplane in $\mathbb{R}^3$ is a Euclidean plane, here the plane satisfying the equation $x - y + 2z = 2$. The vector w is normal to the plane.

As an application one recognizes immediately that XOR is not representable by a perceptron. In fact, the fibers $\text{XOR}^{-1}(1)$ and $\text{XOR}^{-1}(0)$ are two sets of two points each, which both lie on diagonal positions on the rectangle spanned by Q^2, see figure 41.7. Such subsets evidently cannot be separated by a hyperplane $H \subset \mathbb{R}^2$, i.e., a line.

The question is whether one can find a weight w which yields a separation of the two fibers of a logical function f by means of an algorithm. Now, one procedure would be to start from a first weight chosen at random, and then successively approximating a "good" weight. This leads us to the learning process, which has been defined by the Hebb function h above, i.e., to non-constant weight streams.

But there is more to be done in our conceptualization in order to solve the XOR problem and other problems related to the representation of logical functions by use of perceptrons.

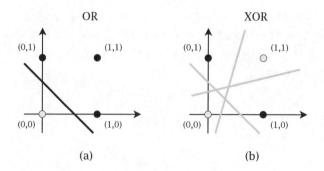

Fig. 41.7. (a) In $\mathbb{R}^2$ there is a hyperplane separating the fibers $OR^{-1}(1)$ and $OR^{-1}(0)$. (b) In the case of XOR, there is no hyperplane which can separate $XOR^{-1}(1)$ and $XOR^{-1}(0)$.

41.3 Neural Networks

Single formal neurons are only the elements of a construction which tends to simulate neural behavior by mathematical objects. The combination of neurons is defined as follows. Let $\mathcal{N}$ be the set of all neurons.

Definition 240 *If N_1 and N_2 are neurons, and $1 \leq i \leq dim(N_2)$, then the triple (N_1, N_2, i) is called an* elementary morphism *from N_1 to N_2, and is denoted by $i : N_1 \to N_2$. Denote by $\mathcal{E}$ the set of all elementary morphisms. Then the* neural category $C\mathcal{N}$ *is the path category of the directed graph $E\mathcal{N} : \mathcal{E} \to \mathcal{N}^2$, where $\Gamma(i : N_1 \to N_2) = (N_1, N_2)$. A* neural network *is a (finite) diagram $\mathcal{D}$ in $C\mathcal{N}$. Except when explicitly stressed, neural networks are assumed to be* elementary *in the sense that its morphisms are all elementary. This means that we are given a morphism $\mathcal{D} : \Delta \to E\mathcal{N}$ of digraphs. An* output neuron *in a neural network is a neuron which is not the domain of a morphism of $\mathcal{D}$.*

Observe that there is no elementary morphism $i : N_1 \to \rhd o$ whose codomain is an input neuron, because $dim(\rhd o) = 0$. An elementary morphism $i : N_1 \to N_2$ gives rise to the following diagram

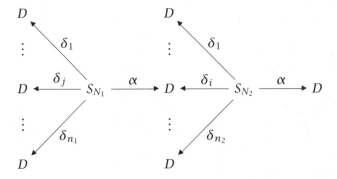

denoted by $\blacktriangleright_{N_1} \xrightarrow{i} \blacktriangleright_{N_2}$ and called *input-output (or IO-) shape of the elementary morphism* $i : N_1 \to N_2$. Such a neural connection is also visualized by triangles, where the morphism index i is visualized as a bullet, see figure 41.8.

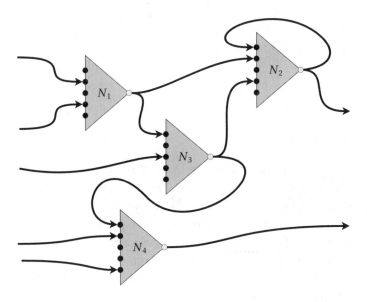

Fig. 41.8. Graphical representation of a neural network showing how the axon output serves as dendritic input for one or more other neurons.

More generally, if we are given any morphism

$$f : N_1 \xrightarrow{i_1} N_2 \xrightarrow{i_2} N_3 \cdots \xrightarrow{i_k} N_k,$$

we obtain a corresponding diagram

$$\blacktriangleright f : \blacktriangleright N_1 \xrightarrow{i_1} \blacktriangleright N_2 \xrightarrow{i_2} \blacktriangleright N_3 \cdots \xrightarrow{i_k} \blacktriangleright N_k$$

also called *input-output (or IO-) shape of f*. In particular, we have $\blacktriangleright_{Id_N} = \blacktriangleright_N$. If one is given a neural network $\mathcal{D}$, then one deduces the diagram of set maps from the IO-shapes of the diagram's morphisms. This one is denoted by $\blacktriangleright_{\mathcal{D}}$ and called the *input-output (or IO-) shape of $\mathcal{D}$*. The limit $S_{\mathcal{D}} = \lim \blacktriangleright_{\mathcal{D}}$ of this diagram is called the *state space of the neural network $\mathcal{D}$*. If one fixes a family $w. = (w_i)_i$ of weight streams w_i, the subspace of $S_{\mathcal{D}}$ with $w.$ as weight coordinate is denoted by $S_{\mathcal{D}}^{w.}$. The subfamily of input neurons of $\mathcal{D}$ is denoted by $\triangleright\mathcal{D}$, whereas the subfamily of output neurons is denoted by $\mathcal{D}\blacktriangleright$. This gives rise to a diagram $\triangleright S_{\mathcal{D}}\blacktriangleright$ of set maps given by the projections for all inputs and outputs:

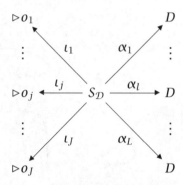

and called *input-output (or IO-) interface of the neural network $\mathcal{D}$*. If one has fixed a weight stream family $w.$, the corresponding diagram is denoted by $\triangleright S_{\mathcal{D}}^{w.}\blacktriangleright$. Here, the α_l-arrow is the projection into the stream domain given by the limit projection to the axonal map codomain of the l-th output neuron. The input map ι_j is the projection onto the unique stream o_j defined by the specific j-th input neuron. If in a neural network $\mathcal{D}$, a neuron has an input index which is not connected to the axonal output of another neuron, it is called a *free input*.

These constructions deserve and need a number of comments and illustrations.

Remark 34 To begin with, if a general neural network is given, it may happen that one would like to have the output of a neuron N_i which is not an output neuron. For example, it may happen that one has a circular neural network, defined by a cyclic digraph, and that one would like to trace the local outputs of the cycle. This is easily realized by adding a neuron $D \xrightarrow{\text{Thru}} D$, which starts at the required local output domain. One may also require that the time shift of the output is not $+1$ as given per default on Thru. This is easily settled in the following exercise 196. By using such Thru pipes, one may suppose, without loss of generality, that the only output of a neural network $\mathcal{D}$ is the information really available on its output $\mathcal{D} \blacktriangleright$.

Exercise 196 The morphism of neurons defined by a path of k concatenated Thru neurons

$$\text{Thru}^k : D \xrightarrow{\text{Thru}} D \xrightarrow{\text{Thru}} D \cdots \xrightarrow{\text{Thru}} D$$

produces a time shift of k units. The morphism of neurons defined by a path of k concatenated Thru neurons in the reversed direction

$$\text{Thru}^{-k} : D \xleftarrow{\text{Thru}} D \xleftarrow{\text{Thru}} D \cdots \xleftarrow{\text{Thru}} D$$

produces a $(-k)$-fold time shift. The concatenation $\text{Thru}^0 : D \xrightarrow{\text{Thru}} D \xleftarrow{\text{Thru}} D$ produces a zero-time shift morphism, i.e., a morphism, whose output is exactly the input. These types of neural morphisms play the role of stream pipes without further functionality, except for a controlled time shift.

Remark 35 The threshold θ in a perceptron $N = \text{MCPerc}_{n,\theta}$ can be externalized and thereby set to 0 by the following trick. We first change to a perceptron $N^\theta = \text{MCPerc}_{n+1,0}$ with the learning rule $h^\theta : \mathbb{R}^{n+1} \times \mathbb{R} \to \mathbb{R}^{n+1} : ((w,z),x) \mapsto (w,\theta)$. This means that the old learning rule is preserved for the first n coordinates (w) and yields the constant value θ for the $(n+1)$-st coordinate. The input is augmented by one dimension, but here, at the $(n+1)$-st input slot, we insert the output of the input neuron $\triangleright_{-1}$ with constant stream -1. This yields the bilinear values $((w,\theta),(x,-1)) = (w,x) - \theta$, and we have recovered the old threshold. This means that the neural network $\triangleright_{-1} \to N^\theta$ simulates $N = \text{MCPerc}_{n,\theta}$ with a specialized input set, but with the zero threshold on N^θ instead of θ. This is a variant of the classical method to turn an affine map into a linear map by homogenization, see also section 22.1 in volume 1.

The following construction of a perceptron learning algorithm is crucial for finding weights that separate two finite sets $X, Y \subset \mathbb{R}^n$ of constant streams by adequate hyperplanes H. Without loss of generality, we may even suppose that X and Y have no points on H (otherwise, shift H a little away from X). By the previous remark 35, we may suppose that the threshold vanishes, i.e., that $H = w^\perp$ is a vector subspace. Then separation means that $X \subset H^+, Y \subset H^-$, both disjoint from H. But this is clearly equivalent to $X \cup -Y \subset (H^n)^o$, or else, $(z, w) > 0$ for all $z \in X \cup -Y$. So we are led to this problem: Given a finite subset $Z \subset \mathbb{R}^n$, find a vector w such that $(z, w) > 0$, all $z \in Z$. The point is that one has to know that such a vector exists. Supposing this is the case, how can one find a candidate? This is the perceptron learning algorithm. The remarkable thing is that it is itself a neuron N, more precisely:

Definition 241 (Perceptron Learning Algorithm) *Suppose that these data are given: A natural number $n > 0$, a constant weight $w_0 \in \mathbb{R}^n$, a finite set $Z \subset \mathbb{R}^n$ of constant n-streams, an input stream p with positive bounds, $0 < e \le p(t) \le f < \infty$, all $t \in \mathbb{Z}$. Let u be in D^n, with values $u(t) \in Z$, and such that for each $z \in Z$, and each time t_0, there is $t \ge t_0$ with $u(t) = z$.*

Let $N_{u,p,w_0} = (a, o, h)$ be the n-dimensional neuron with $a = o = Id_\mathbb{R}$. The learning function $h = (h(t))_t$ reads a follows:

  (i) *$h(t) = const. = w_0$ for all $t \le 0$.*

 (ii) *Suppose that $t > 0$. Then for $(v, \lambda) \in \mathbb{R}^n \times \mathbb{R}$, if $(v, z) > 0$, for all $z \in Z$, then $h(t)(v, \lambda) = v$.*

(iii) *If for $t > 0$ there is a $z \in Z$ with $(v, z) \le 0$, set $h(t)(v, \lambda) = v$ if $\lambda > 0$, else set $h(t)(v, \lambda) = v + u(t - 1)p(t - 1)$.*

The perceptronlearning algorithm for u, p, w_0 *is the neural network*

$$\triangleright u \to N_{u,p,w_0}.$$

Its state space is evidently a singleton and may be identified with the recursively defined weight stream $w(u, p, w_0)$, called the generated weight.

The point of this weight construction is that $w(u, p, w_0)$ is eventually constant and solves the problem of finding a separating vector!

Proposition 356 (Perceptron Convergence Theorem) *For $n > 0$, let $Z \subset \mathbb{R}^n$ be a finite set of constant streams. Suppose there is a constant weight $w^* \in \mathbb{R}^n$ such that $(w^*, z) > 0$, all $z \in Z$. Let u, p, w_0 as in definition 241. Then $w(u, p, w_0)$ is eventually constant and the eventually constant value w_∞ also has the separating property*

$$(w_\infty, z) > 0, \ all \ z \in Z.$$

Proof Let $\alpha = e \cdot \min\{(w^*, z) \mid z \in Z\}$ and $\beta = f \cdot \max\{\|z\| \mid z \in Z\}$. So for all times and all $z \in Z$, $0 < \alpha \leq p(t) \cdot (w^*, z)$. Set $w' = \frac{\beta^2}{\alpha} w^*$. Consider the case where the weight $w = w(u, p, w_0)$ needs a proper update $w(t+1) \neq w(t)$ from time t to time $t + 1$, i.e., $(w(t), u(t)) \leq 0$. Then

$$
\begin{aligned}
\|w(t+1) - w'\|^2 &= \|w(t) + p(t)u(t) - w'\|^2 \\
&= \|w(t) - w'\|^2 + 2p(t)(u(t), (w(t) - w')) + p(t)^2 \|u(t)\|^2 \\
&\leq \|w(t) - w'\|^2 - 2p(t)(u(t), w') + \beta^2 \\
&\leq \|w(t) - w'\|^2 - 2p(t)(u(t), w^*)\frac{\beta^2}{\alpha} + \beta^2 \\
&\leq \|w(t) - w'\|^2 + \beta^2.
\end{aligned}
$$

Now, take an increasing sequence of indexes $t_0 < t_1 < t_2, \ldots$ such $w(t_i - 1) \neq w(t_i)$. Then the weight sequence for these times fulfills the conditions for the above estimation. So we obtain $0 \leq \|w(t_{s+1}) - w'\|^2 \leq \|w(t_0) - w'\|^2 - s\beta^2$, which means $s \leq \frac{\|w(t_0) - w'\|^2}{\beta^2}$. Therefore such changes must stop for large times as the increasing sequences $t_0 < t_1 < t_2 \ldots$ must be finite. □

41.4 Multi-Layered Perceptrons

We have seen that a perceptron is not capable of representing the XOR function. But it suffices to consider very simple neural networks to do so. Here is one construction. To this end, consider the perceptrons $\text{NOT}_{\text{MCPerc}} = \text{MCPerc}_{1,-0.5}^{-1}$, $\text{AND}_{\text{MCPerc}} = \text{MCPerc}_{2,1.5}^{(1,1)}$ and $\text{OR}_{\text{MCPerc}} = \text{MCPerc}_{2,0.5}^{(1,1)}$, which were defined in exercise 195. Then we consider this neural network XOR:

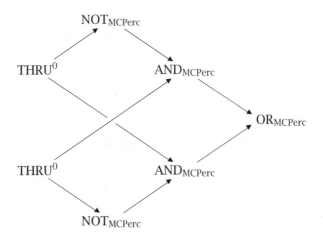

Then the IO-shape of XOR evaluates the free input values (x_1, x_2) to $XOR(x_1, x_2)$. Therefore, one is interested in the construction of neural networks in order to represent logical functions $f : Q^n \to Q$.

The above functionality can however be achieved by a slightly simpler neural network. To this end, we integrate the negation in the conjunction through this change of weights and thresholds: Let $AND^*_{MCPerc} = MCPerc^{(-1,1)}_{2,-0.5}$ and $AND^{**}_{MCPerc} = MCPerc^{(1,-1)}_{2,-0.5}$. Then the neural network XOR^*

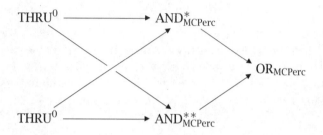

has an IO-shape which also represents the XOR function. It is a network which is of a type called *multi-layered perceptron*. We recognize an input layer consisting of the two $THRU^0$ neurons, a so-called hidden layer consisting of the perceptrons AND^*_{MCPerc} and AND^{**}_{MCPerc}, and the output layer of the OR perceptron.

Definition 242 *Let $n > 0$ be a natural number. A digraph $\Gamma : A \to V$ is called n-layered if $V = \coprod_{i=0,1,...n,n+1} V_i$ is a disjoint union of non-empty subsets, the layers of Γ, all arrows are from V_i to V_{i+1}, for each $v \in V_i, i \le n, deg^+(v) > 0$, for each $v \in V_i, 0 < i, deg^-(v) > 0$, for $v \in V_0, deg^-(v) = 0$, and for $v \in V_{n+1}, deg^+(v) = 0$. Layer V_0 is called the* input layer, *layer V_{n+1} is called the* output layer *of Γ. The other layers are called the* hidden layers.

Definition 243 *For a sigmoid function σ, a neural network $\mathcal{D}$ is called an n-layered (σ-)perceptron iff its graph is n-layered, and each neuron of the network is a (σ-)perceptron (including input neurons $\triangleright o$ for the input layer). Moreover, $\mathcal{D}$ is called* saturated *iff no neuron has a free input, unless it lies in the input layer.*

Exercise 197 Show that every non-saturated n-layered perceptron can be embedded in a saturated $(n + 2)$-layered σ-perceptron, whose input neurons are all one-dimensional by adding an adequate number of THRU^0 neurons, without changing its state space.

Definition 244 *A* feed-forward neural network *is a neural network* $\mathcal{D}$ *whose digraph is acyclic. A neural network whose digraph is cyclic is called* recurrent.

Example 198 An n-layered (σ-)perceptron is a feed-forward neural network.

Example 199 A *Hopfield* network $\mathcal{D}$ is the following type of recurrent neural network. The digraph of $\mathcal{D}$ is a complete digraph with m vertexes, i.e., for any two vertexes $v_i, v_j, i, j = 1, 2, \ldots m$, there is exactly one arrow $v_i \to v_j$ (see also example 23 in volume 1). Its neurons are perceptrons $\mathrm{Perc}^{w_i}_{m, \theta_i}$. Call the $(m \times m)$-matrix W, whose rows are the weight vectors w_i, the *weight matrix*, and write $\theta = (\theta_i)$. If the total output vector is $x \in D^m$, then we have the Hopfield network equation

$$x(t + 1) = \eta(W \cdot x(t) - \theta)$$

for all times $t \in \mathbb{Z}$, with η being applied to each coordinate. It can be shown that for a symmetric weight matrix, the state space contains only constant streams $x_\infty \in \mathbb{R}^m$ or else period-2 streams x_∞, i.e., ${}^2 x_\infty = x_\infty$.

The following proposition shows that 1-layer perceptrons can be used to represent any logical function.

Proposition 357 *Any logical function* $f : Q^n \to Q$ *can be represented by a* 1-*layer saturated perceptron.*

The following proposition shows that 2-layer perceptrons are sufficient for the separation of any two disjoint closed sets, if one of them is bounded.

Proposition 358 *Any two disjoint subsets* $A, B \subset \mathbb{R}^n$, *with A closed and B compact, can be separated by a function* $f : \mathbb{R}^n \to Q$ *which is realized by a* 2-*layer saturated perceptron, i.e.,* $A \subset f^{-1}(0)$ *and* $B \subset f^{-1}(1)$.

For proofs of these propositions, see [42].

41.5 The Back-Propagation Algorithm

The back-propagation algorithm is a procedure for the successive improvement of the weights of a multi-layered σ-perceptron based on given inputs and required corresponding outputs. It was first proposed by Frank Rosenblatt in [33]. In contrast to proposition 358, this algorithm starts with a given neural network and does not vary its vertexes, but only the weights. This procedure does not aim at finding weights which generate exact outputs for given inputs, but just offers a method for approaching the desired output. In general, it is not even guaranteed that this algorithm really produces convergent results, but only a way of "training" the network to work better according to a predefined process. It is however one of the backbones of artificial neural network theory and practice.

The back-propagation algorithm typically uses a sigmoid function

$$\sigma(x) = \frac{1}{1 + e^{-x}}$$

which is a diffeomorphism $\sigma : \mathbb{R} \overset{\sim}{\to}]0, 1[$ onto the open unit interval. It has the property that its derivative is expressed in terms of the function itself, i.e., $\sigma' = \sigma(1 - \sigma)$. Its inverse is $\sigma^{-1}(x) = \log(\frac{x}{1-x})$, see figure 41.9.

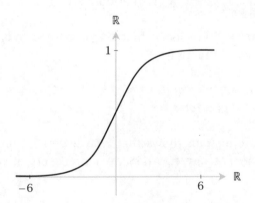

Fig. 41.9. The sigmoid function $\sigma(x) = \frac{1}{1+e^{-x}}$.

The neurons of the multi-layered perceptron $\mathcal{D}$ in this situation are σ-perceptrons $\mathrm{Perc}_{d,0,\sigma}$, where one has set the threshold to 0 as describe in remark 35. We also assume that $\mathcal{D}$ is saturated and one-dimensional. To fix the ideas, we assume that $\mathcal{D}$ is $(M - 1)$-layered, so the output layer is V_M. Within the layer $V_m, m > 0$, of $\mathcal{D}$, the weight of perceptron $P_i^m = \text{Perc}_{d_i,0,\sigma} \in V_m$ at time t is denoted by $w_i^m(t) \in \mathbb{R}^{d_i}$. Its component at time t for the axon connection from perceptron $P_j^{m-1} \in V_{m-1}$ is denoted by $w_{ij}^m(t)$. Denote by $w(t)$ the sequence $(w_{ij}^m(t))_{m,i,j} \in \mathbb{R}^W$ of all weight components of $\mathcal{D}$ at time t. The input vector from the input family $\triangleright \mathcal{D}$ at time t is denoted by $x(t) \in \mathbb{R}^I$, whereas the output vector[2] at time t from the output family $\mathcal{D}\blacktriangleright$ is denoted by $z(t) \in \mathbb{R}^O$. The network is given a target output $\zeta(t) \in \mathbb{R}^O$. The problem is that the output z is different from the target ζ, and that one would like to adjust the weight w such that the output is as near as possible to ζ in the sense that the Euclidean distance measure $E_z(w) = \frac{1}{2}\|\zeta - z\|^2$ is as small as possible. To this end, one offers the system a *training set*, i.e., a sequence $T = (x_1 = x(t_1), \ldots x_N = x(t_N))$ of training inputs for a temporarily constant weight w, make it calculate the corresponding outputs $z_1, \ldots z_N$, and asks for a minimization of the summed distance $E_T(w) = \sum_{i=1}^N E_{z_i}(w)$.

The correction of w is obtained from the map

$$E_T : \mathbb{R}^W \to \mathbb{R}$$

knowing that the value of $E_T(w)$ grows maximally in the direction of the gradient $dE_T(w) = (D_{w_{ij}^m} E_T(w)) \in \mathbb{R}^W$. In fact, if $y(s) \in \mathbb{R}^W$ is a curve with $y(0) = w$, such that $E_T(y(s))$ is constant, then the derivative of $E_T \circ y$ must vanish identically, we therefore have $0 = D(E_T \circ y)(0) = dE_T(w)(Dy(0))$, i.e., the tangent of y at $s = 0$ is orthogonal to the gradient. This means that one has to move along the gradient direction with the argument w in order to reach a weight for a local minimum with maximal efficiency.

This means that $w(t)$ is corrected by the amount $-\phi_t \cdot dE_T(w(t))$ to a new weight

$$w(t + \Delta t) = w(t) - \phi_t \cdot dE_T(w(t)) \qquad (*)$$

with a time delay Δt, which expresses the time needed by the system to calculate the correction. The usually positive coefficient ϕ_t is called the *learning rate at time t*. The back-propagation algorithm manages to calculate the correction term $dE_T(w(t))$ by means of a recursive calculation of correction terms related to the output layer, and then going back to

---

[2] Observe that because of the acyclicity of multi-layered perceptrons, the output is always uniquely determined by the input for given weights.

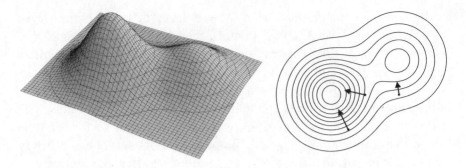

Fig. 41.10. E_T visualized as a hilly landscape. Peaks represent minimal E_T values. On the right side, gradient vectors indicate the fastest path to a peak.

terms for successively earlier layers until it reaches V_1; this is why it is called back-propagation algorithm.

In order to ease notation, we temporarily omit the time reference in the following calculations. Call $_iv_j^m$ the output of perceptron V_j^m for the input x_i. Denote by $_ih_j^m$ the scalar product $\sum_k w_{jk}^m \, _iv_k^{m-1}$. Then the gradient component for the variable w_{jl}^m, i.e., the partial derivative of E_T with respect to w_{jl}^m is

$$
\begin{aligned}
D_{w_{jl}^m} E_T &= \frac{\partial E_T}{\partial w_{jl}^m} \\
&= \sum_{i=1}^{N} \frac{\partial E_T}{\partial _iv_j^m} \cdot \frac{d(_iv_j^m)}{d(_ih_j^m)} \cdot \frac{\partial _ih_j^m}{\partial w_{jl}^m} \\
&= \sum_{i=1}^{N} \frac{\partial E_T}{\partial _iv_j^m} \cdot \sigma(_ih_j^m)(1 - \sigma(_ih_j^m)) \cdot \frac{\partial _ih_j^m}{\partial w_{jl}^m} \\
&= \sum_{i=1}^{N} {_i\epsilon_j^m} \cdot {_iv_l^{m-1}}
\end{aligned}
$$

with the notation

$$
_i\epsilon_j^m = \frac{\partial E_T}{\partial _iv_j^m} \cdot \frac{d(_iv_j^m)}{d(_ih_j^m)} = \frac{\partial E_T}{\partial _iv_j^m} \cdot \sigma(_ih_j^m)(1 - \sigma(_ih_j^m)).
$$

Note that the derivative of σ enters the equation, because $_iv_j^m = \sigma(_ih_j^m)$. Next, the $_i\epsilon_j^m$ must be calculated recursively. For $m = M$, we have

$$
\frac{\partial E_T}{\partial _iv_j^M} = \zeta_j - {_iv_j^M},
$$

where $_iv_j^M$ is a coordinate of the output vector z_i, therefore the numbers $_i\epsilon_j^M$ are given for the start of recursion for the i-th training series pairing (x_i, z_i). The other ϵ numbers for $m < M$ are calculated according to the formula

$$\frac{\partial E_T}{\partial_i v_j^m} = \sum_k \frac{\partial E_T}{\partial_i v_k^{m+1}} \cdot \frac{d(_iv_k^{m+1})}{d(_ih_k^{m+1})} \cdot \frac{d(_ih_k^{m+1})}{d(_iv_k^m)}$$

$$= \sum_k {}_i\epsilon_k^{m+1} \cdot {}_iw_{jk}^{m+1},$$

which implies the desired recursive formula for $_i\epsilon_j^m$ from the values of ϵ on level $m + 1$:

$$_i\epsilon_j^m = \frac{\partial E_T}{\partial_i v_j^m} \cdot \sigma(_ih_j^m)(1 - \sigma(_ih_j^m))$$

$$= \sigma(_ih_j^m)(1 - \sigma(_ih_j^m)) \sum_k {}_i\epsilon_k^{m+1} \cdot {}_iw_{jk}^{m+1}. \qquad (**)$$

When all these numbers have been calculated for each i, the total change is calculated according to the gradient formula $(*)$. This defines the next weight and we may either proceed to another training series T' or else reuse T. *As already mentioned, there is no guarantee that this algorithm eventually yields a stable result and that this result is in fact a local minimum.*

41.5.1 Comments on the Learning Paradigm for Back-Propagation

Learning in neural network theory means that the weights of a network are changed in order to improve the performance of the network. So the previous back-propagation method is used for learning. This type of learning is called *supervised* learning since the change of the weights follows external target, i.e., the given pairings (x_i, z_i) of the training set T. For *unsupervised* learning, the system evolves independently of external targets in the sense of what is called *self-organization.*

If we were to realize the above multi-layered perceptron and the back-propagation rule as a learning artificial brain, this would look like having a training set T and an external control level, where the intermediate results pertaining to the algorithm are stored. At the end of the training, the new weight would be defined from this "deus ex machina" in order to improve the artificial brain's performance. This is far from what is happening in a real environment. There is no such an external level and no intervention into the brain by a "divine" entity.

It is however possible to modify the multi-layered perceptron $\mathcal{D}$ such that an external intervention is no more necessary for improving configuration of the weights. Let us indicate the essential steps needed for rendering the system autonomous. There are several partial problems to solve. We give a sketch of the necessary improvements and temporal refinements and leave the details to the interested reader.

1. The weights on layer $m + 1$ in the recursive formula $(**)$ are not transferred to layer m neurons. One therefore needs to add an elementary morphism from the output of a neuron P_k^{m+1} to an input of a neuron P_l^m if these neurons are connected.

2. If an elementary morphism $P_k^{m+1} \to P_l^m$ is given, we need to generate an output from P_k^{m+1} which represents the weight w_{kl}. Now, if the activation number in P_k^{m+1} is that weight, the inverse sigma function σ^{-1} may convert the output into the weight.

3. Next, we have to look for a method to convert the interior information about a neuron's weight into an output. Here, we need an input vector $v(t)$ at a given time, which has exactly one coordinate 1 at the position l of w_{kl}, and the scalar product $(w_k(t), v(t)) = w_{kl}$ yields the wanted activation value.

4. Conversely to the previous problem, one has to convert the input of the weight information via the morphism $P_k^{m+1} \to P_l^m$ into a weight information of neuron P_l^m. This is achieved by taking a subsidiary (and temporally intermediate) weight vector in P_l^m such that its scalar product with the input yields the input from P_k^{m+1} encoding the weight w_{kl} via its σ-value $\sigma(w_{kl})$. Now, the Hebb learning map can be used to recover w_{kl} from $\sigma(w_{kl})$ by use of σ^{-1}.

5. The training set T should also be realized as a temporal construction in the temporal stream of our system. To this end, each detailed calculation of errors ϵ for a given test index i is realized as a temporal training segment in the whole process. Then, each such training segment is juxtaposed to the previous segment, and at the end of all segments, the calculation of the total change (the gradient) is performed by a Hebb function referring to all previous weight data from the totality of training segments.

The desired output for a given input, which is needed to correct the weights in the neural net can be thought of being hard-coded into the Hebb functions. An alternative would be to extend the input stream by sending the desired outputs along with the inputs, so they are available at the output layer to be used for changing the weights.

Closing this chapter on neural networks, we should add a remarkable theorem by George Cybenko [9]. The question is how well any function may be approximated by a multi-layered network. Here is an answer:

Proposition 359 (Cybenko Approximation Theorem) *Given a sigmoidal function σ, a real number $\varepsilon > 0$, and any continuous function*

$$f : [0, 1]^n \to \mathbb{R},$$

there is a natural number N, two sequences $(\alpha_i)_{i=1,...N}$, $(\theta_i)_{i=1,...N}$ of real numbers, and a sequence of vectors $(y_i)_{i=1,...N}$, $y_i \in \mathbb{R}^n$ such that the function

$$G(x) = \sum_{i=1}^{N} \alpha_i \sigma((y_i, x) - \theta_i)$$

defined on $[0, 1]^n$ approximates f with precision ε, i.e., $|G(x) - f(x)| < \varepsilon$ for all $x \in [0, 1]^n$.

We recognize that this function is the output resulting from a 1-layer perceptron with N hidden neurons, input dimension n and output dimension 1. The α_i are the weights for the output, and y_i and θ_i are the weights and thresholds, respectively, of the N hidden neurons. This means that any approximation is possible by a multi-layered perceptron.

CHAPTER 42

Probability Theory

42.1 Introduction

This short overview of probability theory serves the aim to describe the mathematical framework on which statistical investigations are based. Proper statistics in the sense of inductive statistics, i.e., the scientific theory of testing probability distribution parameters from sample series, must be delegated to more specialized literature, since it is not possible to deal with this science within such a limited space. We shall however give a short sketch of the ideas in inductive statistics, in particular the maximum-likelihood method for guessing probability distribution parameters.

42.2 Event Spaces and Random Variables

Probability deals with events which are observed and which belong to a determined space of possible events. The first definition introduces the type of a mathematical space fundamental to probability theory and statistics.

Definition 245 *Given a set Ω, a σ-algebra over Ω is a subset $\mathfrak{A} \subset 2^\Omega$ such that*

(i) *$\Omega \in \mathfrak{A}$,*

(ii) *If $A \in \mathfrak{A}$, then its complement $A' = \Omega - A$ is an element of $\mathfrak{A}$.*

(iii) *For any countable family $(A_i)_{\in \mathbb{N}}$ of elements $A_i \in \mathfrak{A}$, we have $\bigcup_i A_i \in \mathfrak{A}$.*

In probability theory, a σ-algebra $\mathfrak{A}$ is called an event space, *the elements of $\mathfrak{A}$ are called* events, *whereas the elements $\omega \in \Omega$ are called* outcomes, *and the corresponding singletons $\{\omega\}$, if they are in $\mathfrak{A}$, are called* elementary events. *The maximal element Ω of the event space is called the* certain event, *while its complement $\varnothing$ (which is in $\mathfrak{A}$ by axiom (ii)), is called the* impossible event. *If two events A and B are such that $A \cap B = \varnothing$, they are called* incompatible.

Observe that by (ii) and (iii), the intersection $\bigcap_i A_i = (\bigcup_i A_i')'$ of a countable family $(A_i)_{i \in \mathbb{N}}$ of elements $A_i \in \mathfrak{A}$ is in $\mathfrak{A}$. Further, if $k \geq 0$, taking all members A_i in $(A_i)_{i \in \mathbb{N}}$ constant for $i \geq k$, all finite unions $A_0 \cup A_1 \cup \ldots A_k$ and intersections $A_0 \cap A_1 \cap \ldots A_k$, of elements $A_i \in \mathfrak{A}$ are in $\mathfrak{A}$. Also observe that it is not necessary to denote specifically the set Ω besides the set $\mathfrak{A}$ since we have $\Omega = \bigcup \mathfrak{A}$. We therefore often write $\Omega_{\mathfrak{A}}$ for Ω.

Example 200 The dice[1] space is $\Omega = \{1, 2, 3, 4, 5, 6\}$. Usually one considers the full σ-algebra $\mathfrak{A} = 2^\Omega$. Here, an event is any subset of faces, describing a property of faces, e.g., the subset $E = \{2, 4, 6\}$ of even face numbers.

Fig. 42.1. The outcome space Ω consisting of all possible dice faces, with the event E, subset of Ω, of even face figures.

Example 201 We are checking n devices for their operativeness. Each device $i = 1, 2, \ldots n$, can be either operational or defective. An outcome

---

[1] Henceforth, we assume that dice are of cubic form, their faces bearing the numbers 1 to 6.

is an n-tuple $w = (w_1, w_2, \ldots w_n) \in \Omega = 2^n$, where $w_i = 1$ if device i is operational, and $w_i = 0$ if it is defective . Again, we take $\mathfrak{A} = 2^{\Omega}$. For example, the event $A = \{w \mid \sum_i w_i \geq 3\}$ is the event of all checks of the n devices, such that at least 3 devices are operational.

Example 202 For every subset $E \subset \Omega$, there is a minimal σ-algebra $\langle E \rangle \subset 2^{\Omega}$ containing E. It is called the σ-*algebra generated by E*. For example, the set $\mathfrak{I}$ of all intervals $[a, b]$, $[a, b[$, $]a, b]$, $]a, b[$, be they open, closed, or half-open, with $a, b \in \mathbb{R}$, generates the σ-algebra $\mathfrak{B} = \langle \mathfrak{I} \rangle$ of *Borel sets*. It can be shown that $\mathfrak{B} = \langle \mathfrak{J} \rangle = \langle \mathcal{O} \rangle$, where $\mathfrak{J}$ is the set of all intervals $]-\infty, a]$, $a \in \mathbb{Q}$, and where $\mathcal{O}$ is the set of open sets in the usual topology of $\mathbb{R}$.

The following definition of a morphism of σ-algebras is completely analogous to the definition of continuous maps between topological spaces, which uses the characterization by inverse images of open sets, see definition 183 and lemma 238.

Definition 246 *Given two σ-algebras $\mathfrak{A}$ and $\mathfrak{C}$ over the sets of outcomes $\Omega_{\mathfrak{A}}$ and $\Omega_{\mathfrak{C}}$, a morphism $f : \mathfrak{A} \to \mathfrak{C}$ is a set map $f : \Omega_{\mathfrak{A}} \to \Omega_{\mathfrak{C}}$ such that $f^{-1} : 2^{\Omega_{\mathfrak{C}}} \to 2^{\Omega_{\mathfrak{A}}}$ restricts to a map $\mathfrak{C} \to \mathfrak{A}$, i.e., if $c \in \mathfrak{C}$, then $f^{-1}(c) \in \mathfrak{A}$. Clearly, set-theoretic compositions of morphisms of σ-algebras are again such morphisms, the identity $Id_{\Omega_{\mathfrak{A}}} : \mathfrak{A} \xrightarrow{\sim} \mathfrak{A}$ is a morphism, and composition of morphisms is associative. In other words, we have the category σ-Alg of σ-algebras. Often, if the involved event spaces $\mathfrak{A}$ and $\mathfrak{C}$ are clear, one also writes $f : \Omega_{\mathfrak{A}} \to \Omega_{\mathfrak{C}}$ instead of $f : \mathfrak{A} \to \mathfrak{C}$.*

Exercise 198 Show that a continuous map $f : \mathbb{R} \to \mathbb{R}$ defines a morphism $f : \mathfrak{B} \to \mathfrak{B}$.

Example 203 Given the σ-algebra $\mathfrak{B}$ of Borel sets, consider the n-dimensional real space $\mathbb{R}^n$ and the n projections $pr_i : \mathbb{R}^n \to \mathbb{R}$. Then the inverse image maps $pr_i^{-1} : 2^{\mathbb{R}} \to 2^{\mathbb{R}^n}$ restrict to $pr_i^{-1} : \mathfrak{B} \to 2^{\mathbb{R}^n}$. Their images $pr_i^{-1}(\mathfrak{B})$ by definition generate the σ-algebra $\mathfrak{B}^n$ *of n-dimensional Borel sets*. This algebra is also generated by the set $\mathcal{O}$ of open sets in $\mathbb{R}^n$, and also by the set of inverse images $pr_i^{-1}(]a, b[)$ of open intervals $]a, b[\subset \mathbb{R}$. It follows that a map $f : \Omega_{\mathfrak{A}} \to \mathbb{R}^n$ is a morphism $f : \mathfrak{A} \to \mathfrak{B}^n$ iff all its projections $pr_i \circ f : \Omega_{\mathfrak{A}} \to \mathbb{R}$ are morphisms $pr_i \circ f : \mathfrak{A} \to \mathfrak{B}$.

The most important morphisms of event spaces[2] in probability theory are the random variables:

Definition 247 A random variable *on an event space $\mathfrak{A}$ is a morphism*

$$X : \mathfrak{A} \to \mathfrak{B}, \text{ often denoted by } X : \Omega_{\mathfrak{A}} \to \mathbb{R},$$

of σ-algebras into the algebra $\mathfrak{B}$ of Borel sets. If the event set $\mathfrak{A}$ is finite or countable, then X is called a discrete *random variable. We denote by V_X the image set of X, and call it the* set of values *of X.*

Observe that if the event space is the power set space $\mathfrak{A} = 2^{\Omega_{\mathfrak{A}}}$ of the certain event $\Omega_{\mathfrak{A}}$, then any set map $f : \mathfrak{A} \to \mathfrak{B}$ is a morphism.

Example 204 If we take the dice space as $\mathfrak{A} = 2^{\{1,2,3,4,5,6\}}$, the random variable $X(A) = \sum_{i \in A} i$ adds the values of the faces contributing to the event A.

Example 205 Consider the event space $\mathfrak{A}$ of device operativeness from example 201 over the total event set $\Omega = 2^n$. For $i = 1, 2, \ldots n$, we take the random variable $X_i : \mathfrak{A} \to \mathfrak{B}$ with $X_i(w) = 1$ if the i-th device is working, and $X_i(w) = 0$ else. Another random variable is the sum $X = \sum_i X_i$, whose value is the number $X(w) = \sum_i X_i(w)$ of operational devices at outcome w.

The last example gives rise to an n-dimensional generalization of random variables and leads to the construction of one-dimensional variables deduced from n-dimensional variables.

Definition 248 *Given a positive natural number n, an n-dimensional random variable on an event space $\mathfrak{A}$ is an n-tuple $X = (X_1, X_2, \ldots X_n)$ of random variables. Equivalently, by the universal property of n-dimensional Borel sets discussed in example 203, X is a morphism*

$$X : \mathfrak{A} \to \mathfrak{B}^n, \text{ often denoted by } (X_1, X_2, \ldots X_n) : \Omega_{\mathfrak{A}} \to \mathbb{R}^n,$$

of σ-algebras into the algebra $\mathfrak{B}^n$ of n-dimensional Borel sets.

---

[2] We shall mainly use the term *event space* for σ-algebras in our present context.

If $X = (X_1, X_2, \ldots X_n) : \Omega_{\mathfrak{A}} \to \mathbb{R}^n$ is an n-dimensional random variable, then any morphism of σ-algebras $g : \mathfrak{B}^n \to \mathfrak{B}^m$ induces an m-dimensional random variable $g \circ X : \mathfrak{A} \to \mathfrak{B}^m$. For example, in the above example 205, we took $m = 1$ and the continuous function $g(x_1, \ldots x_n) = \sum_i x_i$, defining the sum $\sum_i X_i$ of the random variables X_i. More generally, any $\mathbb{R}$-linear combination $\sum_i \lambda_i \cdot X_i$ of random variables is again a random variable, i.e., the set of random variables on $\mathfrak{A}$ is a real vector space Rand$(\mathfrak{A})$. An important linear combination of random variables $X_1, X_2, \ldots X_n$ is their *arithmetic mean* or *average*

$$X = \frac{1}{n} \sum_i X_i.$$

More generally, any polynomial function $\sum_{i_1, i_2, \ldots i_n} \lambda_{i_1, i_2, \ldots i_n} X^{i_1} X^{i_2} \ldots X^{i_n}$ of random variables X_i with real coefficients $\lambda_{i_1, i_2, \ldots i_n}$ is a random variable, since polynomial functions are continuous. So Rand$(\mathfrak{A})$ is in fact a commutative ring containing the subring of constant random variables $\Lambda \in \mathbb{R}$.

42.3 Probability Spaces

Probability theory deals with the chance that a random variable takes a given combination of values. In fact, it is less interesting to know an event in the dice event space than the frequency it has when throwing a dice for a number of times. This leads to the second pillar of probability theory: probability functions on event spaces. Only the combination of random variables and probability functions generates relevant probabilistic statements. Here is the axiomatic setup introduced by Andrei Nikolaevich Kolmogorov in 1933 [20].

Definition 249 (Kolmogorov Axioms) *Let $\mathfrak{A}$ be an event space. Then a* probability measure P for $\mathfrak{A}$ *is a map*

$$P : \mathfrak{A} \to \mathbb{R}$$

with the following properties:

(i) $P(A) \geq 0$, *for all $A \in \mathfrak{A}$,*

(ii) $P(\Omega_{\mathfrak{A}}) = 1$,

(iii) *For every family $(A_i)_{i \in \mathbb{N}}$ of pairwise incompatible (i.e., disjoint) events $A_i \in \mathfrak{A}$, we have*

$$P(\bigcup_i A_i) = \sum_i P(A_i).$$

The domain $\mathfrak{A} = \mathfrak{A}_P$ of the probability measure P is also called a probability space over the set $\Omega_{\mathfrak{A}}$ of outcomes, if one wants to stress the space $\mathfrak{A}$ and its set $\Omega_{\mathfrak{A}}$ of outputs.

Observe that for any event $A \in \mathfrak{A}$, we have $P(A \cup A') = P(A) + P(A') = 1$. Therefore $P(A) \leq 1$, and the convergence of the right side of the sum $\sum_i P(A_i)$ is guaranteed, and is also independent of the order of the summation by the "unconditional convergence" theorem for absolutely convergent series, see the observation following proposition 256 in chapter 27.

Sorite 360 Let $P : \mathfrak{A} \to \mathbb{R}$ be a probability measure. Then we have these properties:

  (i) For every event $A \in \mathfrak{A}$, $0 \leq P(A) \leq 1$.

 (ii) $P(\emptyset) = 0$ and $P(\Omega_{\mathfrak{A}}) = 1$.

(iii) For every event $A \in \mathfrak{A}$, $P(A') = 1 - P(A)$.

 (iv) If A and B are events in $\mathfrak{A}$, then $A \subset B$ implies $P(A) \leq P(B)$.

  (v) If A and B are events in $\mathfrak{A}$, then $A \subset B$ implies

$$P(A \cup B) = P(A) + P(B) - P(A \cap B).$$

 (vi) More generally, if $A_1, A_2, \ldots A_n$ are events in $\mathfrak{A}$, then

$$P(A_1 \cup A_2 \cup \ldots A_n) =$$
$$\sum_i P(A_i) - \sum_{i<j} P(A_i \cap A_j) + \sum_{i<j<k} P(A_i \cap A_j \cap A_k) -$$
$$\ldots + (-1)^{n+1} P(A_1 \cap \ldots A_n)$$

Proof The first three claims are immediate from the formula $P(A \cup A') = P(A) + P(A') = 1$.

Claim (iv) follows from $P(B) = P(A \cup (B - A)) = P(A) + P(B - A) \geq P(A)$, since $P(B - A) \geq 0$.

For claim (v), we have the disjoint union $A \cup B = (A - B) \cup (B - A) \cup (A \cap B)$, whence $P(A \cup B) = P(A - B) + P(B - A) + P(A \cap B) = (P(A - B) + P(A \cap B)) + (P(B - A) + P(A \cap B)) - P(A \cap B) = P(A) + P(B) - P(A \cap B)$.

To show claim (vi), we apply statement (v) to $(A_1 \cup \ldots A_{n-1}) \cup A_n$, which yields
$P((A_1 \cup \ldots A_{n-1}) \cup A_n) = P(A_1 \cup \ldots A_{n-1}) + P(A_n) - P((A_1 \cap A_n) \cup \ldots (A_{n-1} \cap A_n))$.
Then induction for the first and last terms yields

$$P(A_1 \cup \ldots A_{n-1}) = \sum_{i<n} P(A_i) - \sum_{i<j<n} P(A_i \cap A_j) + \sum_{i<j<k<n} P(A_i \cap A_j \cap A_k) -$$
$$\ldots + (-1)^n P(A_1 \cap \ldots A_{n-1})$$

and

$$P((A_1 \cap A_n) \cup \ldots (A_{n-1} \cap A_n)) = \sum_{i<n} P(A_i \cap A_n) - \sum_{i<j<n} P(A_i \cap A_j \cap A_n)$$
$$+ \sum_{i<j<k<n} P(A_i \cap A_j \cap A_k \cap A_n) -$$
$$\ldots + (-1)^n P(A_1 \cap \ldots A_{n-1} \cap A_n).$$

This adds up to the correct formula for n. □

Exercise 199 In the dice space $\mathcal{A} = 2^{\{1,2,3,4,5,6\}}$, the probability P may be taken as the relative number of outcomes within an event A, i.e., $P(A) = card(A)/6$. For example, $P(\{2,4,6\}) = \frac{3}{6} = \frac{1}{2}$. Verify that this probability measure conforms to the Kolmogorov axioms.

The previous exercise realizes the so-called *Laplace principle*, which requires that if one has no further knowledge about the elementary events in the powerset algebra $\mathcal{A} = 2^\Omega$ of a finite set Ω, then the probability $P(\{w\})$ of an elementary event associated with an outcome w should be

$$P(\{w\}) = \frac{1}{card(\Omega)},$$

i.e., every elementary event has equal probability, summing up to 1 for the certain event.

Example 206 Let us observe a harddisk regarding its reliability. The outcomes are the sequences s_i of $i \geq 1$ attempts to write data to the harddisk, with the first failure occurring at the i-th attempt. So $\Omega = \{s_i \mid i = 1, 2, \ldots\}$ is a denumerably infinite set. Suppose that there is a chance of $0 < p < 1$ to obtain a failure for one writing process unit. Then the first $i - 1$ attempts each have chance $1 - p$ to happen, while the last has chance p to happen. It is therefore reasonable to define the probabilities for elementary events $P(\{s_i\}) = p \cdot (1 - p)^{i-1}$, and then necessarily $P(A) = \sum_{s_i \in A} p \cdot (1 - p)^{i-1}$. We have $P(\Omega) = \sum_{1 \leq i} p \cdot (1 - p)^{i-1} =$

$p \sum_{0 \le i} (1-p)^i = p \cdot \frac{1}{1-(1-p)} = 1$, the sum of a geometric series, see example 104. The third Kolmogorov axiom is clear from the definition of P as a sum of the values of the outcomes in an event.

Probabilities are often considered as relative values with respect to a given event B, in the sense that one pretends that B is the certain event. This leads to the concept of conditional probability, which we shall motivate after the following definition.

Definition 250 *For a probability space $P : \mathcal{A} \to \mathbb{R}$ and a selected event $B \in \mathcal{A}$, with $P(B) > 0$, the* conditional probability *of an event $A \in \mathcal{A}$ is defined by*

$$P(A|B) = \frac{P(A \cap B)}{P(B)}.$$

Proposition 361 *Let $B_1, B_2, \ldots B_n$ be a sequence of pairwise incompatible events in the probability space $\mathcal{A}$ with probability function P, such that $\bigcup B_i = \Omega_{\mathcal{A}}$, and $P(B_i) > 0$, for all i, then we have*

$$P(A) = \sum_i P(A|B_i) \cdot P(B_i).$$

If, further, $P(A) > 0$, then we have the Bayes formula

$$P(B_i|A) = \frac{P(A|B_i) \cdot P(B_i)}{P(A)}$$

for all i.

Proof These formulas are straightforward applications of the definition of conditional probability.   □

Example 207 The Bayes formula is often useful in experimental setups. Imagine that chemist analyzes a sample x. He already knows, from previous tests, that it contains traces of a substance a with probability $\frac{2}{3}$, i.e., $P(A) = \frac{2}{3}$, or of a substance b with probability $\frac{1}{3}$, i.e., $P(B) = \frac{1}{3}$, but not both, i.e., $P(A \cup B) = P(A) + P(B)$. To test for a specific substance, he uses an indicator that colors green, we call this event G, or remains unchanged. From long experience, he knows that if a sample contains a, then the indicator colors green with probability $\frac{4}{5}$, i.e. $P(G|A) = \frac{4}{5}$, and if it contains b, the indicator becomes green with probability $\frac{1}{5}$, i.e. $P(G|B) = \frac{1}{5}$. With all this information, the chemist wants to know the probability that, given that the indicator turns green, the sample contains traces of a, i.e., he seeks $P(A|G)$. The Bayes formula yields

$$P(A|G) = \frac{P(G|A) \cdot P(A)}{P(G)}.$$

The only number yet to be determined is $P(G)$. But $P(G) = P(G \cap (A \cup B))$, since $A \cup B$ is the certain event. Therefore, by distributivity, $P(G) = P((G \cap A) \cup (G \cap B))$. Since A and B are incompatible, $G \cap A$ and $G \cap B$ are so too and $P(G) = P(G \cap A) + P(G \cap B)$. Thus, converting to conditional probabilities, $P(G) = P(G|A) \cdot P(A) + P(G|B) \cdot P(B)$ and the Bayes formula becomes

$$P(A|G) = \frac{P(G|A) \cdot P(A)}{P(G|A) \cdot P(A) + P(G|B) \cdot P(B)}.$$

All the quantities on right hand side are known to the chemist, therefore

$$P(A|G) = \frac{\frac{4}{5} \cdot \frac{2}{3}}{\frac{4}{5} \cdot \frac{2}{3} + \frac{1}{5} \cdot \frac{1}{3}}$$
$$= \frac{8}{9}.$$

Since the method used in the previous example is very common, we now state the general formula. Let A_i, with $1 \le i \le n$, be mutually incompatible events that sum to A, i.e., $A_i \cap A_j = \varnothing$, for $i \ne j$, and $\bigcup_i A_i = A$. Then

$$P(A_j|A) = \frac{P(A|A_j) \cdot P(A_j)}{\sum_i P(A|A_i) \cdot P(A_i)}.$$

Example 208 (Monty Hall Problem) An amusing application of the Bayes formula is the following problem, whose solution by Marylin vos Savant caused an uproar and fierce discussions among mathematicians and amateurs alike because of its counterintuitiveness.

Suppose game show host Monty Hall gives you the choice of three doors, A, B or C. Behind one door is a car, behind the other two are goats (you definitely don't want to win a goat). Monty lets you pick a door without opening it. He knows what is behind the doors and opens one which has a goat, then asks you if you want to switch to the third door. The problem is: Is it better to switch or not to switch?

An intuitive argument may conclude that it doesn't matter, the probabilities for the two still closed doors to hide the car being equal, i.e., $\frac{1}{2}$.

The a priori probability that door X hides the prize is $P(X) = \frac{1}{3}$. So far, so good. Without loss of generality, assume that you have picked door A and Monty Hall opens door B with a goat behind it. By MB denote the event "Monty Hall opens B", and by X the event "The car is behind X", for

$X \in \{A, B, C\}$. Note that A, B and C are mutually incompatible. Then we have the following conditional probabilities:

$$P(MB|A) = \tfrac{1}{2} \tag{a}$$

$$P(MB|B) = 0 \tag{b}$$

$$P(MB|C) = 1 \tag{c}$$

The probability (a) holds because Monty has equal choice between B and C, (b) holds because Monty will never open the door with the prize and (c) holds because Monty is forced to open B, since C has the prize. Now, we want to compute $P(A|MB)$, i.e., the probability that the first chosen door A contains the prize, after we know that Monty opened B.

$$
\begin{aligned}
P(A|MB) &= \frac{P(MB|A) \cdot P(A)}{P(MB|A) \cdot P(A) + P(MB|B) \cdot P(B) + P(MB|C) \cdot P(C)} \\
&= \frac{\frac{1}{2} \cdot \frac{1}{3}}{\frac{1}{2} \cdot \frac{1}{3} + 0 \cdot \frac{1}{3} + 1 \cdot \frac{1}{3}} \\
&= \frac{1}{3}.
\end{aligned}
$$

Similarly, $P(C|MB) = \tfrac{2}{3}$. Therefore you can double your chance of winning by switching from A to C!

Example 209 From a card game with 32 cards, including 4 aces, the player draws two cards, one after the other, without putting back the first one. How large is the probability to draw an ace in the second turn? The set of outcomes is the set $\Omega = 32 \times 31$, and the event set is $\mathfrak{A} = 2^{\Omega}$. Under the Laplace principle, we have the probability $P(\{w\}) = \frac{1}{32 \cdot 31}$ for each elementary event $w = (0,0), (0,1), \ldots (31,30)$. Let A be the event "ace in the second turn", B_1 the event "ace in the first turn", and B_2 the event "no ace in the first turn", i.e., the complement of B_1. Then we have

$$P(B_1) = \frac{4 \cdot 31}{32 \cdot 31} = \frac{1}{8}, \qquad P(B_2) = \frac{28 \cdot 31}{32 \cdot 31} = \frac{7}{8},$$

$$P(A \cap B_1) = \frac{4 \cdot 3}{32 \cdot 31} = \frac{3}{248}, \qquad P(A \cap B_2) = \frac{28 \cdot 4}{32 \cdot 31} = \frac{7}{62},$$

whence

$$P(A|B_1) = \frac{P(A \cap B_1)}{P(B_1)} = \frac{3}{31}, \qquad P(A|B_2) = \frac{P(A \cap B_2)}{P(B_2)} = \frac{4}{31}.$$

An application of proposition 361 then yields

$$P(A) = P(A|B_1) \cdot P(B_1) + P(A|B_2) \cdot P(B_2) = \frac{3}{31} \cdot \frac{1}{8} + \frac{4}{31} \cdot \frac{7}{8} = \frac{1}{8}.$$

Proposition 362 (Multiplication Formula) *Let $A_1, A_2, \ldots A_n$ be n events in the event space $\mathcal{A}$ with $P(A_1 \cap A_2 \cap \ldots A_n) > 0$. Then we have*

$$P(A_1 \cap A_2 \cap \ldots A_n) =$$
$$P(A_1) \cdot P(A_2|A_1) \cdot P(A_3|A_1 \cap A_2) \cdot \ldots P(A_n|A_1 \cap A_2 \cap \ldots A_{n-1}).$$

Proof This follows by induction on $n \geq 2$. For $n = 2$, it is the definition of conditional probability. The general case results from $P(A_1 \cap A_2 \cap \ldots A_n) = P((A_1 \cap A_2 \cap \ldots A_{n-1}) \cap A_n) = P(A_1 \cap A_2 \cap \ldots A_{n-1}) \cdot P(A_n|(A_1 \cap A_2 \cap \ldots A_{n-1}))$. $\square$

Definition 251 *Two events A and B of a probability space $\mathcal{A}_P$ are called* independent *iff we have*

$$P(A \cap B) = P(A) \cdot P(B).$$

Exercise 200 Show that if two events A and B are independent, then so are the pairs A and B', and therefore, by an exchange of the respective roles of these events, so are also A' and B, and A' and B'.

Example 210 Consider a device T, which is composed of two subdevices T_1 and T_2. They can be defective, and we suppose that they can be operational independently. Call A_i the event "T_i is operational". The subdevices may be combined in a series, T_2 after T_1, say, yielding the combined event G, or in parallel, yielding the combined event G^*. Then the serial combination G is operational iff each of the subdevices is so. Thus, the probability of the event G is

$$P(G) = P(A_1 \cap A_2) = P(A_1) \cdot P(A_2).$$

In the parallel combination G^*, the operativeness is guaranteed if either A_1 or A_2 is operational. Thus, the probability of G^*, using sorite 360 (vi), is

$$P(G^*) = P(A_1 \cup A_2) = P(A_1) + P(A_2) - P(A_1) \cdot P(A_2).$$

If, for example, $P(A_1) = \frac{3}{4}$ and $P(A_2) = \frac{2}{5}$, then we obtain

$$P(G) = \frac{3}{4} \cdot \frac{2}{5} = \frac{3}{10} = 0.3$$

and

$$P(G^*) = \frac{3}{4} + \frac{2}{5} - \frac{3}{10} = \frac{17}{20} = 0.85.$$

The definition of independence of n events is as follows:

Definition 252 *A sequence* $A_1, A_2, \ldots A_n$ *of* n *events is said to be* independent, *if for every subsequence* $A_{i_1}, A_{i_2}, \ldots A_{i_k}$, *we have*

$$P(A_{i_1} \cap A_{i_2} \cap \ldots A_{i_k}) = P(A_{i_1}) \cdot P(A_{i_2}) \cdot \ldots P(A_{i_k}).$$

Equivalently, independence means that

$$P(A_1 \cap A_2 \cap \ldots A_n) = P(A_1) \cdot P(A_2) \cdot \ldots P(A_n)$$

and that any subsequence of $n - 1 \geq 2$ *elements is independent.*

Exercise 201 Prove the equivalence of the two versions of independence in definition 252.

42.4 Distribution Functions

The access to the information encoded in a probability space $P : \mathfrak{A} \to \mathbb{R}$ and a random variable X on the event space $\mathfrak{A}_P$ is often eased by a concept which is more important for calculation purposes: the distribution function. To this end, given a random variable X over $\mathfrak{A}$, probability theory makes use of the notation $P(X|I) = P(X^{-1}(I))$ for the probability of the inverse image under X of a Borel set $I \in \mathfrak{B}$. In a more intuitive notation, for example, if $I = [a, b]$, one writes $P(X|[a, b]) = P(a \leq X \leq b)$; or, if $I =]-\infty, x]$, one writes $P(X|]-\infty, x]) = P(X \leq x)$.

Definition 253 *Given a probability space* $P : \mathfrak{A} \to \mathbb{R}$, *the* distribution *of* P *(or else of* $\mathfrak{A}_P$, *if* P *is clear), is the map*

$$F_{P,X} : \mathbb{R} \to [0, 1] : x \mapsto P(X \leq x).$$

If P *is clear, one writes* F_X, *and if* X *is also clear, one writes* F.

Sorite 363 *Given a probability space* $P : \mathfrak{A} \to \mathbb{R}$ *and a random variable* X *over* $\mathfrak{A}$, *we have these properties of the distribution* $F = F_{P,X}$. *Hereafter, we fix the real numbers* $a < b$ *and an argument* x.

(i) *The distribution* F *is monotonous, i.e.,* $x \leq y$ *implies* $F(x) \leq F(y)$.

(ii) $P(a < X \leq b) = F(b) - F(a)$.

(iii) $P(X > a) = 1 - F(a)$.

(iv) *The following limits exist and are denoted as on the left side of the equations:*

$$F(x - 0) = \lim_{h>0, h\to 0} F(x - h)$$

$$F(x + 0) = \lim_{h>0, h\to 0} F(x + h)$$

$$F(-\infty) = \lim_{y\to-\infty} F(y)$$

$$F(\infty) = \lim_{y\to\infty} F(y)$$

(v) *The distribution F is right continuous, i.e., $F(x) = F(x + 0)$.*

(vi) $F(-\infty) = 0$ *and* $F(\infty) = 1$.

(vii) $P(X = x) = F(x) - F(x - 0)$.

(viii) $P(a \le X \le b) = F(b) - F(a - 0)$.

(ix) $P(a \le X < b) = F(b - 0) - F(a - 0)$.

Proof If $x \le y$, then we have the inclusion $(X \le x) \subset (X \le y)$ of events, whence claim (i).

Since we have $(a < X \le b) = (X \le b) - (X \le a)$, claim (ii) follows.

Further, $\Omega_A - (X \le a) = (X > a)$ implies claim (iii).

Since F is monotonous by (i), and bounded to the range $[0, 1]$, the four limits in claim (iv) do exist.

We have a countable disjoint union

$$\mathbb{R} = \,]-\infty, x] \cup \,]1 + x, \infty[\,\cup \bigcup_{i\in\mathbb{N}}]x + \tfrac{1}{i+2}, x + \tfrac{1}{i+1}].$$

Therefore $1 = P(\mathbb{R}) = P(]-\infty, x]) + P(]1 + x, \infty[) + \sum_{i\in\mathbb{N}} P(]x + \tfrac{1}{i+2}, x + \tfrac{1}{i+1}])$. This implies that the partial sums $P(]1 + x, \infty[) + \sum_{i\le n} P(]x + \tfrac{1}{i+2}, x + \tfrac{1}{x+1}])$ converge to $1 - P(]-\infty, x]) = 1 - F(x)$. In other words, the complementary set $X \le x + 1/x_n$ defines a value $F(x + 1/x_n)$ which tends to $F(x)$, whence claim (v).

We further have the countable disjoint union

$$\mathbb{R} = \,]-\infty, 0] \cup \bigcup_{i\in\mathbb{N}}]i, i + 1],$$

whence $1 = P(\mathbb{R}) = P(]-\infty, 0]) + \sum_i P(]i, i + 1])$. Therefore the partial sums $P(]-\infty, 0]) + \sum_{i\le n} P(]i, i+1])$ converge to 1 as $n \to \infty$, hence $F(\infty) = 1$. Similarly we have the disjoint union $\mathbb{R} = \,]0, \infty[\,\cup \bigcup_{i\in\mathbb{N}}]-i - 1, -i]$, which as before implies that the partial sums $P(]0, \infty[) + \sum_{i\le n}]-i-1, -i]$ converge to 1 as $n \to \infty$. But this means that the probability $P(X \le -n - 1)$ of the complement event $X \le -n - 1$ converges to 0, i.e., $F(-\infty) = 0$, thus claim (vi).

The statements (vii) through (ix) are left as easy exercises to the reader. $\square$

What is the point of this sorite? It allows us to calculate the probability of values of a random variable X in terms of its distribution function. This means that for calculations, one may often stick to the distribution function. In what follows, we describe a number of frequent distribution functions.

42.4.1 Distribution Functions for Discrete Random Variables

In the following definitions, we always select a discrete variable $X : \mathfrak{A}_P \to \mathbb{R}$. Recall that V_X denotes the set of values of X. We shall only define special values $P(X = v_i), v_i \in V_X$, where i is a natural index of the finite or denumerable value set, since the distribution at a value in V_X can be calculated from the summation of these values. The values for other arguments $x \notin V_X$ are set to the value at $\underline{x}$, the largest natural number $\leq x$. This means that the distribution function is a step function.

Definition 254 (Geometric Distribution) *If* $V_X = \mathbb{N} - \{0\}$ *and* $0 < p < 1$, *then the distribution of* X *is called* geometric *iff for every* $i \in V_X$,

$$P(X = i) = p \cdot (1 - p)^{i-1}.$$

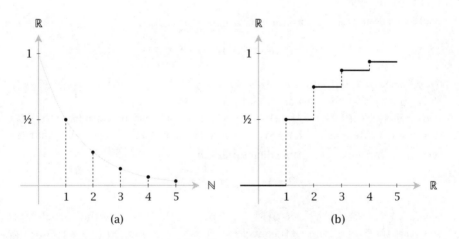

Fig. 42.2. Geometric distribution with $p = \frac{1}{2}$: (a) probability measure, (b) distribution function.

Example 211 The geometric distribution is often used to describe coin flipping. If p is the probability that a head is flipped, then $1 - p$ is the probability that a tail is flipped. The random variable X describes the number of flips until a head results, i.e., the probability $P(X = i)$ is the probability that a tail is flipped $i - 1$ times (i.e., $(1-p)^{i-1}$), times the probability that a head is flipped (i.e., p). The distribution function $P(X \leq i)$ is the probability that a head results after less than or equal to i flips. If the coin is "fair", then $p = \frac{1}{2}$, and the graphs of probability and distribution functions look like those in figure 42.2.

Definition 255 (Binomial Distribution) *If $V_X = \{0, 1, 2, \ldots n\}$ and $0 < p < 1$, the binomial distribution, more precisely, $B(n, p)$-distribution, is defined by*

$$P_{B(n,p)}(X = i) = \binom{n}{i} \cdot p^i \cdot (1 - p)^{n-i}.$$

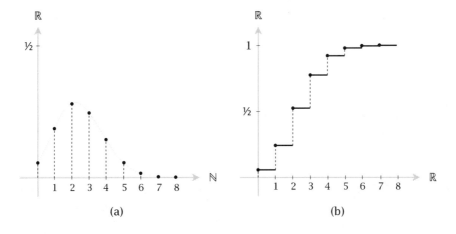

(a) (b)

Fig. 42.3. Binomial distribution with $p = \frac{1}{3}$: (a) probability measure, (b) distribution function.

Example 212 Taking up coin flipping again, consider a series of n flips. Then a certain configuration, for example *HTTTHTHTTT*, where $n = 10$, contains i heads and $n - i$ tails, in this case 3 heads and 7 tails. The probability of occurrence of this particular configuration is $p^i \cdot (1-p)^{n-i}$, if p is the probability of a head flip. Since there are $\binom{n}{i}$ configurations in which exactly i heads occur, the probability that a series of n flips contains exactly i heads is $P(X = i) = \binom{n}{i} \cdot p^i \cdot (1 - p)^{n-i}$. See figure 42.3 for the graphs of the binomial distribution.

Definition 256 (Poisson Distribution) *If* $V_X = \mathbb{N}$ *and* $\lambda > 0$, *the* Poisson distribution with parameter λ *is defined by*

$$P_\lambda(X = i) = \frac{\lambda^i}{i!}e^{-\lambda}.$$

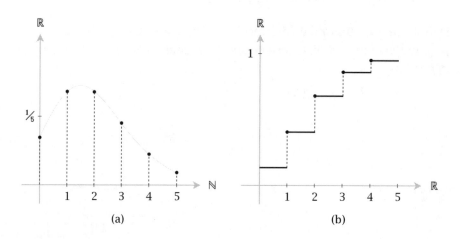

(a)   (b)

Fig. 42.4. Poisson distribution with $\lambda = 2$: (a) probability measure, (b) distribution function.

Example 213 The Poisson distribution arises from the binomial distribution by fixing the expected number $\lambda = np$ and letting n become large. Then

$$P_\lambda(X = i) = \lim_{n\to\infty} P_{B(n,\lambda/n)}(X = i).$$

For example, let $\lambda = 4$ be the average number X of misprints per page of a particular book. Because the number of characters on a page is large compared to the number of expected misprints, X can be modeled using the Poisson distribution P_4. The probability of 5 misprints on a page is

$$P_4(X = 5) = \frac{4^5}{5!}e^{-4} \approx 0.1563.$$

42.4.2 Distribution Functions for Continuous Random Variables

The continuous analogue of a specific function for $P(X = i)$ is a density function, which instead of summing up to the distribution function yields this function by an integral.

Definition 257 *A random variable X is called* continuously distributed *with density f if its distribution function F has the shape*

$$F(x) = \int_{-\infty}^{x} f,$$

where $f \geq 0$ is a non-negative integrable[3] function on $\mathbb{R}$, and where the integral is the limit $\int_{-\infty}^{x} f = \lim_{a \to -\infty} \int_{a}^{x} f$

Sorite 364 *If the random variable X is continuously distributed with density f, then:*

(i) $F(\infty) = \int_{-\infty}^{\infty} f = 1$.

(ii) *The distribution F is continuous on $\mathbb{R}$ and, therefore,*

$$P(X = x) = F(x) - F(x - 0) = 0.$$

(iii) *If f is continuous in x, then F is differentiable in x, and*

$$\frac{dF}{dx}(x) = f(x).$$

Exercise 202 Give a proof of sorite 364. Check the following points: The first claim follows from sorite 363. The second follows from the boundedness of the density. The third follows from the fact that the difference quotient $\frac{1}{h} \cdot \int_{x}^{x+h} f$ is limited by $f(x) \pm \varepsilon$ if $x \in U_\delta(x)$ for an adequate $\delta > 0$ if f is continuous in x.

Definition 258 (Rectangular (or Uniform) Distribution) *Let $a < b$ be real numbers, then the continuously distributed random variable X is $R(a, b)$-distributed (R for "rectangular") iff its density has the form*

$$f(x) = \frac{1}{b - a} \cdot \chi_{]a,b[}(x).$$

The associated distribution function is

---

[3] Bounded by definition.

$$F(x) = \begin{cases} 0 & \text{if } x \le a, \\ \frac{x-a}{b-a} & \text{if } a < x < b, \\ 1 & \text{else.} \end{cases}$$

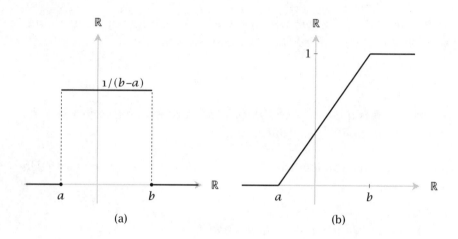

Fig. 42.5. Rectangular distribution: (a) probability measure, (b) distribution function.

Example 214 A random variable X which is uniformly distributed takes values equally likely between a and b. Good pseudo-random number generators (PRNG) implemented by computer algorithms approximate such a distribution for $a = 0$ and $b = 1$.

Definition 259 (Normal Distribution) *Let $\mu, \sigma \in \mathbb{R}, \sigma > 0$. Then the continuously distributed random variable X is $N(\mu, \sigma^2)$-distributed (or normally distributed) iff its density has the form*

$$f(x) = \frac{1}{\sigma\sqrt{2\pi}} \cdot \exp\left(-\frac{1}{2}\left(\frac{x-\mu}{\sigma}\right)^2\right).$$

For $\mu = 0$ and $\sigma = 1$, X is called standard normally distributed. The distribution function associated to the standard normal distribution is

$$\Phi(x) = \frac{1}{\sqrt{2\pi}} \int_{-\infty}^{x} \exp\left(-\frac{1}{2}t^2\right) dt.$$

See figure 42.6 for the graphical representation.

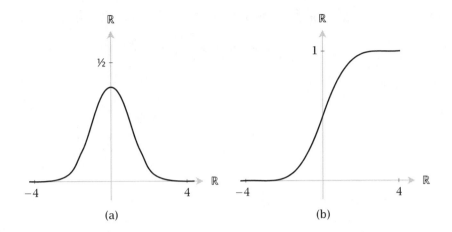

Fig. 42.6. Standard normal distribution: (a) density function, (b) distribution function.

Note that the standard normal distribution function $\Phi(x)$ cannot be expressed using the functions we have seen so far. Therefore a special function has been introduced, the so-called "error" function:

$$\mathrm{Erf}(x) = \frac{2}{\sqrt{\pi}} \int_0^x e^{-t^2}\, dt.$$

Using Erf and the fact that $\Phi(0) = \frac{1}{2}$, the standard normal distribution function can then be written as

$$\Phi(x) = \frac{1}{2}\left(1 + \mathrm{Erf}\left(\frac{x}{\sqrt{2}}\right)\right).$$

It should be observed that the normal distribution density function is an affine deformation of the standard normal density function, similar to deformations of wavelets in section 39.2.

Normal distributions with parameters μ and σ are very common for events whose values concentrate around μ. Here σ is a measure for how far from μ values tend to deviate. The normal distribution fully reveals its importance in the context of the central limit theorem 371.

Proposition 365 *If $a \neq 0$ and b are real numbers, and X is a random variable, then if X is $N(\mu, \sigma^2)$-distributed, the random variable $a \cdot X + b$ is $N(a \cdot \mu + b, a^2 \cdot \sigma^2)$-distributed.*

If for $i = 1, \ldots n$, the random variables X_i are $N(\mu_i, \sigma_i^2)$-distributed, then the $\mathbb{R}$-linear combination $X = \sum_i a_i X_i$ is $N(\sum_i a_i \mu_i, \sum_i a_i^2 \sigma_i^2)$-distributed.

Proof If the random variable X is $N(\mu, \sigma^2)$-distributed, then we have

$$P(X \leq x) = \frac{1}{\sigma\sqrt{2\pi}} \cdot \int_{-\infty}^{t} \exp\left(\frac{-1}{2}\left(\frac{t-\mu}{\sigma}\right)^2\right) dt$$

$$= \frac{1}{\sqrt{2\pi}} \cdot \int_{-\infty}^{\frac{x-\mu}{\sigma}} \exp\left(\frac{-1}{2}s^2\right) ds$$

$$= \Phi\left(\frac{x-\mu}{\sigma}\right).$$

Now, taking the so-called *standardization* $V = \frac{X-\mu}{\sigma}$ of X, the above equation yields

$$P(V \leq x) = P(X \leq \mu + \sigma x) = \Phi(x),$$

which means that V has the standard normal distribution. Let $a > 0$. With $X = \sigma \cdot V + \mu$, the random variable $T = a \cdot X + b$ has $P(T \leq t) = P(a \cdot (\sigma \cdot V + \mu) + b \leq t) = P\left(V \leq \frac{t-(a\mu+b)}{a\sigma}\right) = \Phi\left(\frac{t-(a\mu+b)}{a\sigma}\right)$. So T is a $N(\mu, \sigma^2)$-distributed random variable. Using the symmetry of the normal distribution density, one proves the claim for $a < 0$ by the same method. We omit the proof of the linearity of normal distribution, which is quite technical, and refer to [35]. $\square$

Definition 260 (Exponential Distribution) *Let* $\lambda > 0$ *be a real number. Then the random variable* X *is called* $Ex(\lambda)$-*distributed, or also* exponentially distributed with parameter λ *iff it is defined by the density function*

$$f(x) = \begin{cases} 0 & \text{if } x \leq 0, \\ \lambda e^{-\lambda x} & \text{if } x > 0. \end{cases}$$

The distribution function is

$$F_\lambda(x) = \begin{cases} 0 & \text{if } x \leq 0, \\ 1 - e^{-\lambda x} & \text{if } x > 0. \end{cases}$$

Example 215 The exponential distribution is often used to model situations where an event occurs after a time interval t. An example is the time between successive events in a waiting queue. The probability of the arrival of the next event in a queue between t_0 and t_1 is

$$P(t_0 \leq X \leq t_1) = \int_{t_0}^{t_1} \lambda e^{-\lambda x} dx = F(t_1) - F(t_0) = e^{-\lambda t_0} - e^{-\lambda t_1}.$$

Or, more simply, the probability that the next event occurs before time t is

$$P(X \leq t) = F(t) = 1 - e^{-\lambda t}.$$

Of course, the parameter λ has to be so chosen as to meet the requirements of the situation.

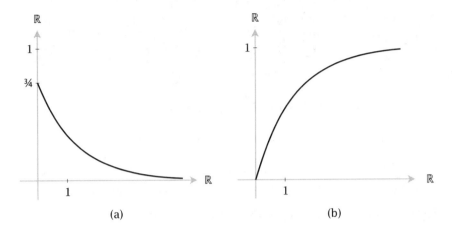

Fig. 42.7. Exponential distribution with $\lambda = \frac{3}{4}$: (a) density function, (b) distribution function.

There are many more types of distributions, but those that we have introduced are the most common and are suitable for modeling a large variety of situations.

42.5 Expectation and Variance

The theory of random variables is the theoretical packaging of what in descriptive statistics is calculated from sample series. We therefore give a short overview of quantities usually deduced from sample series. We consider a one-dimensional sample series, i.e., a sequence $x. = (x_1, x_2, \ldots x_n)$ of real numbers acquired from the measurement of $n > 0$ events. The sequence $x.$ is also called the *data vector*. For such a sample series, one has these common quantities:

Definition 261 *Given a sample series* $x. = (x_1, x_2, \ldots x_n)$, *their arithmetic mean is*

$$\overline{x} = \frac{1}{n} \sum_i x_i.$$

If the sample series is permuted by π *to* $x_{\pi(.)} = (x_{\pi(1)}, x_{\pi(2)}, \ldots x_{\pi(n)})$ *such that the permuted sequence is ordered, i.e.,* $x_{\pi(i)} \leq x_{\pi(j)}$ *if* $i < j$, *then the* median *is the value*[4]

$$\tilde{x} = \begin{cases} x_{\pi(n/2)} & \text{if } n \equiv 0 \ (2), \\ x_{\pi((n+1)/2)} & \text{else.} \end{cases}$$

The range *of the sample series is the difference*

$$v = x_{\pi(n)} - x_{\pi(1)}$$

of an ordered permuted sequence, i.e., the difference between the largest and the smallest value in the sequence x. of measurements.

For $n > 1$, the standard deviation *of x. is the number*

$$s = \sqrt{\frac{1}{n-1} \sum_i (x_i - \overline{x})^2},$$

and its square s^2 is called the empirical variance.

The data vector x. is called *deterministic* - if it is constant, i.e, $x. = c \cdot \Delta$, where $\Delta = (1, 1, \dots 1)$. The above quantities are involved in solving the problem of finding a deterministic data vector $c \cdot \Delta$ such that the difference $x - c \cdot \Delta$ has minimal length, i.e., such that $(x - c \cdot \Delta) \perp \Delta$.

Exercise 203 Show that the unique solution of $(x - c \cdot \Delta) \perp \Delta$ is the arithmetic mean $\overline{x} = c$. We have $x. = \overline{x} \cdot \Delta + (x. - \overline{x} \cdot \Delta)$, where the summands are orthogonal. Therefore, by the theorem of Pythagoras, $\|x.\|^2 = \|\overline{x} \cdot \Delta\|^2 + \|x. - \overline{x} \cdot \Delta\|^2$. This means

$$\sum_i x_i^2 = n\overline{x}^2 + \sum_i (x_i - \overline{x})^2$$

or else

$$s^2 = \frac{1}{n-1} \sum_i (x_i^2 - n\overline{x}^2),$$

a common expression for the calculation of the variance.

With this background, random variables are seen as the mathematical counterparts of empirical measurements. Here is the concept representing the arithmetic mean of random variables:

---

[4] Observe that this value is independent of the non-uniquely determined permutation π.

Definition 262 *Let X be a random variable. The expectation $E(X)$ is defined as follows, depending on whether X is discrete or not:*

(i) *(Discrete Case) If X is discrete, then, if $\sum_{x \in V_X} |x| P(X = x) < \infty$, the expectation of X is the number*[5]

$$E(X) = \sum_{x \in V_X} x \cdot P(X = x).$$

(ii) *(Continuous Case) If X is continuously distributed with density f, then if $\int_{-\infty}^{\infty} |x| \cdot f(x)\, dx < \infty$ (the limit $\lim_{a \to \infty} \int_{-a}^{a} |x| \cdot f(x)\, dx$), the expectation of X is the number*

$$E(X) = \int_{-\infty}^{\infty} x \cdot f(x)\, dx.$$

Example 216 A random variable X with $B(n, p)$-distribution has expectation

$$E(X) = \sum_{x=0}^{n} x \cdot \binom{n}{x} \cdot p^x \cdot (1 - p)^{n-x}$$
$$= np$$

For our example of n flippings of a fair coin ($p = \frac{1}{2}$), the expected number of heads is $\frac{n}{2}$.

The expectation of a random variable Y, which is uniformly distributed between a and b, is easily calculated:

$$E(Y) = \int_{-\infty}^{\infty} x \cdot \frac{1}{b - a} \cdot \chi_{]a,b[}(x)\, dx$$
$$= \int_{a}^{b} x \cdot \frac{1}{b - a}\, dx$$
$$= \left(\frac{1}{b - a} \frac{x^2}{2} \right) \Big|_{a}^{b}$$
$$= \frac{b^2 - a^2}{2(b - a)}$$
$$= \frac{a + b}{2}.$$

The expected value of a PRNG will therefore approximately be $\frac{1}{2}$.

---

[5] Observe that the absolute convergence of the sum guarantees that the order of summation is irrelevant in this definition.

A strong criterion for the calculation of expectation values is the symmetric case:

Definition 263 *For a real number z, a random variable X is called symmetric around a real number z iff*

$$P(X \leq z - x) = P(X \geq z + x)$$

for all real $x \geq 0$.

Exercise 204 Show that if a continuously distributed random variable X has a density f such that $f(z - x) = f(z + x)$ for all $x \in \mathbb{R}$, then X is symmetric.

Proposition 366 *If X is symmetric around z, and if the expectation of X exists, then we have*

$$E(X) = z.$$

Proof Let us prove the continuous case for a symmetric density $f(z-x) = f(z+x)$, the other cases are similar, we have to replace the integrals by corresponding sums.

$$
\begin{aligned}
E(X) &= \int_{-\infty}^{\infty} x \cdot f(x)\, dx \\
&= \int_{-\infty}^{z} x \cdot f(x)\, dx + \int_{z}^{\infty} x \cdot f(x)\, dx \\
&= \int_{0}^{\infty} (z - \xi) \cdot f(z - \xi)\, d\xi + \int_{0}^{\infty} (z + \xi) \cdot f(z + \xi)\, d\xi \\
&= \int_{0}^{\infty} z(f(z - \xi) + f(z + \xi))\, d\xi - \int_{0}^{\infty} \xi(f(z - \xi) - f(z + \xi))\, d\xi \\
&= \int_{0}^{\infty} z(f(z - \xi) + f(z + \xi))\, d\xi \\
&= z \cdot \int_{-\infty}^{z} f(\xi)\, d\xi + z \cdot \int_{z}^{\infty} f(\xi)\, d\xi = z \cdot 1 = z.
\end{aligned}
$$

$\square$

Example 217 If a random variable X is $N(\mu, \sigma^2)$-distributed, then its expectation is $E(X) = \mu$.

Often, the expectation $E(X)$ can be calculated directly from the distribution function of X:

Proposition 367 *If $X \geq 0$, then*

$$E(X) = \int_0^\infty 1 - F_X$$

if this improper integral exists.

We then also have

$$E(X) = \int_0^\infty P(X \geq x)\, dx = \int_0^\infty P(X > x)\, dx.$$

In particular, if X is discrete with non-negative integer values, then

$$E(X) = \sum_{n=0}^\infty P(X > n)$$

if this sum converges.

Proof Let us prove the claim for a continuously distributed variable defined by a density function $f(x)$, the case of a discrete variable is similar. Since X has no negative values, we have $\int_{-\infty}^x f = F(x) = P(X \leq x) = 0$ for $x < 0$. We therefore may suppose $f(x) = 0$ for $x < 0$. We have

$$\int_0^\infty (1 - F(t))dt = \int_0^\infty \left(\int_t^\infty f(x)dx \right) dt$$

$$= \int_\Delta f \quad \text{with } \Delta = \{(x, t) \mid x \geq t \geq 0\}$$

$$= \int_0^\infty \left(\int_0^t dt \right) f(x)\, dx$$

$$= \int_0^\infty x f(x)\, dx$$

$$= E(X).$$

In the case of a continuously distributed variable, the distribution function $F(x)$ is continuous, therefore $P(X > x) = 1 - F(x) = 1 - F(x - 0) = P(X \geq x)$, and the second claim follows. For a discrete variable, the distribution function is only non-continuous for an at most countable set, i.e., the functions $P(X > x)$ and $P(X \geq x)$ differ at most for a countable set of points, which has measure zero. Therefore (see the criterion 283 for integrability) the integrals also coincide. $\square$

Example 218 If the random variable X has a geometric distribution with parameter p, then

$$P(X > n) = \sum_{i=n}^\infty p \cdot (1 - p)^i = (1 - p)^n.$$

Thus, proposition 367 yields

$$E(X) = \sum_{n=0}^{\infty} (1-p)^n = \frac{1}{p}.$$

The probability of throwing the number 6 with a fair dice is $\frac{1}{6}$. This means that the expected number of rollings before throwing a 6 is $1/(1/6) = 6$.

For an exponentially distributed Y with parameter λ

$$P(Y \leq x) = 1 - e^{-\lambda x},$$

thus

$$P(Y > x) = e^{-\lambda x}.$$

As $Y \geq 0$, the expectation of Y is then

$$\begin{aligned}
E(Y) &= \int_0^{\infty} P(|Y| > x)\, dx \\
&= \int_0^{\infty} P(Y > x)\, dx \\
&= \int_0^{\infty} e^{-\lambda x}\, dx \\
&= \left(-\frac{1}{\lambda} e^{-\lambda x} \right) \Big|_0^{\infty} \\
&= \frac{1}{\lambda}.
\end{aligned}$$

The next result states that the expectation operator is linear:

Proposition 368 *If $a_i \in \mathbb{R}$, for $i = 0, \ldots n$, and if $X_1, \ldots X_n$ are random variables having expectations $E(X_i)$, then*

$$E\left(a_0 + \sum_{i=1}^{n} a_i \cdot X_i \right) = a_0 + \sum_{i=1}^{n} a_i \cdot E(X_i).$$

Proof This claim follows from the special cases $E(X + Y) = E(X) + E(Y)$ and $E(a \cdot X + b) = a \cdot E(X) + b$ for real numbers a and b. This requires the theory of probability functions for several random variables, which we will not develop in sufficient details, we shall only give a short introduction in section 42.6. We refer to [22] for this subject.   □

The next empirical quantity, which we want to adopt for random variables, is variance.

Definition 264 *If X is a random variable such that $E(X)$ and $E([X - E(X)]^2)$ exist, then the number*

$$Var(X) = E([X - E(X)]^2)$$

is called the variance *of X. Its square-root $\sigma = \sqrt{Var(X)}$ is called the* standard deviation *of X.*

Example 219 If a random variable X is $N(\mu, \sigma^2)$-distributed, then its variance is $Var(X) = \sigma^2$. This, together with example 217, provides an explanation for the parameters μ and σ of normal distributions.

Here is a useful result for squares of random variables:

Proposition 369 *Let X be a random variable such that $E(X^2)$ exists, and let $a, b \in \mathbb{R}$. Then:*

   (i) *The expectation $E(X)$ and variance $Var(X)$ both exist.*
   (ii) *$Var(X) = E(X^2) - (E(X))^2$.*
   (iii) *$Var(a \cdot X + b) = a^2 \cdot Var(X)$.*

Proof If $E(X^2) = \int_x^\infty P(X^2 > x)\,dx$ exists, then this integral is also equal to $\int_x^\infty P(|X| > \sqrt{x})\,dx$. Since for $x \geq 1$, we have $(|X| > x) \subset (|X| > \sqrt{x})$, and therefore $P(|X| > x) \leq P(|X| > \sqrt{x})$, the integral $\int_x^\infty P(|X| > x)\,dx$ is also finite, but $(X > x) \subset (|X| > x)$, and therefore $E(X) = \int_x^\infty P(X > x)\,dx$ is also finite. Using proposition 368, this yields

$$\begin{aligned} Var(X) &= E([X - E(X)]^2) \\ &= E(X^2) - 2E(X)^2 + E(X)^2 \\ &= E(X^2) - E(X)^2. \end{aligned}$$

Finally,

$$\begin{aligned} Var(a \cdot X + b) &= E((a \cdot X + b - a \cdot E(X) - b)^2) \\ &= E(a^2 \cdot (X - E(X))^2) \\ &= a^2 \cdot Var(X). \end{aligned}$$

$\square$

Example 220 The variance of a uniformly distributed random variable X is

$$Var(X) = E(X^2) - (E(X))^2.$$

We already know that $(E(X))^2 = \left(\frac{a+b}{2}\right)^2$. It remains to calculate $E(X^2)$:

$$E(X^2) = \int_a^b x^2 \cdot \frac{1}{b-a}\, dx$$

$$= \frac{1}{b-a} \cdot \left(\frac{x^3}{3}\right)\bigg|_a^b$$

$$= \frac{b^3 - a^3}{3 \cdot (b-a)}$$

$$= \frac{a^2 + ab + b^2}{3}$$

Therefore

$$Var(X) = \frac{a^2 + ab + b^2}{3} - \left(\frac{a+b}{2}\right)^2$$

$$= \frac{(a-b)^2}{12}$$

The variance of the PRNG is therefore $\frac{1}{12}$. The linearity of expectation allows us to start from a PRNG X delivering values between 0 and 1 and build a new one, Y, which uniformly delivers values between a and b, i.e., $Y = (b-a) \cdot X + a$. Then

$$E(Y) = E((b-a) \cdot X + a)$$

$$= (b-a) \cdot E(X) + a$$

$$= \frac{1}{2}(b-a) + a$$

$$= \frac{a+b}{2}$$

and

$$Var(Y) = Var((b-a) \cdot X + a)$$

$$= (b-a)^2 \cdot Var(X)$$

$$= \frac{(b-a)^2}{12}$$

as expected.

42.6 Independence and the Central Limit Theorem

We now generalize distribution functions to n-dimensional random variables. Given an n-dimensional random variable $X = (X_1, \ldots X_n)$ on the probability space $\mathfrak{A}_P$, and a vector $x = (x_1, \ldots x_n) \in \mathbb{R}^n$, each Borel set of form $I = \,]-\infty, x_1] \times \ldots]-\infty, x_n] \in \mathfrak{B}^n$ gives rise to the Borel set $X^{-1}(I)$ and thereby to the n-*dimensional distribution function* $F_{P,X} : \mathbb{R}^n \to [0, 1]$ via

$$F_{P,X}(x_1, \ldots x_n) = P(X^{-1}(I)).$$

Observe that $X^{-1}(I)$ is the intersection of the Borel sets defined by the condition "$X_1 \leq x_1$", … "$X_n \leq x_n$". We again use the intuitive notation for the values of F and write

$$F_{P,X}(x_1, \ldots x_n) = P(X_1 \leq x_1, \ldots X_n \leq x_n).$$

This suggests the definition of independent random variables:

Definition 265 *The n random variables $X_1, \ldots X_n$ on the probability space $\mathfrak{A}_P$, are said to be* independent *iff*

$$F_{P,X}(x_1, \ldots x_n) = F_{P,X_1}(x_1) \cdot \ldots F_{P,X_n}(x_n)$$

or, restated using the probability measure,

$$P(X_1 \leq x_1, \ldots X_n \leq x_n) = P(X_1 \leq x_1) \cdot \ldots P(X_n \leq x_n)$$

for all $x_1, \ldots x_n \in \mathbb{R}$.

Here is a useful theorem concerning the density function of a sum of independent continuously distributed random variables (see [35]):

Proposition 370 *If $X, Y : \mathfrak{A}_P \to \mathbb{R}$ are independent, continuously distributed random variables with densities f_X, f_Y, respectively, then their sum $X + Y$ is continuously distributed with the density*

$$f_{X+Y}(z) = \int_{-\infty}^{\infty} f_X(\xi) \cdot f_Y(z - \xi) \, d\xi.$$

The integral on the right hand side is also called the convolution *of f_X and f_Y, and written $f_X * f_Y$. Thus we can write more concisely*

$$f_{X+Y}(z) = (f_X * f_Y)(z).$$

Example 221 If a random variable $Z \in Rand(\mathfrak{A})$ has the distribution function

$$F_Z(x) = P(Z_1^2 + \ldots Z_s^2 \leq x), \text{ for all } x \in \mathbb{R},$$

with all the $Z_i \in Rand(\mathfrak{A})$ being independent and having the standard normal distribution, then Z is called χ_s^2-*distributed*.

The following is a fundamental theorem of probability theory. It justifies the importance of normal distributions.

Proposition 371 (Central Limit Theorem)
Let the random variables $X_1, \ldots X_n : \mathfrak{A}_P \to \mathbb{R}$ be independent and have the same distribution. Suppose that their common expectation is $E(X_i) = \mu$ and that their common variance is $Var(X_i) = \sigma^2 > 0$ and consider their sum $S_n = \sum_{i=1}^{n} X_i$.

Then the distribution of the random variable

$$Z_n = \frac{1}{\sigma \sqrt{n}} \left(S_n - n\mu \right)$$

converges to the standard normal distribution in each point, i.e., for all $x \in \mathbb{R}$,

$$\lim_{n \to \infty} F_{Z_n}(x) = \Phi(x).$$

This means effectively that the distribution of S_n tends to $N(n\mu, n\sigma^2)$, thus $E(S_n) \to n\mu$ and $Var(S_n) \to n\sigma^2$, as $n \to \infty$.

Intuitively, this means that linear combinations of a large number of identically distributed independent variables tend towards a normal distribution.

Example 222 Figure 42.8 shows Galton's board, also known as quincunx. Marbles roll from top to bottom through a grid of pegs to land in one of n compartment, here $n = 15$. At each peg, a marble may fall to left or to the right with equal probability $p = \frac{1}{2}$. Since there are $n - 1$ levels, we can describe a path by a sequence of random variables X_i taking the value 0 for left and 1 for right. The number of 1 in the sequence, or simply $\sum_i X_i$, is the number of the compartment that a marble following this path reaches.

We know that $S_{n-1} = \sum_{i=0}^{n-1} X_i$ follows a binomial distribution $B(n - 1, \frac{1}{2})$, i.e. the probability that a marble lands in case i, for $0 \le i < n$, is $P_{B(n-1,1/2)}(S_{n-1} = i)$ (it is revealing to compare this with example 212).

Since the random variables X_i are independent and have the same distribution, namely $P(X_i = 0) = \frac{1}{2}$ and $P(X_i = 1) = \frac{1}{2}$, we can apply the central limit theorem to get an approximation to the binomial distribution. We have $\mu = E(X_i) = 0 \cdot \frac{1}{2} + 1 \cdot \frac{1}{2} = \frac{1}{2}$ and $\sigma^2 = Var = E(X_i^2) - (E(X_i))^2 = \frac{1}{4}$. The central limit theorem now states that the distribution of S_{n-1} is approximatively $N((n - 1)\mu, (n - 1)\sigma^2) = N(\frac{1}{2}(n - 1), \frac{1}{4}(n - 1))$, in our

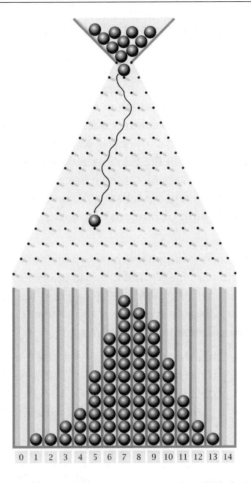

Fig. 42.8. Marbles rolling down Galton's board.

concrete case this means that S_{14} approaches a random variable Y with distribution $N(7, \frac{7}{2})$. This means that the probability that a marble lands in case i is approximatively $P(Y = i)$. In an experimental setting such as the one shown in the figure, after m marbles have been rolled, each compartment i will contain approximatively $m \cdot P(Y = i)$ marbles. The columns built by the marbles will tend in shape towards the density function of a normal distribution. Of course, more compartments and more marbles will improve the approximation.

42.7 A Remark on Inferential Statistics

The preceding considerations look like nice abstract mathematics and somewhat resemble the situation in quantum mechanics, where the common physical reality is paralleled by a "fictitious" reality of observables, whose eigenvalues are the measured quantities. However, in statistics, one is given sample series $x. = (x_1, \ldots x_n)$ of numbers, and not random variables. What, then, is the relation between sample series and the abstract level of random variables, which are never observed as a such? This is the subject of inferential statistics: To investigate the relations between the "Platonic" world of random variables and the empirical world of sample series.

To get off the ground, one supposes that each measured quantity x_i is the value of one random variable X_i, and one also assumes that the n random variables X_i are independent and realize one and the same distribution $F_{X_i} = F$. The event space is given from the outcome set of all possible sequences of n measurements of the system.

Next, one supposes that the distribution F is defined by a parameter vector δ, such as the parameter pair $\delta = (\mu, \sigma^2)$ of a normal distribution. Let $\phi(x, \delta)$ be the probability $P(X = x)$ for a discrete distribution or the density $f(x, \delta)$ for a continuously distributed variable. Then the number

$$L(x., \delta) = \prod_i \phi(x_i, \delta)$$

describes the probability of a sample series $x.$ to be realized for the system parameter δ. Under these assumptions, one wants to find a parameter δ such that this probability $L(x., \delta)$ is maximal. This leads to the definition

Definition 266 *With the above assumptions and notations, a distribution parameter δ is called a* maximum likelihood estimate *iff*

$$L(x., \delta) \geq L(x., \delta')$$

for every possible δ'.

Example 223 The following is an example of the maximum likelihood method in the case of a continuously distributed variable. Suppose that the distribution is normal, i.e.,

$$\phi(x, \delta) = f(x, \mu, \sigma^2)$$
$$= \frac{1}{\sigma\sqrt{2\pi}} \cdot \exp\left(\frac{-1}{2}\left(\frac{x - \mu}{\sigma}\right)^2\right).$$

Then the likelihood function is

$$L(x., \delta) = L(x., \mu, \sigma^2)$$
$$= \left(\frac{1}{\sigma\sqrt{2\pi}}\right)^n \cdot \prod_i \exp\left(\frac{-1}{2}\left(\frac{x_i - \mu}{\sigma}\right)^2\right).$$

To maximize this function, we maximize its logarithm:

$$l(x., \mu, \sigma^2) = \log(L(x., \mu, \sigma^2))$$
$$= -n \cdot (\log(\sqrt{2\pi}) + \log(\sigma)) - \frac{1}{2}\sum_i \left(\frac{x_i - \mu}{\sigma}\right)^2.$$

A necessary condition for a local maximum of $l(x., \mu, \sigma^2)$ is that $D_\mu l = D_\sigma l = 0$ vanishes for its partial derivatives. This means

$$D_\mu l(x., \mu, \sigma^2) = \sum_i \frac{x_i - \mu}{\sigma} = 0$$

$$D_\sigma l(x., \mu, \sigma^2) = -\frac{n}{\sigma} + \sum_i \frac{(x_i - \mu)^2}{\sigma^3} = 0.$$

These equations yield $\mu = \frac{1}{n}\sum_i x_i = \bar{x}$ and $\sigma^2 = \frac{1}{n}\sum_i (x_i - \mu)^2$. This means that the maximum likelihood estimate is the arithmetic mean plus nearly the empirical variance s. In fact, we have the denominator n instead of $n - 1$ as required for s. See [35] for a detailed discussion of the maximum likelihood method.

CHAPTER 43

Lambda Calculus

43.1 Introduction

The two volumes of this book introduce a wide range of mathematical objects and functions. Since the second half of the nineteenth century, mathematicians have been concerned about how these fit into a framework of effective computation. The development of mathematical logic has been the main outcome of this quest. From the 1930's on mechanical computation has become real, and these questions have attained practical relevance. Here the link to effective computation is the *Church-Turing thesis* which says that any real-world computation can be translated into a computation on a Turing machine or in Alonzo Church's (1903–1995) own invention, the λ-calculus. Church devised the λ-formalism as a kind of computational logic and showed that it is equivalent to a universal[1] Turing machine.

The λ-calculus has become a major tool for describing computations and is the foundation of many real-world programming languages, of which functional programming languages like LISP, Scheme, SML or Haskell come closest to its spirit. We may even say that functional languages are more or less direct implementations of the λ-calculus. Stated concisely, one can say that the λ-calculus is the theory of abstraction and application of functions.

---

[1] A Turing machine that can emulate any specific Turing machine.

In this chapter we formally describe the λ-language and introduce its computation model. In the end we will be able to use the formalism just like a real programming language.

Towards the end of the chapter we describe the kind of functions which the Church-Turing thesis originally envisioned.

43.2 The Lambda Language

The λ-calculus is based on a formal language L as discussed in section 19.1 ff. of volume 1. The language $L = L(V, C)$ is built from a denumerable set $V = \{x, y, z \ldots\}$ of *variables* and a second, disjoint, set $C = \{c, d, \ldots\}$ of *constants*. We always suppose that V and C are given a well-ordering. If C is empty, L is called *pure*. It finally contains the symbols $(,), \lambda$ and the dot $(.)$. The language L is by definition the smallest language over the alphabet $\mathcal{A} = V \cup C \cup \{(,), \lambda, .\}$ with the properties 1–3 listed below. Its elements are called λ-*terms*. The length of a term t, being composed of variables, constants, parentheses, or the dot, is denoted by $l(t)$, and counts by definition the occurrences of variables and constants. Observe that this differs from the length of t as a word: the parentheses, the symbol λ and the dot are ignored.

1. $V \cup C \subset L$, i.e., variables or constants are the λ-terms of length one; they are called *atoms of L*.

2. If $N, M \in L$, then $(NM) \in L$; the λ-term (NM) is called *application (of N to M)*; we have $l((NM)) = l(N) + l(M)$.

3. If $x \in V$ and M is a λ-term of L, then $(\lambda x.M)$ is a λ-term in L; it is called *abstraction*. We have $l((\lambda x.M)) = l(M) + 1$.

Exercise 205 Give a definition of the syntax of λ-language $L(V, C)$ as a context free grammar and cast it in BNF notation (in particular, the alphabet of the grammar must be finite).

In practice, the large number of parentheses in λ-terms is irritating, one therefore adopts the convention to write the typical terms

- $PWQR$ for $(((PW)Q)R)$ (association to the left),

- $\lambda x.PQR$ for $(\lambda x.((PQ)R))$,

- $\lambda xy \dots z.P$ for $(\lambda x.(\lambda y \dots (\lambda z.P) \dots)))$.

- Moreover, if one writes $\lambda x_1 \dots x_n.M$, then the special case $n = 0$ means M. In the same vein, if one writes $MN_1 \dots N_k$, then the special case $k = 0$ means M. Observe that the empty word is not a term in the language of the λ-calculus.

In a term t, the forms of application or abstraction are mutually exclusive, and in either case, the subterms M and N of $t = (MN)$, or x and M of $t = (\lambda x.M)$ are uniquely determined. If a term t is part of a larger term s, one says that t *occurs in s*. Moreover, if a term $t = (\lambda x.M)$ occurs in s, then the term M is called the *scope of the variable x in t*. Attention, the scope of a variable is strictly tied to the λ which precedes the variable.

Example 224 The completely parenthesized λ-term

$$(\lambda x.(x(\lambda y.(\lambda x.(xy)))))$$

is written according to the convention

$$\lambda x.x\lambda y.\lambda x.xy.$$

In each of following terms, the scope of boldfaced variables is underlined:

$$\lambda \boldsymbol{x}.\underline{x\lambda y.\lambda x.xy}, \qquad \lambda x.x\lambda \boldsymbol{y}.\underline{\lambda x.xy}, \qquad \lambda x.x\lambda y.\lambda \boldsymbol{x}.\underline{xy}.$$

Definition 267 *Let $t \in L(V,C)$ be a term. Then $V(t)$ denotes the set of variables occurring in t. The set $FV(t)$ of free variables of t is defined by recursion on $l(t)$:*

- (i) *If $t = x$ is a variable ($l(t) = 1$), then $FV(t) = \{x\}$.*
- (ii) *If $t = (MN)$, then $FV(t) = FV(M) \cup FV(N)$.*
- (iii) *If $t = (\lambda x.M)$, then $FV(t) = FV(M) - \{x\}$.*

If $FV(t) = \varnothing$, then t is called a closed *term. A variable in $V(t) - FV(t)$ is called a* bound *variable of t.*

A word of caution: A variable may be free in a term, but the same variable may be bound in a larger scope by a λ abstraction. Similarly a variable may be bound in a term, but free in a larger context. Thus, for example $FV(xy) = \{x,y\}$, but $FV(\lambda x.xy) = \{y\}$ and $FV(\lambda y.\lambda x.xy) = \varnothing$. But also, $FV(x\lambda x.\lambda y.\lambda x.xy) = \{x\}$. Therefore it is important to always specify the term, relative to which a variable is considered to be either free or bound.

43.3 Substitution

In the present state of our theory, the calculus is very rigid. To begin with, this functional formalism should be extended to make it possible to replace bound variables by specific terms in order to evaluate the abstraction forms. For that purpose the concept of substitution is introduced. Observe that the following notation for substitutions is *not* part *of* the language L, but is a notation for operations *on* the language.

Definition 268 *Suppose that u and t are terms in L, and that $x \in V$ is a variable. Then the term $[t/x]u$ of substitution is defined as follows by recursion on the length of u:*

(i) *If $u = x$, then $[t/x]u = t$.*

(ii) *If u is a constant or a variable different from x, then $[t/x]u = u$.*

(iii) *If $u = (MN)$, then $[t/x]u = ([t/x]M\,[t/x]N)$.*

(iv) *If $u = \lambda x.v$, then $[t/x]u = u$.*

(v) *If $u = \lambda y.v$ with $y \neq x$ and $y \notin FV(t)$, or $x \notin FV(v)$ then*

$$[t/x]u = \lambda y.[t/x]v.$$

(vi) *If $u = \lambda y.v$ with $y \neq x$, and $y \in FV(t)$, and $x \in FV(v)$, then choose the first variable z greater than all variables in the set $FV(t) \cup FV(v)$. Then*

$$[t/x]u = \lambda z.[t/x]([z/y]v).$$

For the expression $[t/x]u$ we say "substitute x by t in u".

Again, pay attention to the notation $[t/x]u$: this is not a term in L, it is an expression denoting the action of the map $[x/t] : L \to L$ on the term u, and the result $[t/x]u$ as a term in L looks completely different from "$[t/x]u$".

The following sorites state some useful properties of substitution.

Sorite 372 *If $x, y \in V$ and $t, u \in L$, then*

(i) (Identity action) *We have $[x/x]u = u$.*

(ii) (Ineffectiveness of bound variables) *If $x \notin FV(u)$, then $[t/x]u = u$.*

(iii) (Invariance of length) *We have $l([y/x]u) = l(u)$.*

(iv) (Free variables of substitution) *If $x \in FV(u)$, then*

$$FV([t/x]u) = FV(t) \cup (FV(u) - \{x\}).$$

Proof The proof strategy is obvious: We have to check all six cases in the definition of a substitution, and in each case, if reasonable, proceed by induction on the word length of a term.

Let us prove claim (i). For cases (ii) and (iv) in definition 268, the action is the identity by definition. In case (i), $t = x = u$ yields $[t/x]u = t = x = u$. Case (iii) is immediate by induction. Case (vi) cannot occur since $y \neq x$ excludes $y \in FV(t) = FV(x) = \{x\}$.

For claim (ii), case (i) is excluded, case (ii) yields the right result anyway, case (iii) inherits the condition $x \notin FV(u) = FV(M) \cup FV(N)$ on free variables to the terms M and N, and induction applies. Case (iv) works by definition, case (v) inherits $x \notin FV(u)$ to v since the formula $FV(u) = FV(v) - \{y\}$ does not omit the variable x from $FV(v)$, so $x \notin FV(v)$, and induction applies. Case (vi) is impossible since $x \in FV(v)$ implies $x \in FV(u)$ in this case.

For claim (iii), cases (i), (ii) and (iv) are trivial. Case (iii) can be shown by induction, since $l(M) < l(u)$, and $l(N) < l(u)$. For case (v), we have $[t/x]u = \lambda y.[t/x]v$ and therefore $l([t/x]u) = l(\lambda y.[t/x]v) = 1 + l([t/x]v)$. But because $l(v) < l(u)$, we have $l([t/x]v) = l(v)$ by induction, and we are done. Case (vi) can be shown in a similar way.

Claim (iv) is immediate for all cases. $\square$

Sorite 373 *Let $x, y, z \in V$ be three distinct variables, and let $u, t, q \in L$ be terms such that no bound variable in u is free in ztq. Then*

(i) (Canceling) *If $z \notin FV(u)$, then $[t/z][z/x]u = [t/x]u$.*

(ii) (Inversion) *If $z \notin FV(u)$, then $[x/z][z/x]u = u$.*

(iii) (Composition) *If $z \notin FV(t)$, then*

$$[t/x][q/y]u = [([t/x]q)/y][t/x]u.$$

(iv) (Commutation) *If $y \notin FV(t)$ and $x \notin FV(q)$, then*

$$[t/x][q/y]u = [q/y][t/x]u.$$

(v) (Repeated substitution) *We have*

$$[t/x][q/x]u = [([t/x]q)/x]u.$$

Proof Claim (ii) follows from claim (i) and statement (i) in sorite 372.

Claim (iv) follows from claim (iii) and statement (ii) in sorite 372.

For claim (i), we go through all six cases:

Case (i): $u = x$, we suppose $z \neq u = x$, then $[t/z][z/x]x = [t/z]z = t = [t/x]x$.

Case (ii): the atom u is $\neq x$ and we suppose $z \neq u$: $[t/z][z/x]u = [t/z]u = u = [t/x]u$.

Case (iii): this is evident by induction.

Case(iv): $u = \lambda x.p$, then $[t/z][z/x]u = [t/z][z/x](\lambda x.p) = [t/z]\lambda x.p = \lambda x.p$ since z is not free in $u = \lambda x.p$. This coincides with $\lambda x.p = [t/x]\lambda x.p$.

Case (v): again $u = \lambda y.p$ with $y \neq x$ and $y \neq z$ by the hypothesis on free and bound variables. So by induction, $[t/z][z/x]\lambda y.p = [t/z]\lambda y.[z/x]p = \lambda y.[t/z][z/x]p = \lambda y.[t/x]p = [t/x]\lambda y.p$.

Case (vi): again $u = \lambda y.p$ with $y \neq x$ and $y = z$, $x \in FV(p)$, but $y = z$ contradicts the hypothesis on free and bound variables, so the case is impossible.

The proofs of points (iii) and (v) work along the same mechanical lines, we leave them as easy exercises to the reader. □

Despite their formalistic shape, these statements are completely natural. They express in a precise way when and how we may substitute free variables.

Exercise 206 Perform the substitution $[t/x]u$ for the following t and u:

1. $t = x$ and $u = \lambda y.x$,

2. $t = y$ and $u = \lambda zx.x$,

3. $t = \lambda z.zy$ and $u = \lambda y.xy$,

4. $t = \lambda v.vz$ and $u = \lambda y.x\lambda z.x$.

43.4 Alpha-Equivalence

In the following discourse, we shall be interested in certain (equivalence) relations among λ-terms which are compatible with the constructors of application and abstraction.

Definition 269 A relation $R \subset L \times L$ is called λ-compatible *iff*

   (i) *whenever tRs, then also $(\lambda x.t)R(\lambda x.s)$, for all variables $x \in V$;*

   (ii) *whenever tRs, uRv, then $(tu)R(sv)$.*

This type of relation is desirable to hold between terms that are the "same up to bound variables", which means that a change of bound variables should be irrelevant to the meaning of a term. Here is the typical λ-compatible relation of congruence:

Definition 270 *If in a term t, the term $\lambda x.u$ occurs, and if for a variable y, we have $y \notin FV(u)$, then one can replace t by the term, where the given occurrence of $\lambda x.u$ is replaced by $\lambda y.[y/x]u$. The new term t' is said to be obtained from t by a change of a bound variable in t. One says that t is α-congruent to s, or also that t α-converts to s, in signs $t \equiv_\alpha s$, iff it is obtained by a sequence of $0 \leq k < \infty$ changes of bound variables from t.*

Example 225 If x, y, u, v, z are variables, then we have

$$\lambda xy.xzxy\lambda x.x = (\lambda x.(\lambda y.((((xz)x)y)(\lambda x.x))))$$
$$\equiv_\alpha (\lambda x.(\lambda v.((((xz)x)v)(\lambda x.x))))$$
$$\equiv_\alpha (\lambda u.(\lambda v.((((uz)u)v)(\lambda x.x))))$$
$$\equiv_\alpha (\lambda u.(\lambda v.((((uz)u)v)(\lambda w.w))))$$
$$= \lambda uv.uzuv\lambda w.w.$$

The congruence relation is exactly what one expects from identification of terms if they play the same role. This is made explicit in the sorite about α-congruence:

Sorite 374 *let u and v be terms in L, and let $x_1, \ldots x_k$ be variables in V. Then*

(i) *If $u \equiv_\alpha v$, then $FV(u) = FV(v)$.*

(ii) *There exists a term u' in L such that $u \equiv_\alpha u'$ and none of the x_i is bound in u'.*

(iii) *The congruence relation $\equiv_\alpha$ on L is an equivalence relation. The set of equivalence classes is denoted by $\Lambda = L/\equiv_\alpha$.*

(iv) *The congruence relation is λ-compatible.*

Proof The proof of claim (i) is again by induction on the term length and on the length of the sequence of changes of bound variables. We only need to deal with sequences of length one. For $u = \lambda x.w$, we use statement (iv) in sorite 372 and get $FV(\lambda y.[y/x]w) = (\{y\} \cup (FV(w) - \{x\})) - \{y\} = FV(w) - \{x\}$, whereas $FV(\lambda x.w) = FV(w) - \{x\}$. If $\lambda x.w$ is a proper subterm of u, then it may be a subterm in an application $u = (qt)$, i.e., $\lambda x.w$ occurs in q or in t, and induction applies with the subterm q or t. If $\lambda x.w$ occurs in q for $u = \lambda y.q$, then we again apply induction since the free variables of u are those of q minus y. This does not change if q is replaced by a congruent term, and we are done.

Claim (ii) is evident since there are infinitely many variables in V, and we may produce congruences by eliminating all the x_i.

For claim (iii), it is evident that congruence is reflexive and transitive. Let us show that it is also symmetric. It must be shown that for $y \neq x$, if $y \notin FV(u)$, then $\lambda y.[y/x]u \equiv_\alpha \lambda x.u$. Suppose that $x \notin FV(u)$. Then $[y/x]u = u$. So $\lambda y.[y/x]u = \lambda y.u$, and the latter yields $\lambda x.u = \lambda x.[x/y]u$ by a change of the bound variable y. If $x \in FV(u)$, then by sorite 372, statement (iv), we have $x \notin FV([y/x]u)$, therefore $\lambda y.[y/x]u \equiv_\alpha \lambda x.[x/y][y/x]u$.

We have to show $[x/y][y/x]u \equiv_\alpha u$ for $x \neq y$ and $y \notin FV(u)$. But by sorite 373, (ii), $[x/y][y/x]u = u$ if $y \notin FV(u)$ and no bound variable in u is x or y. Else, if x is bound in u, then $[y/x]u = u$, and then $[x/y]u = u$. If x is free in u, but y is bound in u, then $[x/y][y/x]u \equiv_\alpha u$ by a mechanical case-by-case inductive proof on the length of u, which we delegate to the reader.

for claim (iv), suppose that $t \equiv_\alpha s$ by a change of a bound variable, then this operation also α-converts $\lambda x.t$ to $\lambda x.s$, since the operation take place within the t term. If $t \equiv_\alpha s, u \equiv_\alpha v$, then we have this chain of α-conversions: $(tu) \equiv_\alpha (su) \equiv_\alpha (sv)$, and we are done. $\square$

The last statement in sorite 374 enables the *definition of application and abstraction on the set Λ of term congruence classes*. Moreover, substitution is also defined on congruence classes in the following sense:

Proposition 375 *If $s \equiv_\alpha t$ and $u \equiv_\alpha v$, then $[s/x]u \equiv_\alpha [t/x]v$. This implies that substitution is well-defined on congruence classes in Λ.*

Proof This proposition is again proved by a case-by-case verification for the intermediate relations $[s/x]u \equiv_\alpha [t/x]u \equiv_\alpha [t/x]v$. $\square$

43.5 Beta-Reduction

Whereas α-congruence is an equivalence relation, which essentially reduces to a change of bound variables, β-reduction has a semantic background, which has much more dramatic implications for the form of terms. The essence is that the abstraction $\lambda x.u$ formally captures the idea of turning u into a function in the variable x, which can be replaced by a concrete value in an "abstract" formula, and the application (ut)

captures the idea of evaluating u at the value t. And in the case of an abstraction, $(\lambda x.u)t$, this means evaluating the abstract formula $\lambda x.u$ at the value t. This means replacing all occurrences of x in u by t. In other words, we are "reducing" the term $(\lambda x.u)t$ to the term $[t/x]u$, which results from u by replacing all occurrences of x by t in the way the operation of substitutions is defined in definition 268.

Definition 271 *If a term $(\lambda x.u)t$ occurs in a term v, then its replacement by $[t/x]u$ in v defines a new term w. We say that w is a β-contraction of v, or that v contracts to w if w results from v by such a replacement operation, in symbols: $v \triangleright_{1\beta} w$. A term $(\lambda x.u)t$ is called a β-redex, while the term resulting from the operation $[t/x]u$ is called* its contractum. *Clearly, the contractum of a β-redex does not uniquely determine the β-redex in general.*

We say that a term v β-reduces to w, in signs: $v \triangleright_{\beta} w$ iff there is a sequence of terms $v = v_1, \ldots v_k = w$, with $k \geq 1$, such that for each $i = 1, 2, \ldots k - 1$, we have either $v_i \equiv_\alpha v_{i+1}$ or $v_i \triangleright_{1\beta} v_{i+1}$. In other words, if w results from v by a finite (possibly empty) sequence of changes of bound variables or contractions.

Sorite 376 *With the notations of definition 271, we have these facts:*

(i) *If $v \triangleright_{\beta} w$, and $v \equiv_\alpha v'$ and $w \equiv_\alpha w'$, then $v' \triangleright_{\beta} w'$, in other words: β-reduction induces a (homonymous) relation $\triangleright_{\beta}$ on the set Λ of congruence classes.*

(ii) *The relation $\triangleright_{\beta}$ is reflexive and transitive. In other words, the collection of objects in L (or else in Λ), together with the arrows $u \triangleright_{\beta} v$, defines the structure of a category.*

(iii) *(Substitution is $\triangleright_{\beta}$-order preserving) If $v \triangleright_{\beta} w$, then for any term t and variable x, we have $[t/x]v \triangleright_{\beta} [t/x]w$.*

(iv) *(Substitution decreases free variables) If $v \triangleright_{\beta} w$, then $FV(v) \supseteq FV(w)$.*

(v) *If $v \triangleright_{\beta} w$, then for any term t, we have $[v/x]t \triangleright_{\beta} [w/x]t$.*

Proof Claim (i) is immediate since the chain $v \triangleright_{\beta} w$ can prolonged by the prepended relation $v' \equiv_\alpha v$ (recall that $\equiv_\alpha$ is an equivalence relation) and the appended relation $w \equiv_\alpha w'$ to yield a chain $v' \triangleright_{\beta} w'$.

Claim (ii) is obvious: the identity is the identity relation, which is included in $\equiv_\alpha$.

We leave the proof of claims (iii)–(v) to the reader. □

Example 226 Let x, y, z, w be four variables, then

$$(\lambda x.(\lambda y.xy)z)w \triangleright_{1\beta} [w/x](\lambda y.xy)z$$
$$= (\lambda y.wy)z$$
$$\triangleright_{1\beta} [z/y]wy$$
$$= wz.$$

Example 227 An example showing that β-reduction need not simplify terms and may even expand them is the following case:

$$(\lambda x.xxy)(\lambda x.xxy) \triangleright_{1\beta} (\lambda x.xxy)(\lambda x.xxy)y$$
$$\triangleright_{1\beta} (\lambda x.xxy)(\lambda x.xxy)yy$$
$$\triangleright_{1\beta} (\lambda x.xxy)(\lambda x.xxy)yyy$$
$$\triangleright_{1\beta} \ldots$$

In this case, we say that the term $(\lambda x.xxy)(\lambda x.xxy)$ *diverges*.

The following, similar, term remains the same after β-contraction, but since there is always a β-redex, β-reduction does not terminate.

$$(\lambda x.xx)(\lambda x.xx) \triangleright_{1\beta} (\lambda x.xx)(\lambda x.xx)$$
$$\triangleright_{1\beta} (\lambda x.xx)(\lambda x.xx)$$
$$\triangleright_{1\beta} (\lambda x.xx)(\lambda x.xx)$$
$$\triangleright_{1\beta} \ldots$$

The termination of a sequence of β-contraction is by definition guaranteed iff at a certain stage, no β-redex term is left.

Definition 272 *A term that contains no β-redex is called a β-normal form or β-normal. The set of β-normal forms is denoted by $\beta-NF$, and for a β-reduction $v \triangleright_\beta w$ with $w \in \beta-NF$, one says that v has the β-normal form w.*

The term $(\lambda x.(\lambda y.xy)z)w$ from example 226, has a β-normal form wz, while neither terms from example 227 has no β-normal form. But there are even worse cases: A term may be β-contracted in different ways such that in one way, the result never β-normalizes, while in a second way, β-normalization is reached.

Example 228 Let x, y, z, w be four variables, then the term $q = (\lambda z.w)r$ with the diverging term $r = (\lambda x.xxy)(\lambda x.xxy)$, has these β-reductions:

$$q = (\lambda z.w)r$$
$$\triangleright_{1\beta} [r/z]w$$
$$= w.$$

by performing a β-contraction first on the outer β-redex. If β-contraction is applied first to the r-term at each step in the derivation, then

$$q = (\lambda z.w)r$$
$$\triangleright_{1\beta} (\lambda z.w)(ry)$$
$$\triangleright_{1\beta} (\lambda z.w)(ryy)$$
$$\triangleright_{1\beta} \dots,$$

thus, in this case, q diverges. The reason this happens is, of course, that in $\lambda z.w$, the "argument" z is not needed for the evaluation of the "body" w. Therefore there is no need to reduce r first and get stuck in an infinite process.

It has become customary to distinguish two concrete strategies for reducing λ-terms: *applicative order* and *normal order*. Applicative order can be stated concisely as *leftmost innermost*, i.e., the next redex to consider is the one at the deepest level. If there are several of them at the same level, ties are broken by choosing the leftmost one. Normal order reduction in contrast is *leftmost outermost*. Here the next redex to choose for reduction is the one at the highest level, ties being broken again by choosing the leftmost. The distinguishing property is that, if a λ-term has a normal form, normal order will find it (hence the name), applicative order, in contrast, may diverge for the same term. The disadvantage of normal-order is that the terms in the derivation may grow fast in size. Applicative order reduction, if it doesn't diverge, generally performs better in this regard. In Example 228, the first reduction uses normal order, the second applicative order. Some programming languages, like Haskell, implement a version of normal order called *lazy evaluation*, however, most, like SML and derivates of C implement applicative order, known as *strict evaluation* in this context. See the excellent [1] for further details.

Despite these subtle variants of reduction, the normal form of a term, if it exists, is unique in Λ, i.e., two normal forms of the same term are α-

congruent. This is what one would expect from a fairly reasonable theory of functional expressions. Here is one of the main results of λ-calculus:

Proposition 377 (Church-Rosser Theorem for β-Reduction) *If for three terms* $t, u, v \in L$, *we have* $t \triangleright_\beta u$ *and* $t \triangleright_\beta v$, *then there is a term* s *such that* $u \triangleright_\beta s$ *and* $v \triangleright_\beta s$. *In other words, each diagram*

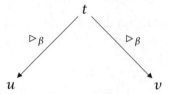

in the category L *can be completed to a square*

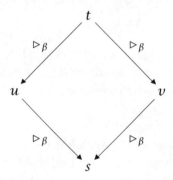

of $\triangleright_\beta$*-relations.*

It is however not claimed that s *is the colimit of the diagram. But it can be shown[2] that* s *can be so chosen as to be the colimit of the diagram, which means that the category* L *(and* Λ*) has pushouts.[3]*

Proof The proof of this classical result is quite involved, splitting in a number of subcases, we cannot give it here and refer to the proof given in appendix 1 of [16]. The existence of a colimit s is proved in Exercise 12.4.4 in [4].   □

This theorem implies what we were looking for:

Corollary 378 *If the term* u *has two β-normal forms* t *and* s, *then they are α-congruent. In other words: In the category* Λ, *any term* t *has at most one arrow* $t \xrightarrow{\triangleright_\beta} w$ *with β-normal codomain* w, *which is then called the β-normal form of* t; *we denote it by* $\beta(t)$.

---

[2] Exercise 12.4.4 in [4].

[3] The pushout is the colimit of the dual diagram (reversed arrows) to a fiber product diagram. Our diagram is precisely of this type.

Proof If we apply the Church-Rosser theorem 377 to the two arrows $u \triangleright_\beta t$ and $u \triangleright_\beta s$, then we obtain a pair of arrows $t \triangleright_\beta w$ and $s \triangleright_\beta w$. But then each of these arrows can only contain $\equiv_\alpha$ relations, and therefore $s \equiv_\alpha t$. □

An even stronger form of the Church-Rosser theorem is true. It is about terms which are connected by any sequence of β-reduction arrows, where the direction may be from a term to its contractum or vice-versa.

Definition 273 *The relation of β-equality is the equivalence relation $=_\beta$ on L (or on Λ) generated by the relation $\triangleright_\beta$. If $t =_\beta s$, we say that t is β-convertible to s.*

The following is immediate:

Sorite 379 *If t, v, w are terms, then $v =_\beta w$ implies $[v/x]t =_\beta [w/x]t$ and $[t/x]v =_\beta [t/x]w$.*

Proposition 380 (Church-Rosser Theorem for β-Conversion) *For terms $v, w \in L$, we have $v =_\beta w$ iff there is a term t such that $v \triangleright_\beta t$, and $w \triangleright_\beta t$. So the horizontal line in the following diagram can always be completed to a triangle with $\triangleright_\beta$ arrows (which realize a particularly short instance of their $=_\beta$-relation):*

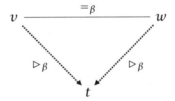

Proof We prove the theorem by induction on the minimal length of a chain of $\equiv_\alpha$ or $\triangleright_{1\beta}$ relations defining $v =_\beta w$. If only $\equiv_\alpha$ are needed, the claim is trivial. We may suppose that only $\triangleright_{1\beta}$ relations intervene, and that there are $n + 1$ such relations connecting v and w, say by $v = v_0 \triangleright_{1\beta} v_1 \ldots v_n = w$ or $w = v_0 \ldots v_{n-1} \triangleright_{1\beta} v_n = v$, depending on whether there is an arrow starting at v or terminating at v. In the first case, by induction, there is a t such that $v_1 \triangleright_\beta t$ and $w \triangleright_\beta t$, and the composition $v \triangleright_{1\beta} v_1 \triangleright_\beta t$, together with $w \triangleright_\beta t$ does the job. For the second case, by induction, there are two arrows $w \triangleright_\beta s$ and $v_{n-1} \triangleright_\beta s$ and, according the the Church-Rosser theorem 377 for the pair $v_{n-1} \triangleright_\beta s$ and $v_{n-1} \triangleright_\beta v$, two arrows $s \triangleright_\beta t$ and $v \triangleright_\beta t$, so that the composed arrow $w \triangleright_\beta s \triangleright_\beta t$ does the job for w. □

Corollary 381 *If we have $v =_\beta t$ where t is β-normal, then $v \rhd_\beta t$. If both, v and t, are β-normal, then $v =_\beta t$ implies $v \equiv_\alpha t$, i.e., on normal terms in Λ, β-conversion classes reduce to equality (congruence classes).*

Proof This is immediate from proposition 380 and from the fact that normality excludes contractions.   □

43.6 The λ-Calculus as a Programming Language

The λ-calculus, even its pure form, is a powerful formalism to describe actual functions. In this section we define a few necessary constructs within the λ-calculus, such as natural numbers with the corresponding arithmetic, Booleans, conditionals (if ... then ... else), and recursion.

We begin with definition of the Booleans *true* and *false*.

$$\textbf{true} = \lambda x.\lambda y.x,$$
$$\textbf{false} = \lambda x.\lambda y.y.$$

Informally **true** and **false** are functions taking two arguments, **true** returns its first argument, **false** its second argument. This is exactly what we need to define the conditional:

$$\textbf{test} = \lambda x.\lambda y.\lambda z.xyz.$$

Here is an example of its use:

$$
\begin{aligned}
\textbf{test true } uv &= (\lambda x.\lambda y.\lambda z.xyz) \textbf{ true } uv \\
&\rhd_\beta (\lambda y.\lambda z.\textbf{true } yz)uv \\
&\rhd_\beta (\lambda z.\textbf{true } uz)v \\
&\rhd_\beta \textbf{true } uv \\
&= (\lambda x.\lambda y.x)uv \\
&\rhd_\beta (\lambda y.u)v \\
&\rhd_\beta u.
\end{aligned}
$$

Likewise **test false** $uv = v$.

The definition of natural numbers follows the Peano axioms. Since in pure λ-calculus there are no constants, the natural number 0 is defined to be the λ-term

$$\underline{0} = \lambda x.\lambda y.y$$
$$= \textbf{false}.$$

More generally, if for two λ-terms x and y and a natural k, we define

$$x^0 y = y,$$
$$x^{k+1} y = x(x^k y),$$

then we have $\underline{k} = \lambda x.\lambda y.x^k y$ for every natural number k. This $\underline{k}$ is called the *Church numeral* associated with the natural number k.

The successor operation is

$$\textbf{succ} = \lambda n.\lambda x.\lambda y.x(nxy).$$

This looks rather abstract, but let us see what **succ 0** reduces to:

$$\textbf{succ } \underline{0} = (\lambda n.\lambda x.\lambda y.x(nxy))(\lambda x.\lambda y.y)$$
$$\triangleright_{1\beta} \lambda x.\lambda y.x((\lambda x.\lambda y.y)xy)$$
$$\triangleright_{1\beta} \lambda x.\lambda y.x((\lambda y.y)y)$$
$$\triangleright_{1\beta} \lambda x.\lambda y.xy$$
$$= \underline{1}.$$

In the same way, we have $\underline{2} = \lambda x.\lambda y.x(xy) = \textbf{succ } \underline{1}$. A useful operation is

$$\textbf{iszero} = \lambda x.((x(\lambda y.\textbf{false}))\,\textbf{true})$$

which applied to $\underline{0}$ reduces to **true** and to **false** for $\underline{k}$ with $k > 0$.

Next we define addition:

$$\textbf{plus} = \lambda u.\lambda v.\lambda x.\lambda y.ux(vxy).$$

Exercise 207 Perform complete reductions of **iszero** $\underline{0}$ and **iszero** $\underline{2}$. Reduce the λ-term **plus** $\underline{3}$ $\underline{2}$ and check that the result is effectively $\underline{5}$. Use **plus** to define **times**.

Using similar principles, it is possible, but a little more intricate, to define **pred**, such that **pred** $\underline{x} = \underline{x-1}$, if $x > 0$, and **pred** $\underline{x} = \underline{0}$, if $x = 0$. See [31] for details.

There is no explicit provision in λ-calculus for recursion. Interestingly, there is a λ-term that effectively computes the fixpoint of a function:

$$Y = \lambda f.(\lambda x.f(\lambda y.xxy))(\lambda x.f(\lambda y.xxy)).$$

This so-called *Y-combinator* is hard to read, and even harder to understand. However, it is sufficient here to note that its essential property is that $g = Yf$ computes the fixpoint g of f, i.e., $g =_\beta fg$ or $Yf =_\beta f(Yf)$.

Example 229 Let us construct the λ-term for the factorial function using the Y-combinator. The factorial contains a recursive call to itself, so the trick is to define a function that not only takes a numerical argument, but as a further argument the function that is to be called in recursion.

$$g = \lambda f.\lambda x.(\textbf{test } (\textbf{iszero } x) \ 1 \ (\textbf{times } x \ (f \ (\textbf{pred } x))))).$$

It is clear that if we had a "real" factorial *fact*, then g *fact* would again be the "real" factorial. This, however, just means that we look for the solution *fact* of *fact* $= g$ *fact*, and this is just what the Y-combinator provides. Thus:

$$\textbf{factorial} = Yg.$$

Exercise 208 Calculate a reduction of **factorial** $\underline{2}$. Be warned: This is a long and wearisome exercise. You may replace terms of the form **iszero** $\underline{x}$, **times** $\underline{x} \ \underline{y}$ and **pred** $\underline{x}$ immediately by their results.

43.7 Recursive Functions

The lambda calculus provides us with a formalism for representing all partial recursive functions, "representation" being understood in a precise technical sense to be explained in section 43.8. Recursive functions are a type of functions which can be defined using a recursive rule, applying the recursion theorem (proposition 55 in volume 1), and restricting the domains and codomains to powers $\mathbb{N}^d$ of the set of natural numbers. In other words, it is all what you might expect from doing "strictly natural" arithmetic. To this end, we consider the set $\mathcal{NF} = \bigcup_{d \in \mathbb{N}} \textbf{Sets}(\mathbb{N}^d, \mathbb{N})$. The sets of recursive functions we shall consider are subsets of $\mathcal{NF}$. Here is their definition.

Definition 274 *The set* $\mathcal{PF}$ *of* primitive recursive functions *is the smallest subset of* $\mathcal{NF}$ *containing:*

(i) *the elementary functions, i.e., the*

Zero function:
$$z : \mathbb{N} \to \mathbb{N} : n \mapsto 0,$$

Successor function:
$$s : \mathbb{N} \to \mathbb{N} : n \mapsto n + 1.$$

For each positive power $\mathbb{N}^k$ and each index $1 \le i \le k$,
the projection
$$p_i^k : \mathbb{N}^k \to \mathbb{N} : (x_1, \ldots x_k) \mapsto x_i.$$

(ii) *(Composition) If $f_i : \mathbb{N}^n \to \mathbb{N}$, for $i = 1, 2, \ldots m$, and $g : \mathbb{N}^m \to \mathbb{N}$ are in $\mathcal{PF}$, then so is*

$$g \circ (f_1, f_2, \ldots f_m) : \mathbb{N}^n \to \mathbb{N} : x \mapsto g(f_1(x), f_2(x), \ldots f_m(x)),$$

where $(f_1, f_2, \ldots f_m) : \mathbb{N}^n \to \mathbb{N}^m$ is the universal arrow guaranteed by the universal property of the Cartesian product.

(iii) *(Primitive recursion) If $g : \mathbb{N}^n \to \mathbb{N}$ and $h : \mathbb{N}^{n+2} \to \mathbb{N}$ are in $\mathcal{PF}$, then the function $r : \mathbb{N}^{n+1} \to \mathbb{N}$ is also in $\mathcal{PF}$, where r is defined by*

$$r(x_1, \ldots x_n, 0) = g(x_1, \ldots x_n),$$
$$r(x_1, \ldots x_n, y + 1) = h(x_1, y, r(x_1, \ldots x_n, y)).$$

In the special case $n = 0$, $g : \mathbb{N}^0 \to \mathbb{N}$ is a constant value in $g \in \mathbb{N}$ and we have $h : \mathbb{N}^2 \to \mathbb{N}$, and $r : \mathbb{N} \to \mathbb{N}$ is defined by

$$r(0) = g,$$
$$r(y + 1) = h(y, r(y)).$$

Example 230 Using the constructs for primitive recursion, we can now define the sum *plus* of natural numbers:

$$plus(x, 0) = g(x),$$
$$plus(x, y + 1) = h(y, plus(x, y)).$$

This is an instance of the schema (iii) of primitive recursion with $n = 1$. We still have to define g and h:

$$g(x) = p_1^1(x), \quad h(y, z) = s(p_2^2(y, z)).$$

Exercise 209 Show that the following functions are primitive recursive: natural multiplication $times : \mathbb{N} \times \mathbb{N} \to \mathbb{N} : (x, y) \mapsto x \cdot y$, and the factorial $fact : \mathbb{N} \to \mathbb{N} : x \mapsto x!$.

Example 231 The function $P : \mathbb{N} \to \mathbb{N}$ with $P(n) = 1$ iff n is prime, and $P(n) = 0$ else (the characteristic function of the set of natural primes) is in $\mathcal{PF}$.

Despite their simple structure, primitive recursive functions are already quite powerful, as the examples suggest. There are, however, functions which are not in $\mathcal{PF}$. One such function is the *Ackermann function* $A : \mathbb{N}^2 \to \mathbb{N}$. It is recursively defined as follows:

$$A(0,0) = 1,$$
$$A(0,1) = 2,$$
$$A(0, y + 2) = y + 4,$$
$$A(x + 1, 0) = A(x, 1),$$
$$A(x + 1, y + 1) = A(x, A(x + 1, y)).$$

This is a recursive definition which is not of the primitive recursive form. This does *not* mean, of course, that no primitive recursive definition for A can be found. It can however be shown that $A \notin \mathcal{PF}$ and that, therefore, there is no primitive recursive form for A. The larger set $\mathcal{PRF}$ of *partial recursive* functions encompasses $\mathcal{PF}$ and additionally includes this type of functions. Its definition requires a slightly more general set of domains, i.e., we admit that a function $f : D \to \mathbb{N}$ has a domain D which is a proper subset $D \subset \mathbb{N}^d$. This makes things a bit more complicated, since composition of only partially defined functions (not for all tuples) is not always defined. We work around this complication by introducing a special sign $\bot \notin \mathbb{N}$ (also called *bottom*). Then we consider the larger set $\mathbb{N}_\bot = \{\bot\} \cup \mathbb{N}$. We now look at the subset $\mathcal{NF}_\bot \subset \bigcup_{d \in \mathbb{N}} \mathbf{Sets}(\mathbb{N}_\bot^d, \mathbb{N}_\bot)$ consisting of all set functions $f : \mathbb{N}_\bot^d \to \mathbb{N}_\bot$ such that $f(z_1, \ldots z_d) = \bot$ if at least one of the arguments z_i is $\bot$. Then for such an f, its domain D is the complement in $\mathbb{N}_\bot^d$ of the fiber $f^{-1}(\bot)$ over $\bot$. Observe that all tuples with at least one $\bot$ are contained in this fiber. It also guarantees that if the image $f(x_1, \ldots x_n)$ of an argument $(x_1, \ldots x_n) \in \mathbb{N}^n$ is not in the domain of a second function g, then the argument is also in the fiber of $(g \circ f)^{-1}(\bot)$, i.e., the composition is not defined in $x_1, \ldots x_n$.

Definition 275 *The set $\mathcal{PRF}$ of* partial recursive functions *is the smallest subset $\mathcal{PRF} \subset \mathcal{NF}_\perp$ which contains these functions:*

(i) *(Elementary functions) It contains all elementary functions introduced in definition 274.*

(ii) *(Composition) If $f_i : \mathbb{N}_\perp^n \to \mathbb{N}_\perp$, for $i = 1, 2, \ldots m$, and $g : \mathbb{N}_\perp^m \to \mathbb{N}_\perp$ are in $\mathcal{PRF}$, then so is*

$$g \circ (f_1, f_2, \ldots f_m) : \mathbb{N}_\perp^m \to \mathbb{N}_\perp : x \mapsto g(f_1(x), f_2(x), \ldots f_m(x)),$$

where $(f_1, f_2, \ldots f_m) : \mathbb{N}_\perp^n \to \mathbb{N}_\perp^m$ is the universal arrow guaranteed by the universal property of the Cartesian product.

(iii) *(Primitive recursion) If $g : \mathbb{N}_\perp^n \to \mathbb{N}_\perp$ and $h : \mathbb{N}_\perp^{n+2} \to \mathbb{N}_\perp$ are in $\mathcal{PRF}$, then the function $r : \mathbb{N}_\perp^{n+1} \to \mathbb{N}_\perp$ is also in $\mathcal{PRF}$, where r is defined as follows:*

$$r(x_1, \ldots x_n, 0) = g(x_1, \ldots x_n),$$
$$r(x_1, \ldots x_n, y + 1) = h(x_1, y, r(x_1, \ldots x_n, y)).$$

(iv) *(Minimization) If $f : \mathbb{N}_\perp^{n+1} \to \mathbb{N}_\perp$ is partially recursive, then so is*

$$\mu(f) : \mathbb{N}_\perp^n \to \mathbb{N}_\perp,$$

$$\mu(f)(x_1, \ldots x_n) = \begin{cases} x & \text{if } f(x_1, \ldots x_n, x) = 0 \\ & \text{and } f(x_1, \ldots x_n, y) \neq \perp, 0 \text{ for all } u < x, \\ \perp & \text{else.} \end{cases}$$

Exercise 210 Show that there is a canonical embedding $\mathcal{PF} \subset \mathcal{PRF}$.

Exercise 211 Show that the Ackermann function is in $\mathcal{PRF}$.

43.8 Representation of Partial Recursive Functions

This section merges the theory of partial recursive functions and lambda calculus in the sense that one can characterize partial recursive functions by the properties of β-normalization of λ-terms.

We have already defined Church numerals $\underline{k}$, where k is a natural number. If Ψ is any term, and if $\underline{k_1}, \underline{k_2}, \ldots \underline{k_r}$ are Church numerals, then we have the term $\Psi \underline{k_1}, \underline{k_2}, \ldots \underline{k_r}$ which by the left association convention equals the application chain $(\ldots ((\Psi \underline{k_1}) \underline{k_2}) \ldots \underline{k_r})$.

Definition 276 *Let* $\psi : \mathbb{N}^m_\perp \to \mathbb{N}_\perp$ *be a partial recursive function. Then a term* $\Psi \in L$ *is said to* represent ψ *iff for all* $(k_1, k_2, \ldots k_m) \in \mathbb{N}^m$,

- *if* $\psi(k_1, k_2, \ldots k_m) = \perp$ *(i.e.,* ψ *is undefined for* $k_1, k_2, \ldots k_m$*), then* $\Psi \underline{k_1}, \underline{k_2}, \ldots \underline{k_m}$ *has no normal form;*
- *if* $\psi(k_1, k_2, \ldots k_m) = k \in \mathbb{N}$ *(i.e.,* ψ *is defined for* $k_1, k_2, \ldots k_m$*), then* $\beta(\Psi \underline{k_1}, \underline{k_2}, \ldots \underline{k_m})$ *exists and is equal to* $\underline{k}$.

Finally we have the following significant connection between partial recursive functions and the power of lambda calculus:

Proposition 382 *Every partial recursive function* ψ *can be represented by a* λ*-term* Ψ.

Proof For a proof, we refer to [21].   □

Appendix

Further Reading

Calculus. Both books *Calculus* [39] and *Calculus on Manifolds* [38] by Michael Spivak are among the most brilliantly written books on modern calculus, the latter being a concise and substantial introduction to the essential theorems of calculus. *Advanced Calculus* [23] by Lynn Loomis and Shlomo Sternberg is a comprehensive treatment of calculus, ODEs and calculus of manifolds, including classical mechanics.

Numerics. The book on numerical mathematics [36] by Hans Rudolf Schwarz is a comprehensive and computer oriented reference. *Ordinary Differential Equations* by Fred Brauer and John Nohel is not only a reliable reference to the theory of ODEs, but also includes some chapters on numerics of ODEs, such as Runge-Kutta methods.

Categories. Saunders Mac Lane's *Categories for the Working Mathematician* [26] is a standard reference written by one of the fathers of category theory. Benjamin Pierce's little book *Basic Category Theory for Computer Scientists* [30] is especially tailored to computer scientists.

Splines in all variants are a workhorse in computer graphics, especially CAD (computer aided design) applications. Many books on numerical mathematics include a discussion of splines. The encyclopedic volume *Computer Graphics: Principles and Practice* by James Foley and colleagues [11] provides many details on theoretical and implementation issues.

Fourier Theory. Discussions of Fourier theory is generally found in any book on numerical mathematics (see above). Since especially the Fast

Fourier Transform appears throughout digital signal processing, a book on DSP like Richard Lyons' [25] will be useful for the technical details.

Wavelets. *Wavelet Transforms* [32] by Raghuveer Rao and Ajit Bopardikar is a good reference to the theory of wavelets and includes many algorithms and examples. The article *Fourier Analysis and Wavelet Analysis* [41] by James Walker provides an excellent comparison of Fourier against Wavelets algorithms.

Fractals. The approach used in this book is comprehensively developed in Michael Barnsley's *Fractals Everywhere* [5]. It is mathematically complete and beautifully presented. For a wider scope see [29].

Neural Nets. *Introduction to the Theory of Neural Computation* [13] by John Hertz, Andreas Krogh and Richard Palmer offers a thorough treatment of a large variety of neural networks.

Probability Theory. The book on discrete structures [35] by Thomas Schickinger and Angelika Steger is a good reference focused on discrete probability and statistics, including many useful exercises. For a thoroughly mathematical treatment, refer to Kai Lai Chung's *A Course in Probability Theory* [8].

Lambda Calculus. Any text on functional programming worth its subject features an introduction to the λ-calculus, for example the one by Field and Harrison [10]. An accessible dedicated work is Hindley's and Seldin's *Introduction to combinators and λ-calculus* [16]. The classical treatise is Barendregt's comprehensive *The Lambda Calculus* [4].

APPENDIX **B**

Bibliography

[1] Abelson, Harold & Sussman, Gerald Jay. *Structure and Interpretation of Computer Programs*. MIT Press, Cambridge 1996.

[2] Abraham, Ralph & Marsden, Jerrold E. *Foundations of Modern Mechanics*. Addison Wesley, Reading 1967.

[3] Amann, Herbert. *Analysis III*. Birkhäuser, Basel 2001.

[4] Barendregt, Henk P. *The Lambda Calculus*. North Holland, Amsterdam 1984.

[5] Barnsley, Michael F. *Fractals Everywhere*. Morgan Kaufmann, San Francisco et al. 1993.

[6] Brauer, Fred & Nohel, John A. *Ordinary Differential Equations*. Benjamin, New York 1967.

[7] Bronstein, Ilya N. & Semendyayev, Konstantin A. *Handbook of Mathematics*, Thomson Learning, 1991.

[8] Chung, Kai Lai. *A Course in Probability Theory*. Academic Press, London 2000.

[9] Cybenko, George. "Approximation by Superpositions of a Sigmoidal Function." *Mathematics of Control, Signals, and Systems*, 2 (1989) pp. 303–314.

[10] Field, Anthony J. & Harrison, Peter. *Functional Programming*. Addison Wesley, Reading 1988.

[11] Foley, James D. et al. *Computer Graphics: Principles and Practice in C*. Addison Wesley, Reading 1995.

[12] Harary, Frank. *Graph Theory*. Addison Wesley, Reading 1972.

[13] Hertz, John, Krogh, Andreas & Palmer, Richard. *Introduction to the Theory of Neural Computation*. Addison-Wesley, Reading 1991.

[14] Hildebrandt, Stefan. *Analysis I*. Springer, Heidelberg et al. 2002.

[15] Hildebrandt, Stefan. *Analysis II*. Springer, Heidelberg et al. 2003.

[16] Hindley, J. Roger & Seldin, Jonathan P. *Introduction to combinators and λ-calculus*. Cambridge University Press, New York 1986.

[17] John, Fritz. *Partial Differential Equations*. Springer, Heidelberg et al. 1978.

[18] Johnstone, Peter. *Sketches of an Elephant: A Topos Theory Compendium*. Oxford University Press, Oxford 2002.

[19] Jones, Huw. *Computer Graphics through Key Mathematics*. Springer, Heidelberg et al. 2001.

[20] Kolmogorov, Andrei Nikolaevich. *Grundbegriffe der Wahrscheinlichkeitsrechnung*. English Translation: *Foundations of the Theory of Probability*. Chelsea, New York 1950.

[21] Krivine, Jean-Louis. *Lambda-calcul, types et modèles*. Masson, Paris 1990.

[22] Lehn, Jürgen & Wegmann, Helmut. *Einführung in die Statistik*. Teubner, Stuttgart 1992.

[23] Loomis, Lynn H. & Sternberg, Shlomo. *Advanced Calculus*. Addison Wesley, Reading 1968.

[24] Louis, Alfred K. et al. *Wavelets*. Teubner, Stuttgart 1994.

[25] Lyons, Richard G. *Understanding Digital Signal Processing*. Prentice Hall, Upper Saddle River 1996.

[26] Mac Lane, Saunders. *Categories for the Working Mathematician*. Springer, Heidelberg et al. 1971.

[27] Mandelbrot, Benoit B. *The Fractal Geometry of Nature*. W H Freeman & Co., New York 1982.

[28] Mazzola, Guerino. *The Topos of Music*. Birkhäuser, Basel 2002.

[29] Peitgen, Heinz-Otto, Jürgens, Hartmut & Saupe, Dietmar. *Chaos and Fractals: New Frontiers of Science*. Springer, Heidelberg et al. 1992.

[30] Pierce, Benjamin C. *Basic Category Theory for Computer Scientists.* MIT Press, Cambridge 1991.

[31] Pierce, Benjamin C. *Types and Programming Languages.* MIT Press, Cambridge 2002.

[32] Rao, Raghuveer M. & Bopardikar, Ajit S. *Wavelet Transforms.* Addison Wesley, Reading 1998.

[33] Rosenblatt, Frank. *Principles of Neurodynamics.* Spartan Books, Washington DC 1962.

[34] Rudin, Walter. *Functional Analysis.* McGraw-Hill, New York 1991.

[35] Schickinger, Thomas & Steger, Angelika. *Diskrete Strukturen II.* Springer, Heidelberg et al. 2002.

[36] Schwarz, Hans Rudolf. *Numerische Mathematik.* Teubner, Stuttgart 1977.

[37] Scott, Dana. "Symbolic Computation and Teaching." *Artificial Intelligence and Symbolic Mathematical Computation,* AISMC-3, vol. 1138 pp. 1–20, 1996.

[38] Spivak, Michael. *Calculus.* Publish or Perish, Houston 1994.

[39] Spivak, Michael. *Calculus on Manifolds.* Westview Press, Boulder 1971.

[40] Stein, Elias M. & Shakarchi, Rami. *Fourier Analysis.* Princeton University Press, Princeton & Oxford 2003.

[41] Walker, James S. "Fourier Analysis and Wavelet Analysis." *Notices of the AMS,* Volume 44, Number 6, pp. 658–670, 1997.

[42] Zerz, Eva & Helmke, Uwe & Prätzel-Wolters. *Mathematical Theory of Neural Networks.* Berichte der AG Technomathematik 238, Kaiserslautern 2001.

Index

Printing: Krips bv, Meppel
Binding: Litges & Dopf, Heppenheim